Computer-Aided Simulation in Railway Dynamics

MECHANICAL ENGINEERING

A Series of Textbooks and Reference Books

EDITORS

L. L. FAULKNER

Columbus Division
Battelle Memorial Institute

and

Department of Mechanical Engineering
The Ohio State University
Columbus, Ohio

S. B. MENKES

Department of Mechanical Engineering
The City College of the
City University of New York
New York, New York

1. Spring Designer's Handbook, *by Harold Carlson*
2. Computer-Aided Graphics and Design, *by Daniel L. Ryan*
3. Lubrication Fundamentals, *by J. George Wills*
4. Solar Engineering for Domestic Buildings, *by William A. Himmelman*
5. Applied Engineering Mechanics: Statics and Dynamics, *by G. Boothroyd and C. Poli*
6. Centrifugal Pump Clinic, *by Igor J. Karassik*
7. Computer-Aided Kinetics for Machine Design, *by Daniel L. Ryan*
8. Plastics Products Design Handbook, Part A: Materials and Components; Part B: Processes and Design for Processes, *edited by Edward Miller*
9. Turbomachinery: Basic Theory and Applications, *by Earl Logan, Jr.*
10. Vibrations of Shells and Plates, *by Werner Soedel*
11. Flat and Corrugated Diaphragm Design Handbook, *by Mario Di Giovanni*
12. Practical Stress Analysis in Engineering Design, *by Alexander Blake*
13. An Introduction to the Design and Behavior of Bolted Joints, *by John H. Bickford*
14. Optimal Engineering Design: Principles and Applications, *by James N. Siddall*
15. Spring Manufacturing Handbook, *by Harold Carlson*
16. Industrial Noise Control: Fundamentals and Applications, *edited by Lewis H. Bell*
17. Gears and Their Vibration: A Basic Approach to Understanding Gear Noise, *by J. Derek Smith*

18. Chains for Power Transmission and Material Handling: Design and Applications Handbook, *by the American Chain Association*
19. Corrosion and Corrosion Protection Handbook, *edited by Philip A. Schweitzer*
20. Gear Drive Systems: Design and Application, *by Peter Lynwander*
21. Controlling In-Plant Airborne Contaminants: Systems Design and Calculations, *by John D. Constance*
22. CAD/CAM Systems Planning and Implementation, *by Charles S. Knox*
23. Probabilistic Engineering Design: Principles and Applications, *by James N. Siddall*
24. Traction Drives: Selection and Application, *by Frederick W. Heilich III and Eugene E. Shube*
25. Finite Element Methods: An Introduction, *by Ronald L. Huston and Chris E. Passerello*
26. Mechanical Fastening of Plastics: An Engineering Handbook, *by Brayton Lincoln, Kenneth J. Gomes, and James F. Braden*
27. Lubrication in Practice, Second Edition, *edited by W. S. Robertson*
28. Principles of Automated Drafting, *by Daniel L. Ryan*
29. Practical Seal Design, *edited by Leonard J. Martini*
30. Engineering Documentation for CAD/CAM Applications, *by Charles S. Knox*
31. Design Dimensioning with Computer Graphics Applications, *by Jerome C. Lange*
32. Mechanism Analysis: Simplified Graphical and Analytical Techniques, *by Lyndon O. Barton*
33. CAD/CAM Systems: Justification, Implementation, Productivity Measurement, *by Edward J. Preston, George W. Crawford, and Mark E. Coticchia*
34. Steam Plant Calculations Manual, *by V. Ganapathy*
35. Design Assurance for Engineers and Managers, *by John A. Burgess*
36. Heat Transfer Fluids and Systems for Process and Energy Applications, *by Jasbir Singh*
37. Potential Flows: Computer Graphic Solutions, *by Robert H. Kirchhoff*
38. Computer-Aided Graphics and Design, Second Edition, *by Daniel L. Ryan*
39. Electronically Controlled Proportional Valves: Selection and Application, *by Michael J. Tonyan, edited by Tobi Goldoftas*
40. Pressure Gauge Handbook, *by AMETEK, U.S. Gauge Division, edited by Philip W. Harland*
41. Fabric Filtration for Combustion Sources: Fundamentals and Basic Technology, *by R. P. Donovan*
42. Design of Mechanical Joints, *by Alexander Blake*
43. CAD/CAM Dictionary, *by Edward J. Preston, George W. Crawford, and Mark E. Coticchia*

44. Machinery Adhesives for Locking, Retaining, and Sealing, *by Girard S. Haviland*
45. Couplings and Joints: Design, Selection, and Application, *by Jon R. Mancuso*
46. Shaft Alignment Handbook, *by John Piotrowski*
47. BASIC Programs for Steam Plant Engineers: Boilers, Combustion, Fluid Flow, and Heat Transfer, *by V. Ganapathy*
48. Solving Mechanical Design Problems with Computer Graphics, *by Jerome C. Lange*
49. Plastics Gearing: Selection and Application, *by Clifford E. Adams*
50. Clutches and Brakes: Design and Selection, *by William C. Orthwein*
51. Transducers in Mechanical and Electronic Design, *by Harry L. Trietley*
52. Metallurgical Applications of Shock-Wave and High-Strain-Rate Phenomena, *edited by Lawrence E. Murr, Karl P. Staudhammer, and Marc A. Meyers*
53. Magnesium Products Design, *by Robert S. Busk*
54. How To Integrate CAD/CAM Systems: Management and Technology, *by William D. Engelke*
55. Cam Design and Manufacture, Second Edition; with cam design software for the IBM PC and compatibles, disk included, *by Preben W. Jensen*
56. Solid-State AC Motor Controls: Selection and Application, *by Sylvester Campbell*
57. Fundamentals of Robotics, *by David D. Ardayfio*
58. Belt Selection and Application for Engineers, *edited by Wallace D. Erickson*
59. Developing Three-Dimensional CAD Software with the IBM PC, *by C. Stan Wei*
60. Organizing Data for CIM Applications, *by Charles S. Knox, with contributions by Thomas C. Boos, Ross S. Culverhouse, and Paul F. Muchnicki*
61. Computer-Aided Simulation in Railway Dynamics, *by Rao V. Dukkipati and Joseph R. Amyot*
62. Fiber-Reinforced Composites: Materials, Manufacturing, and Design, *by P. K. Mallick*

Mechanical Engineering Software

Spring Design with an IBM PC, *by Al Dietrich*

Mechanical Design Failure Analysis: With Failure Analysis System Software for the IBM PC, *by David G. Ullman*

Computer-Aided Simulation in Railway Dynamics

Rao V. Dukkipati Joseph R. Amyot

Division of Mechanical Engineering
National Research Council of Canada
Ottawa, Ontario, Canada

Marcel Dekker, Inc. ▪ New York and Basel

Library of Congress Cataloging in Publication Data

Dukkipati, Rao V.
Computer-aided simulation in railway dynamics / Rao V. Dukkipati and Joseph R. Amyot.
p. cm. -- (Mechanical engineering ; 61)
Includes bibliographies.
ISBN 0-8247-7787-5
1. Railroads--Trains--Dynamics--Mathematical models.
2. Railroads--Trains--Dynamics--Data processing. I. Amyot, Joseph R. II. Title. III. Series: Mechanical engineering (Marcel Dekker, Inc.) ; 61.
TF550.D84 1988 87-21286
625.1'0072'4--dc19 CIP

MARCEL DEKKER, INC.
270 Madison Avenue, New York, New York 10016

Current Printing (last digit):
10 9 8 7 6 5 4 3 2 1

PRINTED IN THE UNITED STATES OF AMERICA

PREFACE

The growing populations worldwide since the mid-twentieth century have required expanded and improved rail transportation in the form of faster, safer and more comfortable passenger and freight service under heavier loads. This has spawned considerable experimental and analytical research into railway dynamics. Much of this research has been made possible by the advent of computers. Although the research and development of analytical and experimental tools is still in progress, there now exists an important body of knowledge in this area and a need to consolidate it for practical use. This book does this consolidation for simulation in railway dynamics.

Simulation is the act of assuming the behaviour of reality in its absence. The dynamical behaviour of railway vehicles, either individually or in trains, is the reality dealt with in this book. Simulation may be carried out by means of a computer model or a physical model such as a track simulator using rollers to simulate receding track relative to a longitudinally-constrained vehicle placed on them. Simulation results may then be analyzed to predict the behaviour of reality according to certain well-defined performance criteria such as stability and ride quality. The accuracy and reliability of prediction depend on the extent to which the models have been validated. Validation is the process of determining agreement, or lack thereof, between simulation results and real measurements.

The usefulness of simulation stems from the controlled environment in which it is performed. A computer model may be exercised without danger and the cost of varying parameters and obtaining results is usually small compared with the cost of implementing such changes and making measurements in reality. Measurements are also easier, for example, in a roller-based physical track simulation than on a moving vehicle because of a fixed reference frame and, as an advantage over computer simulation, there is no need to identify the vehicle's parameters since it is mounted physically onto the rollers.

The objective of this book is to present a coherent treatment of the state of the simulation art in railway dynamics for practical use. It provides a working understanding of the scope of simulation in railway dynamics, the effective use of simulation facilities, and concepts for the development of computer simulation software and physical track simulators. General methods, procedures and concepts are emphasized while special attention is given to clear presentation of equations, methods of solution, numerical illustrative examples and case studies. Whereas application is the dominant theme, numerous references are provided for the theoretically minded. The book is appropriate for engineers, researchers, system analysts and managers in the railway dynamics field who are familiar with elementary matrix algebra, differential equations, mechanics and computer programming.

Chapter 1 describes the computer modeling and simulation process, defines concepts such as conceptual model, computerized model and validation, discusses the scope of application in railway dynàmics, introduces the five phases of a modeling and simulation project from a planning and management perspective, and raises the make-or-buy question relative to software acquisition. The practical tone of the book is established by selected descriptions, using photographs of a cross-section of railway rolling stock.

The emphasis on application is continued in Chapter 2 with physical descriptions, illustrated by diagrams, of physical aspects relevant to simulation in railway dynamics. A building-block order of description is adopted, starting with tracks and followed by the wheel-rail interaction, multi-wheelset trucks which support the load-carrying body of vehicles, entire vehicles including freight cars, passenger cars and locomotives, and finally systems of interconnected vehicles called trains.

The above railway components must be conceptualized by mathematical models before they can be programmed into computers as computerized models for simulation purposes. Chapter 3 presents these mathematical models, discusses the significance of pertinent terms and parameters, and refers the reader to the literature for detailed derivations. The analysis of performance, either directly through the reduction of simulation results or indirectly by frequency domain methods, is treated similarly.

Parameter identification and model validation are the bridges which connect computer models with reality. Chapter 4 discusses how the numerical values of mathematical model parameters of system components can be determined in practice and how the computer model of a whole system can be validated by comparisons of simulation results with real measurements.

Chapter 5 presents pertinent case studies to illustrate the application of computer modeling and simulation introduced in the preceding chapters. Examples of existing computer programs applied to wheel-rail interaction, passenger car, freight car, locomotive and train dynamics studies are given.

The make-or-buy decision hinges on the availability of software. Therefore, Chapter 6 gives an extensive up-to-date survey of existing railway dynamics software.

Chapter 7 broadens the simulation scenario by extending the computer modeling and simulation presentation of Chapters 2 to 6 to the very current subject of physical track simulators. Advantages and limitations, conceptual bases, design factors and validation are discussed through the medium of a case study. A comparative survey of existing and planned simulators is given along with a range of actual applications.

Our sincere acknowledgement is extended to all who have had a direct impact on this work. Foremost are the numerous contributors to the field of simulation in railway dynamics. Special thanks go to the Division of Mechanical Engineering, National Research Council of Canada, for its wholehearted support of this project. The figures are due to the draughting skills of Mr. W. Brien. We were continually impressed by the expeditiousness and expertise of the Stenographic Services staff, under the management of Mrs. H.S. Cuccaro.

Our parents, wives and children have had a significant influence on our lives and, therefore, this work; so we would like to include them here. For the first author, there is Annapurnamma, Nagabhushanam, Sudha, Ravi and Raghu; for the second, Florence, Oscar, Morlene, Marc, Philip and Paul.

Rao V. Dukkipati
Joseph R. Amyot

CONTENTS

Computer-Aided Simulation in Railway Dynamics

1

INTRODUCTION

1.1 INTRODUCTION

You are driving leisurely on a highway. Suddenly, up ahead, the flashing red signals and bells of a railway crossing spring to life and the gates go down, announcing the approach of a train. As your car comes to a complete stop, the front light of a locomotive can be seen and the sound of its horn can be heard in the distance. Just a few short moments of anticipation later, the clatter of wheels of a seemingly never-ending train is before you. Then, it is all over. More than a hundred vehicles of every description must have passed prior to the train's disappearance. Considering the amount of kinetic energy involved and recalling the ground vibrations, this was indeed an awe-inspiring experience. Fortunately, for practical and economic reasons, it is possible to study the dynamics of such a complex system by simulation methods applied in the relative calm and safety of a computer environment.

Modern freight vehicles are required to carry more load per axle while trains consist of a greater number of vehicles than in the past. Furthermore, in competition with air travel, passenger trains have to go faster and still provide comfort. Thus, modern railway vehicles and trains tend to be more complex and sophisticated. Their design is helped by computer-aided methods such as simulation.

Computer simulation lets a designer focus his attention on one or more of a train's components for purposes of predicting its dynamic behaviour under specified operating conditions. He can repeat a simulation run with different parameter values and iterate until a satisfactory performance is determined. This iteration procedure can be much more costly and time-consuming if done in hardware. Simulation results, however, are useful only as a step in a design procedure which must necessarily culminate in hardware construction and testing. Nevertheless, they can provide insight, reduce costs and accelerate the design process.

During the last twenty-five years, a considerable body of literature and computer programs on the analysis of railway dynamics has emerged. The purpose of this book is to review the capabilities and limitations of computer modeling and simulation as applied to railway dynamics. The emphasis is on the use of: (1) existing computer models for the analysis and design of railway rolling stock, and (2) available mathematical models for the development of new computer models and physical track simulators. These considerations have a direct bearing on the make-or-buy decision regarding computer models.

1.2 SIMULATION CONCEPTS

A useful set of definitions pertaining to simulation has been proposed by the Society for Computer Simulation (SCS) Technical

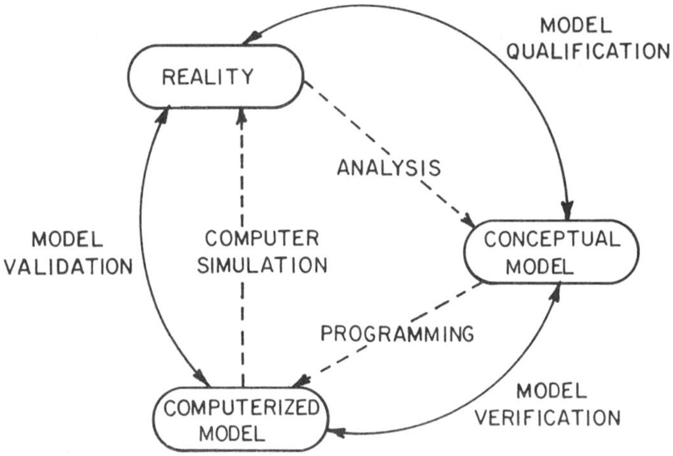

Fig. 1.1 Computer modeling and simulation process. (Adapted from reference [5].)

Committee on Model Credibility [5]. As shown in Fig. 1.1, three basic elements can be defined in the computer modeling and simulation process: *reality, conceptual model and computerized model.* The processes which relate the elements to each other are described by the inner arrows. The outer arrows refer to the procedures whereby the credibility of these processes is evaluated. The following definitions deal with each of the basic elements and their relationships. They are general in that they can be applied to simulations of measurable physical systems as well as ill-defined social and biological systems. Some of the definitions are illustrated by pertinent railway examples.

Reality is an entity, situation, or system which has been selected for analysis. For example, it could be a railway train, vehicle, truck (bogie) or wheelset operating under specified conditions.

The *conceptual model* is a verbal description, equations, governing relationships or *natural laws* that purport to describe reality. Newton's laws of motion are natural laws used in deriving the differential equations of railway components or systems.

The *domain of intended application* of a conceptual model is a set of prescribed conditions for which the conceptual model is intended to match reality. For example, should a set of equations of motion for a railway vehicle be limited to conical wheel-tread profiles and tangent-track operation at sub-critical speeds, these limitations would define part of the domain of intended application of that specific conceptual model. Other limitations (e.g., vehicles with only double-wheelset trucks) would enter into its completely-defined domain of intended application.

The *level of agreement* of a conceptual model is the expected agreement between itself and reality, consistent with the domain of intended application and the purpose for which the model was built. Thus, if a given conceptual model is expected to predict the critical speed of a specific type of vehicle within ten percent, this defines its level of agreement as far as critical speed is concerned.

Model qualification is the determination of adequacy of a conceptual model to provide an acceptable level of agreement for its domain of intended application. This activity may not be very quantitative in nature since it does not involve computer implementation of the model. Rather, it tends to be more concerned with the correctness of assumptions and derivations.

The *computerized model* (sometimes called *computer model)* is an operational computer program which implements a conceptual model. It is the result of an activity called *programming*. A survey of approximately 150 computerized models, ranging from track structure and wheel-rail interaction programs to train dynamics programs, is contained in Chapter 6.

Model verification is the substantiation that a computerized model represents a conceptual model within specified limits of accuracy. It usually involves comparisons of independent calculations with computerized model results for very specific situations, e.g., one point in the time trajectory of a variable.

The *domain of applicability* of a computerized model consists of the prescribed conditions for which it has been tested, compared against reality to the extent possible, and judged suitable for use. Thorough testing is necessary if the domain of applicability of a computerized model is to coincide with the domain of intended application of its conceptual model.

The *range of accuracy* of a computerized model is the demonstrated agreement between itself and reality within a stipulated domain of applicability. It may or may not coincide exactly with the level of agreement of the conceptual model. Taking the above example, whereas a given conceptual model was expected to predict the critical speed of a specific type of vehicle within ten percent, it may be found that the computerized model can predict it only within fifteen percent, or perhaps it can predict it consistently within five percent.

Model validation is the substantiation that, within its domain of accuracy, a computerized model possesses a satisfactory range of accuracy consistent with the intended application of the model. It is in this phase of the modeling and simulation process that the accumulated errors are discovered. If the computer model is found to be inadequate, its errors may be traced back either to programming errors which were overlooked or to a lack of understanding of reality.

Certification documentation is the documentation required to communicate information about a model's credibility and applicability. As a minimum, it should contain the following basic elements: (a) a statement of purpose for which the model was built, (b) a verbal and/or analytical description of the conceptual and computerized models, (c) a specification of the domain of applicability and range of accuracy related to the purpose of the model, and (d) a description of tests used for model verification and validation, and a discussion of their adequacy.

Model certification consists of a user's acceptance of the certification documentation as adequate evidence that the computerized model can be effectively utilized for a specific application.

Computer simulation is the exercise of a certified computerized model to gain insight about reality. It is an experiment using a computerized model and should not be confused with the model itself.

Further definitions describing properties of a computerized model were proposed by Amyot [1]. Let $V_R(t)$ be the output variable vector of a real system, let $V_M(t)$ be the corresponding model vector, and let t be clock time. We may write

$$V_M(t/\beta) = V_R(t) \tag{1.1}$$

where

$$\beta = f(t) \tag{1.2}$$

is a time-scale factor.

If the function f(t) is known, the model may be called *synchronous*; otherwise, it is *asynchronous*. The function f(t) may be a constant when a model is synchronous; however, this is not a necessary constraint. Furthermore, a synchronous model may be defined as being real time if $\beta = 1$, *slower than real time* if $\beta > 1$, and *faster than real time* if $0 < \beta < 1$.

Thus, the output variable vector of a perfectly accurate real time model would coincide with the corresponding real system vector at every clock-time instant. Models need to be in real time for certain applications. For example, real time models can be used in training simulators or to provide test bed environments for hardware controllers. For human observation purposes, to get a *feel* for a system's dynamics at speeds which can be followed naturally, it is sometimes desirable to run a model slower or faster than real time, depending on whether the real system is too fast or too slow, respectively. For most design activities, it is usually sufficient to run a model asynchronously and study the results off-line.

A *physical track simulator* usually replaces a test track by rollers which simulate the receding speed of rails relative to a vehicle which remains stationary except for motions associated with guidance. Inertial moments and forces, such as centrifugal forces, which occur naturally on curved track need to be applied to the vehicle masses in a simulator and controlled according to calculations based on masses, inertias, velocity and track layout. The hardware includes roller drive motors and actuators to control roller yaw angles and apply inertial forces and moments. In the case of a curved track simulator with general velocity mission capabilities, an on-line computer system is required to control the motors and actuators in real time.

1.3 SCOPE

A photograph of a long freight train drawn by five diesel locomotives through a tight curve is shown in Fig. 1.2. A computer model of this reality can be very complex or relatively simple, depending on the simulation objectives. To determine the forces and relative displacements between all rigid bodies during speed variations in the curve would require a very complex model. However, a much simpler model would suffice to predict only the inter-vehicle forces during constant speed travel through the

Fig. 1.2 Long freight train drawn by five diesel locomotives from the Electro-Motive Division of General Motors. (Reproduced from Ellsworth et al. [3] with the permission of General Motors.)

curve. Thus, for a given reality, the complexity of a model can vary considerably as a function of the simulation objectives.

Model complexity also depends on the real system. For example, the freight train with its various types of interconnected vehicles is inherently more complex than a self-propelled electrical rail vehicle in urban traffic. The transit car tandem of Fig. 1.3, consisting of two types of cars, is somewhat more complex. Type A is a self-sufficient electrically-driven car with a mass of approximately 40,370 kg (89,000 lb) and a 70-passenger seating capacity. Type B has a 38,555 kg (85,000 lb) mass, seats 76 passengers and cannot function without being mated to a type A car. Each car is equipped with air suspension trucks and twin 9,072 kg (20,000 lb) air conditioning units.

Fig. 1.3 Transit cars built for the New York City Transit Authority by Pullman Technology. Inc. (Reproduced from Ellsworth et al. [3] with the permission of Pullman Technology Inc.)

Let us continue our exploration of the complexity and diversity of railway systems by means of a few more examples. Fig. 1.4 shows a 24,494 kg (54,000 lb) coal hauler with a 94,347 kg (208,000 lb) rated capacity. Its two transverse tubs are designed to provide greater capacity and a low center of gravity. A two-axle truck and coupler are clearly visible at one end. On the other hand, a three-axle high-carrying-capacity truck, designed for equalized distribution of wheel loads, is seen in Fig. 1.5. Four-axle and six-axle trucks are also available [3].

When the locomotive in Fig. 1.6 is fully-loaded, its mass is approximately 63,500 kg (140,000 lb). One of its two-motor, two-axle trucks appears in Fig. 1.7. The truck's frame is spring supported by equalizers resting upon journal boxes. The supporting helical coil springs are carried in spring seats on the equalizers.

Fig. 1.4 Twin-tub coal hauler developed by the Greenville Steel Car Company. (Reproduced from Ellsworth et al. [3] with the permission of Greenville Steel Car Company.)

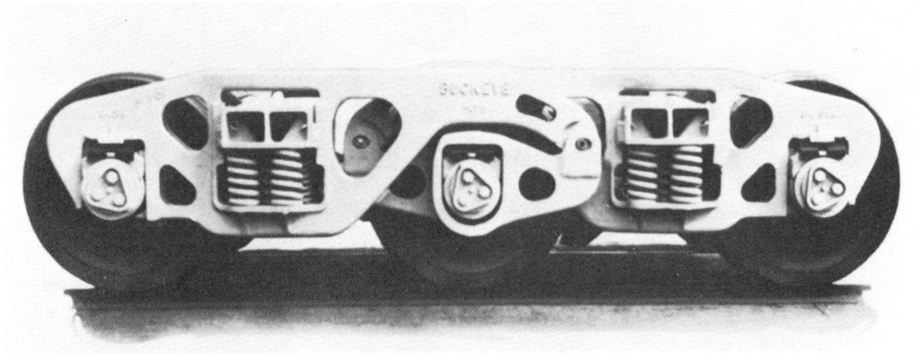

Fig. 1.5 Buckeye six-wheel elasto-cushion truck. (Reproduced from Ellsworth et al. [3] with the permission of Buckeye Steel Castings.)

Fig. 1.6 Diesel locomotive by the Rail and Diesel Products Division of Bombardier. Inc. (Reproduced from Ellsworth et al. [3] with permission of Bombardier. Inc.)

Fig. 1.7 Two-motor and two-axle truck used on the locomotive of Fig. 1.7. (Reproduced from Ellsworth et al. [3] with permission of Bombardier. Inc.)

Friction snubbers are installed in each set of springs. The tractive effort is transmitted to the centre post through rubber elements, called *traction pads*. Truck swivelling is possible due to the elasticity of the sidebearers and traction pads. A lateral shock absorber between the centre post and truck frame is used for directional stability. Resilient lateral stops limit the relative lateral movements of the centre post to ±1.91 cm (0.75 in). The outboard-journal wheelsets are supported by tapered roller bearings. The lateral movement between bearing boxes and pedestal is limited to 0.64 cm (0.25 in). A single cast-iron brake shoe at each wheel is actuated by one cylinder on each side of the truck.

It is evident from the above examples that the complexity of a model must be directly related to the level of real detail introduced into it. If there is too much detail, development costs may soar and the model may be inaccurate or unusable. Conversely, with insufficient detail, it will not meet the objectives. Therefore, the minimum amount of detail which will satisfy the simulation objectives should be incorporated in a model. Starting from this principle, a computer modeling and simulation project may be planned and managed according to the guidelines given in Section 1.4.

In the ensuing chapters, we examine the capabilities and limitations of simulation methodology as applied to physical track simulators and computer models of railway dynamics. Thereby, it should be possible to (a) determine if simulation would be useful in a given railway design project, (b) embark on a survey of available simulation software, starting from the fairly extensive survey of Chapter 6, (c) participate in the make-or-buy decision process regarding an appropriate computer model, and (d) participate in the development or use of a computer model or physical track simulator.

1.4 MODEL DEVELOPMENT

Assuming a decision has been made that a railway dynamics computer model would be useful for a given design project, should it be developed in-house, bought or rented? The answer does not involve only a consideration of direct costs. Other factors should also be examined: (a) considerable real-system insight can be gained from the model development process itself, (b) it is easier to maintain and modify an in-house product, and (c) the product may be marketable. Nevertheless, since the direct costs of in-house development usually exceed purchase or rental costs, a survey of available models is highly recommended, perhaps using Chapter 6 as a starting point.

The purchase or rental costs are straightforward. Unfortunately, the costs of development are not very easy to predict without a knowledge of what is involved in the planning and management of a computer modeling and simulation project [2,4]. Five distinct phases can be identified in such a project; definition, design, implementation, operation, and conclusion. Although there may be some overlapping and iteration, they are separately definable activities executed essentially in the order listed.

Documentation is involved in the successful completion of a computer modeling and simulation project. It can consist of many separate documents or a single document evolving with the project. However, good management practice usually requires a minimum number of separate documents, e.g., project proposal, conceptual model, and computer model. For clarity in the following descriptions of project phases, various *documents* will be introduced as separate entities although they may be partly or wholly integrated in practice.

1.4.1 Definition Phase

The purpose of this first phase is to propose a computer modeling and simulation project. The real system is analyzed to produce a conceptual model to a level of detail commensurate with the project objectives. The architecture of the computer model is determined so that a model-building cost estimate can be made. Finally, the objectives, value and cost of the project are defined.

Depending on the project's scope, it may be advisible to produce a *model feasibility* document before a decision is made to proceed with the development of a conceptual model. Such a document contains a broad description of the real system and specifies the expected contribution of the computer model to the railway design project. It also estimates the manpower, equipment and financial resources required to produce a *conceptual model* document. The conceptual model is deemed adequate when there is a concensus among its contributors of analysis and data that it is capable of satisfying the stated objectives. The definition phase should culminate in a *project proposal* document defining the objectives and cost of the proposed computer modeling and simulation project, and its value to the railway design project. The project approval decision is based on this document.

1.4.2 Design Phase

The computer model is designed in the second phase. The program modules, data sets, the relationship between them, and module verification tests are all specified and presented in a *computer model specification* document. The programming activity performed in the ensuing implementation phase is based on this document and a *programmer's handbook.* The latter document defines programming guidelines and constraints, such as program documentation requirements and variable-naming conventions. A mechanism for controlling the implementation of changes is required. One method is by the use of formalized *change memos* for programming changes. Such memos may also be used to modify the system model or computer model specifications, provided the proposed changes are adequately reviewed before implementation.

1.4.3 Implementation Phase

The implementation of the computer model is the third phase of a computer modeling and simulation project. It encompasses programming, verification and validation. The program modules are coded, debugged, verified and integrated into the model. The complete working model is then verified and validated. Computer listings, pertinent flowcharts, verification data, simulation results and any other pertinent information are accumulated in a *computer model* document. A *user's handbook* containing complete operating instructions is also required. The model's usefulness is determined largely by the clarity and simplicity with which this document is written.

1.4.4 Operation Phase

The use of the computer model constitutes the fourth phase. It is not uncommon for a computer model to be applied in many ways which were unforeseen at the time of its conception. Correspondingly, it may be appropriate to produce a *simulation document* with many parts or, alternatively, a set of simulation documents.

Simulation experiments can be categorized as *unstructured* or *structured*. An unstructured experiment is one in which the user exercises the model interactively in various unplanned ways to get a feel for its characteristics and ideas for structured experiments. On the other hand, structured experiments require organization and planning so that appropriate data can be acquired, reduced and presented in a suitable form for the simulation document(s).

1.4.5 Conclusion Phase

In this concluding phase, the experience gained from the project is summarized in a *project summary* document. Estimated and actual task times are compared. Modeling procedure variations, lessons learned and avoidable mistakes are recorded. Contributions to the railway design project(s) are listed. Based on the above and considering possible future uses, a decision is made regarding the relegation of the model to inactive storage or its continuing maintenance as an active program.

REFERENCES

1. AMYOT, J.R. (1982): Defining a Real-Time Model; *Simulation,* Vol. 38, No. 1, 27, January 1982.
2. BRUCE, P. and PEDERSON, S.M. (1982): The Software Development Project; *John Wiley and Sons*, New York.
3. ELLSWORTH, K.G., HOUSER, F.N. and WHITNEY, D.E. (1980): The Car and Locomotive Cyclopedia, 4th edn; *Simmons-Boardman Publishing Corporation*, Omaha, Nebraska.

4. GAGNE, R.E. (1976): Computer Modeling and Simulation Handbook; *Simulation*, Vol. 26, No. 5, 147-154, May 1976.
5. SCS Technical Committee on Model Credibility (1979): Terminology for Model Credibility; *Simulation*, Vol. 32, No. 3, 103-104, March 1979.

2

PHYSICAL ASPECTS

2.1 INTRODUCTION

In this chapter, we will examine some physical aspects of relevance to simulation in railway dynamics. A building-block approach will be taken, beginning with tracks and the wheel-rail interface, and ending with trucks, vehicles and whole trains.

As an overview, let us consider Fig. 2.1, a diagram illustrating the force paths of a train. The train is assumed to consist of N vehicles, the first of which would normally be a locomotive. The force path between vehicles is through couplers. As an example, the last vehicle is shown to be composed of a carbody supported by two trucks, the rear one consisting of wheelsets, sideframes and a bolster, such as the typical two-axle freight truck of Fig. 2.15. The roller bearing and adaptor constitute the primary suspension. The secondary suspension consists of the springs and snubbing friction shoes located between the sideframes and bolster. The carbody rests directly on the center plate and its weight is transmitted downwards to the track through the bolster, secondary suspension, sideframes, primary suspension, wheelsets and wheel-rail interface. Conversely, forces arising from the track's nominal geometry and irregularities are transmitted upwards through the same components. The resulting carbody accelerations (hence, ride quality) are functions of the parameter values designed into the vehicle.

Good truck design is crucial to the system and compromises are required between the following truck performance objectives: (a) to provide guidance with adequate dynamic stability and good response characteristics on straight and curved track, (b) to provide effective vibration isolation in response to track irregu-

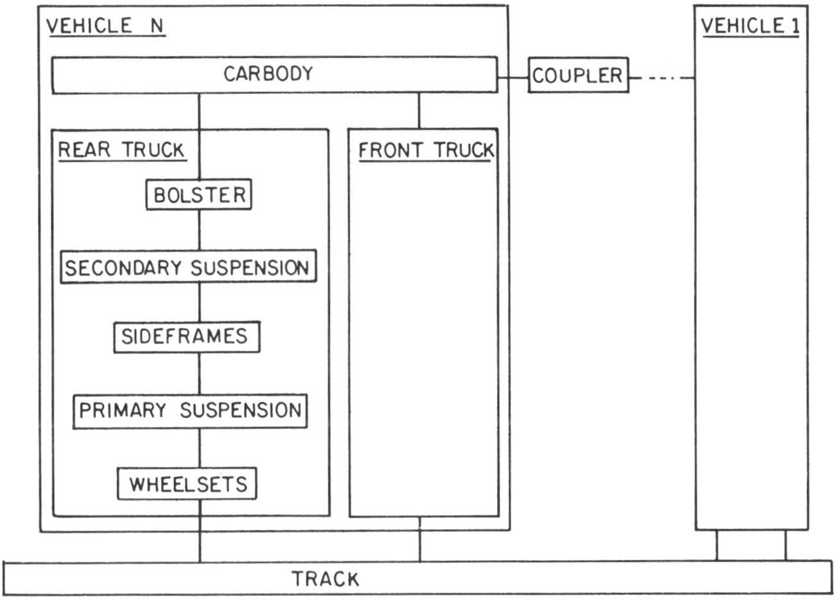

Fig. 2.1 Force Paths of a Train

larities over a given speed range, (c) to support the vehicle under various operating loads, (d) to provide sufficient traction and braking capabilities for all operating conditions, and (e) to produce wheel-rail forces which will minimize wear and damage to vehicle components and tracks [1,12]. To meet these objectives, many different types and configurations of trucks have been developed.

2.2 TRACKS

Train operations and performance are dependent to a great extent on the track condition since train equipment and lading damage can result from improperly constructed or deteriorated track. The criteria for the use of different weights of rail are based on maximum axle loads and tonnage moved over a particular track section. The rails transmit the train loads to the roadbed via small wheel-rail contact surfaces, resulting in high compressive stresses there. These high pressures can, and often do, cause corrugation, shelling and progressive fractures of rails. Wheel diameter, axle load and train speed have a greater effect on contact pressures than wheel and rail head profiles.

Rail lengths are no longer usually joined with joint bars, since welded rail is now much more common. Also, high dynamic forces occur at engine burns, not-so-smooth welds, switch points and frogs, crossings, etc.

Tie plates distribute the rail loads over a greater tie area, thereby protecting ties against mechanical wear and, in turn, helping the track to maintain its nominal gauge. Spikes are used to hold the rails and tie plates in proper alignment and gauge. The spiking patterns employed depend on the types of rail, tie plates, track alignment (tangent or curved) and tonnage handled. Compression fasteners are sometimes used to resist rail turnover.

The longitudinal movement of rails over the tie plates, is known as rail creep. Rail anchors are used to control rail creep, which may be caused by temperature-induced rail expansions and contractions, track grades, rail traffic, and braking action.

Ties, made of wood or concrete, serve to maintain the track gauge and distribute the loads on the ballast. The ballast provides uniform support for the ties, absorbs vibrations and shock, and helps to anchor the track in place by resisting lateral and longitudinal movement.

A moving train is forced to change directions due to the curvature of the track. A curved section of track, following a tangent (i.e., straight) section, usually consists of a constant radius circular curve and a transitional curve, also known as a spiral, inserted between the tangent section and the circular curve. The outside rail constrains the wheelsets to follow the curved path.

2.2.1 Superelevation

In curves, trains are influenced by centrifugal forces which act away from the center of the curve and tend to overturn the cars. The sum of a car's weight and its centrifugal force produces a resultant force directed toward the outer rail, unless the outer rail is raised, or superelevated, as in Fig. 2.2.

A curve is described as being balanced at the equilibrium speed which produces a resultant force trough the center line of the track. In this condition, the vertical rail forces are equal, so that maximum utilization of tractive effort and minimum wear on wheels and rails can be realized.

2.2.2 Spirals

When trains are operated at other than low speeds, it is necessary to insert a spiral between tangent and curve segments. The spiral is a curve of continuously-changing radius, decreasing from an infinite radius at the tangent end to a radius equal to that of the circular curve at the other end.

It provides a smooth transition from tangent to curve segments, and vice versa, allowing the superelevation to change gradually over its length.

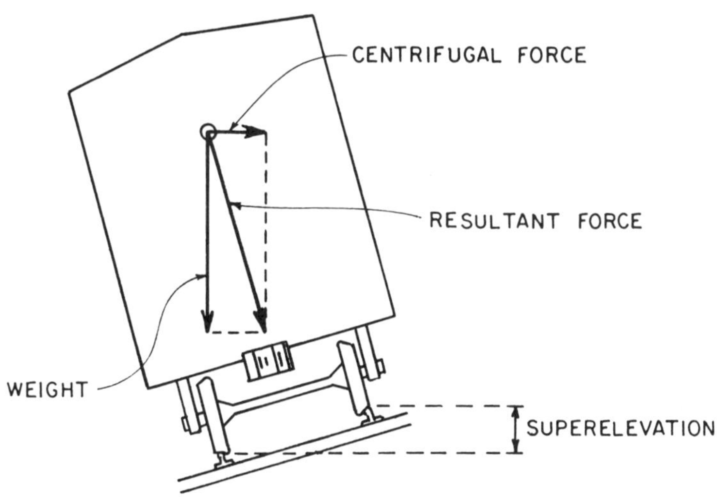

Fig. 2.2 Superelevated Car in Equlibrium

The length of spiral varies directly with the amount of superelevation of the curve. The maximum allowable rate of change of superelevation determines the minimum length of spiral for a given vehicle speed and curve superelevation. The transition between curves of different degrees and superelevations in a compound curve

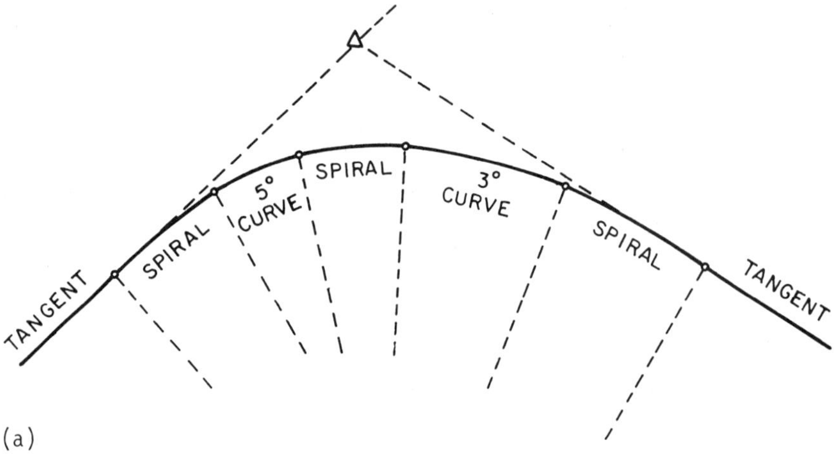

(a)

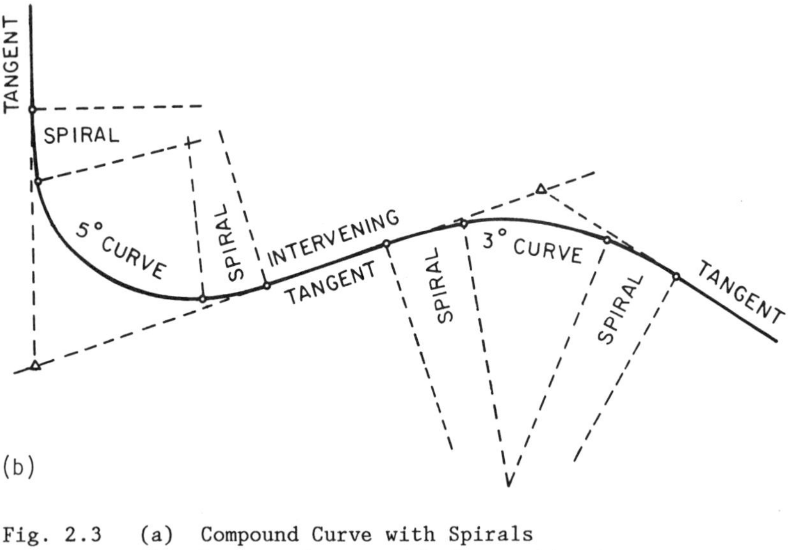

(b)

Fig. 2.3 (a) Compound Curve with Spirals
(b) Reverse Curves with Intervening Tangent

is accomplished by means of a spiral with similar limits on rate of change of superelevation (Fig. 2.3(a)). However, studies have shown that substantially shorter spiral lengths than those normally allowed in practice provide for adequate safety of train operation.

2.2.3 Reverse Curves

An intervening segment of tangent track, at least equal in length to the longest car permitted on the track, is usually inserted between reverse curves to allow the trucks of a car to recover from a curve before entering the next one (Fig. 2.3(b)).

The required length of tangent segment is a function of train speed, degree of curve, spiral lengths, superelevations, and car lengths. At larger speeds, there is less time for the trucks of a car to recover. Thus, even though proper spirals have been provided, it is desirable that the tangent segment between curves be greater in length than one car; otherwise, the authorized speed should not exceed the equilibrium speed of the adjacent curves.

2.2.4 Grade

The slope of the track (ascending or descending) in the longitudinal (forward) direction is known as its grade. The grade is normally expressed as a percentage, indicating the number of metres (feet) by which the track rises, or falls, in a longitudinal distance of 100 metres (feet). Typical grade territories encountered in practice are:

(a) Light Ascending Grade: An ascending grade of 1.0% or less is known as a light ascending grade (Fig. 2.4(a)). Train handling procedures for such a grade assume it to be long enough so that a balance can be established between decelerating gravitational and train resistance retarding influence forces and accelerating locomotive forces.

(b) Light Descending Grade: A descending grade of 1.0% or less is termed a light descending grade (Fig. 2.4(b)). Train handling procedures for such a grade assume it to be long enough so that a balance can be reached between accelerating gravitational forces

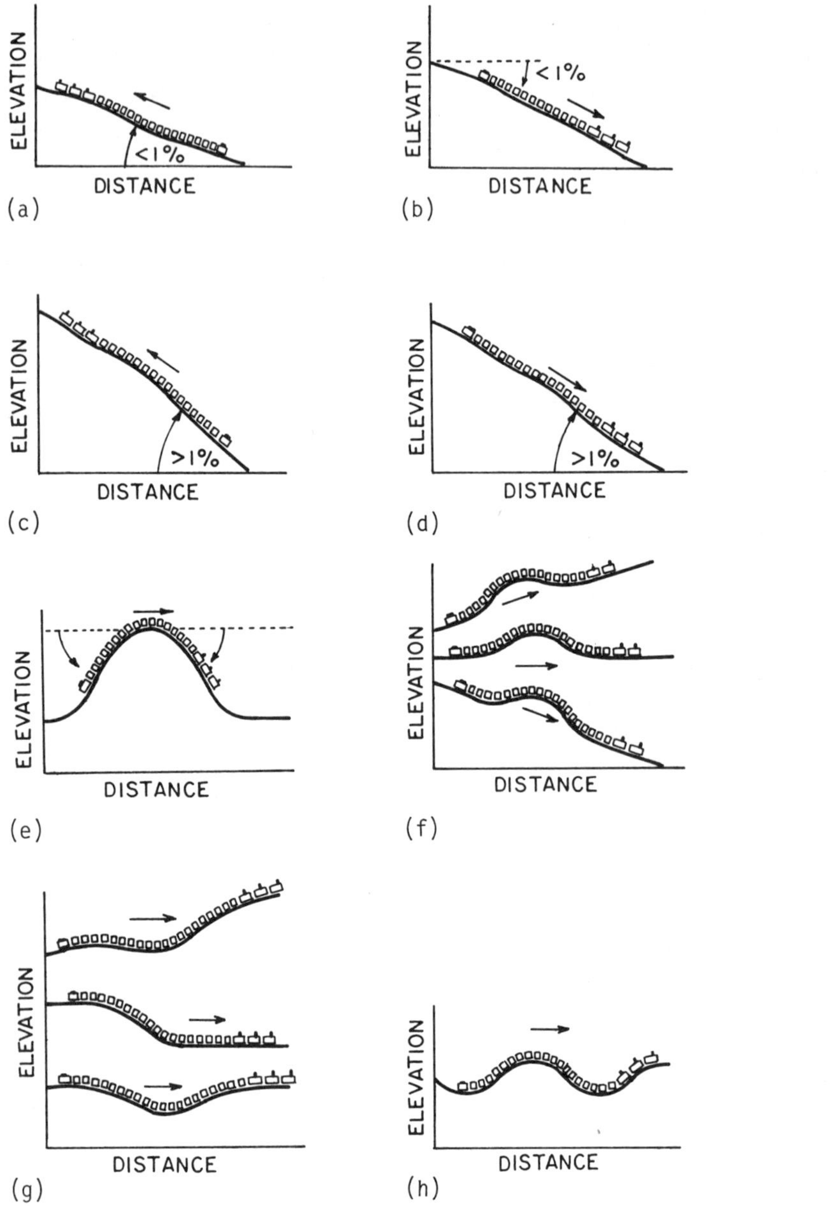

Fig. 2.4 (a) Light Ascending Grade, (b) Light Descending Grade, (c) Heavy Ascending Grade, (d) Heavy Descending Grade, (e) Cresting Grade, (f) Hump, Knoll or Hogback, (g) Sag or Dip, (h) Undulating Grade

due to the grade and decelerating forces due to train resistance and braking.

(c) Heavy Ascending Grade: An ascending grade greater than 1.0% is known as a heavy ascending grade (Fig. 2.4(c)). Train handling procedures are based on the assumption that such grades are long enough so that a balance can be established between decelerating and accelerating forces, as in light ascending grades.

(d) Heavy Descending Grade: A descending grade greater than 1.0% is called a heavy descending grade (Fig. 2.4(d)). Train handling procedures assume that such grades are long enough so that a balance can be reached between accelerating and decelerating forces, as in light descending grades.

(e) Cresting Grade: A long ascending grade which changes to a long descending grade is called a cresting grade, provided both grades are of sufficient magnitude to require a change in train handling procedures at the crest (Fig. 2.4(e)).

(f) Hump, Knoll or Hogback: These terms are used to describe a rapid increase in grade, followed by a decrease in grade sufficient to result in considerable coupler slack adjustment (Fig. 2.4(f)).

(g) Sag or Dip: A sag, or dip, is a descending grade followed by an ascending, level, or nearly level grade, the combination of which is sufficient to result in significantly more coupler slack adjustment than in most other grade territories (Fig. 2.4(g)).

(h) Undulating Grade: An undulating grade defines a track profile where grade changes are so frequent that an average train passing over it covers three or more alternating ascending and descending grades (Fig. 2.4(h)). The train slack is continually adjusting as cars on descending grades tend to move faster than those on ascending grades.

2.2.5 Track Irregularities

Four commonly used parameters to define track irregularities are the alignment, gauge, cross-level and vertical profile (Fig. 3.5). The alignment is the average of two lateral rail positions. The gauge is the lateral distance between the rails. The cross-level is the difference in elevation between two lateral rail positions. The vertical profile is the average elevation of two lateral rail positions. Track irregularities are defined as deviations of these parameters from nominal values.

In hot weather, expansion of continuous welded rail increases its tendency to buckle; therefore, lateral track instability due to thermal stressing is possible. A similar condition may occur in jointed rail when there is inadequate allowance for expansion. In cold weather, the continuous welded rail will contract, which tends to increase the lateral stability of the track but could result in pull-aparts. Lateral forces due to dynamic loading will sometimes aggravate track irregularities. Longitudinal forces in track may also result from rail, or ties, "running" due to insufficient rail anchors, or ballast, and indirectly contribute to lateral instability.

2.3 WHEEL-RAIL INTERACTION

The guidance of a railway vehicle is determined by the interaction between the wheelsets and rails. Thus, the mechanics of wheel-rail interaction is fundamental to the dynamics of railway vehicles.

Since the wheelsets are rigid bodies in solid contact with the rails, the kinematics of guidance is based on the geometry of wheelsets and rails. On the other hand, the dynamics of guidance depend on the wheel-rail interaction forces and moments developed as a result of the vehicle's forward velocity.

Here, we examine the kinematics of guidance, the steady state components of forces tending to cause derailment in curves, and the general wheel-rail interaction forces and moments, also called creep forces and moments.

The wheelset is the fundamental component common to all conventional rail vehicles. It normally consists of two wheels pressed firmly on a common axle, which is then fitted with two plain, or roller, bearings on the outboard side of the wheels (Fig. 2.5).

Since the wheels are not free to rotate independently, their treads are coned, or otherwise profiled, to allow them to negotiate

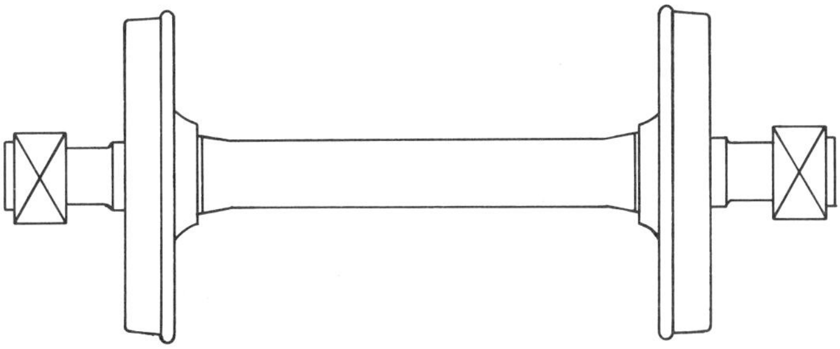

Fig. 2.5 Standard Wheelset

gentle curves without slipping. Large tangential forces, called creep forces, can be generated at the wheel-rail interface in tight curves, or whenever a wheelset is disturbed from its centered position on tangent track. These forces tend to steer and center the wheelset; but, they are also responsible for increased wear, energy loss and tractive effort, particularly on curves.

The creep forces can cause a wheelset, or truck, to describe a cyclic path on tangent track (Fig. 2.6). The frequency of this motion for a single wheelset is given by $\omega = V\sqrt{\lambda/(ar)}$, where V is the forward speed, λ is the coning angle (also called the conicity or taper ratio), r is the nominal rolling radius, and a is the semi-gauge of the rail. If there are no primary suspension

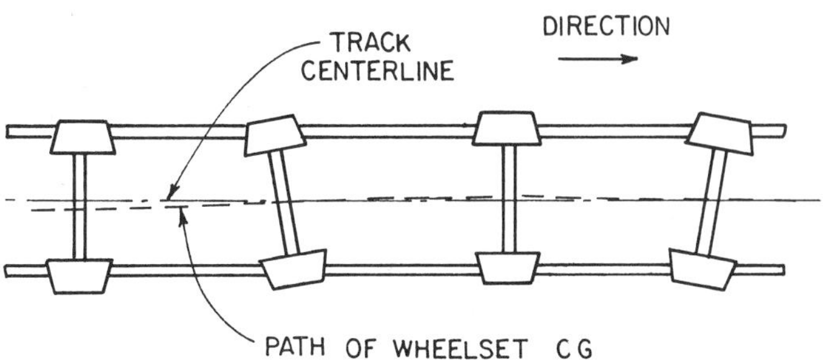

Fig. 2.6 Wheelset Kinematic Mode

elements connecting the wheelset to a truck frame, this oscillation can be self-sustained at relatively low speeds.

This phenomenon, commonly known as hunting, manifests itself first as a lightly damped response to track irregularities. Then, the magnitude of oscillation increases dramatically as a threshold speed, called the critical speed, is approached. The critical speed can be as low as 55 km/h (34.18 mph) with empty or light cars, and the resulting limit cycle (i.e., self-sustained oscillation) contributes to wear and fatigue of parts. The critical speed depends on several parameters, including suspension characteristics, weight distribution, rail surface contamination, effective conicity of wheelsets, and tractive or braking effort. In practice, track class and track geometry are often the dominant operational speed-limiting factors on tangent track. If it is good track, the engine crew will run the limit, oblivious to most truck hunting in the train's vehicles, except the lead unit.

In curve negotiation the leading wheel bears against the outside rail of the curve. The angle between the leading outer wheel and the outside rail is referred to as the "angle of attack". Figure 2.7(a) illustrates the relative position of a three-axle rigid-frame locomotive truck while traversing a curve. In practice, however, the axles can shift relative to one another due to the lateral freedom of motion built into the trucks. This allows the trailing axles to shift slightly toward the outer rail and reduce the angle of attack. As an illustration, Fig. 2.7(b) shows the relative position of the wheelsets of a standard two-axle freight car truck, with and without built-in lateral freedom, while traversing a curve.

Usually, the lengths of cars and locomotives in a train govern the maximum curvature that can be negotiated, but an additional constraint may exist with cars or locomotives having rigid-frame trucks of three or more axles. The maximum curvature that can be negotiated by such trucks, without excessive friction and binding, depends on their wheel base, the lateral freedom of their wheelsets and the track gauge.

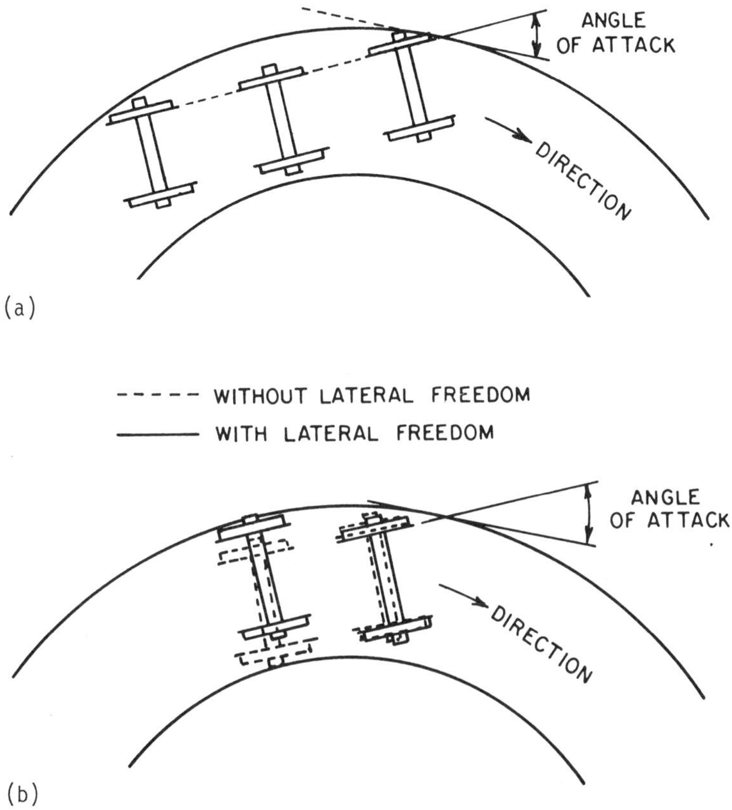

Fig. 2.7 (a) Three-Axle Rigid-Frame Locomotive Truck in Free Curving Position
(b) Standard Two-Axle Freight Car Truck

As a matter of interest, in passing, a wheelset in which at least one of the wheels is not rigidly fixed to the axle is known as an independently rotating wheelset. Studies have shown [4] that this type greatly reduces wheel tread and rail wear and virtually eliminates hunting. However, since the guidance mechanism in a truck using such wheelsets is reduced to only the gravitational restoring force of the coned wheels, bounded by the flanges, a small centrifugal force can make it run continuously against one rail in curves, thereby, increasing rail and flange wear and the proneness for derailment. Thus, they are not used in normal practice.

2.3.2 Curving Forces

The focus of this section is on the steady state frictional and centrifugal force components which tend to cause derailment in constant radius curves, at constant speeds, if the speeds are too large for the radii of curvature. The dynamically-varying components of wheel-rail forces, which can arise from track irregularities, hunting and train handling procedures, are considered in Sections 2.3.3 and 2.6.2.

Frictional forces occur between wheels and rails in curves, even when the centrifugal forces are exactly balanced by track superelevation. Longitudinal slip occurs because the outer wheels have a greater distance to travel around curves than do the inner wheels. Lateral slip occurs because the wheels of a truck may be travelling at a slight angle to the rails at the points of contact, thereby attempting to run off the rails, but are restrained by the wheel flanges which force them to slip sideways. This relative slip between wheels and rails, arising from kinematics, results in

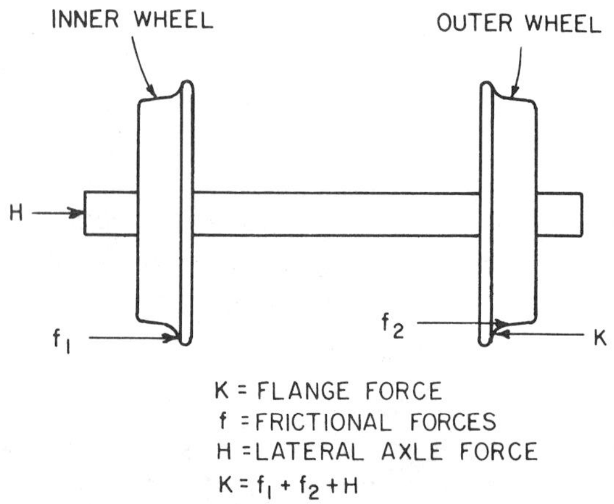

Fig. 2.8 Frictional Forces on a Wheelset

frictional forces which together with the dynamic lateral axle forces, are balanced by the flange forces, as shown in Fig. 2.8.

The relative magnitude of the lateral frictional forces developed for various degrees of curvature for a typical locomotive is shown in Fig. 2.9. It is seen that frictional forces increase sharply with curvature. Frictional forces are significantly increased when engine sand is used because the sand increases the coefficient of friction between wheel and rail.

The greater the angle of attack, the greater are the lateral frictional forces, and the more "scuffing" and wear there is between wheel flanges and rails. As frictional forces increase, the probability of wheel climb and lateral buckling or turnover of rails also increases. Although these kinematic forces would not normally cause excessive lateral forces, by themselves, critical force levels may be reached in the presence of additional dynamic forces such as those resulting from centrifugal accelerations, and train action at the couplers.

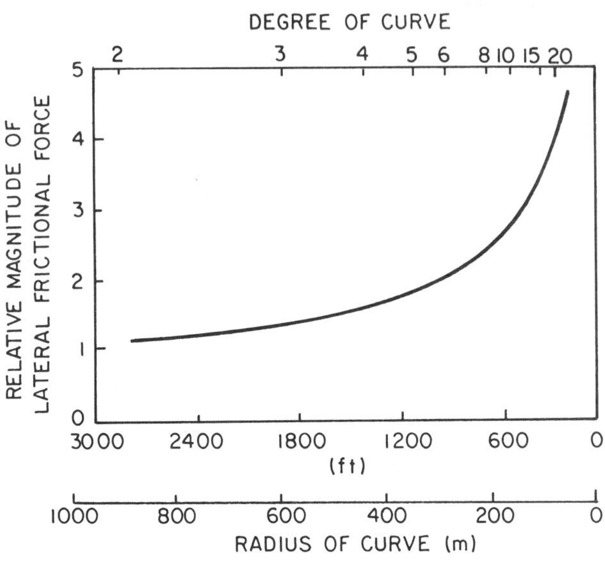

Fig. 2.9 Relationship of Lateral Frictional Forces to Curvature

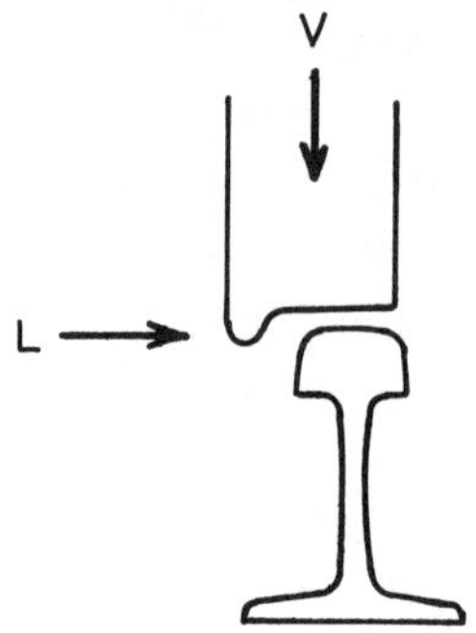

Fig. 2.10 Forces on Rail

The net force acting at a wheel-rail contact can be decomposed into longitudinal, lateral and vertical components, and the ratio of the latter two (Fig. 2.10) is referred to as the L/V ratio. In practice, large steady-state lateral forces are usually accompanied by large vertical loads tending to keep the L/V ratios below critical levels. However, large L/V ratios can occur transiently as a result of sudden, dynamic reductions in vertical load.

The L/V ratio is especially important in predicting wheel climb and rail turnover. The criticalness of a given L/V ratio value is a function of the prevailing dynamic conditions. For example, L/V ratio values that represent a problem in low-speed draft situations may not be the same as in high-speed buff situations. An L/V ratio value in the range (0.8, 0.9) is generally considered a minimum for wheel climb to be likely. Values in excess of double these have been observed but, because of their short duration, wheel climb did not occur. The duration of the occurrence is important. An accepted duration for the onset of wheel climb or rail turnover is approximately 0.3 seconds.

Large L/V ratio value of significant duration can occur when locomotives or cars bounce, pitch or roll. All vehicles have natural frequencies which, when excited by track irregularities, can cause large amplitude motions or instability at various speeds.

Bounce and pitch yield vertical motion whereas yaw and roll result in a swaying and rocking motion, sometimes called 'rock and roll'. Although harmonic rock and roll can occur at relatively low speeds, harmonic bounce and pitch are usually associated with speeds in excess of 80 km/h (50 mph).

Vertical bouncing is initiated by abrupt sags or humps, such as they occur at bridge ends, railroad crossings or soft spots in a track. The dynamic decreases in vertical wheel loading due to bounce may result in large L/V ratio values whereas the dynamic increases in loading cause wheel and rail wear. To minimize slack and prevent possible uncoupling or binding of cars, short, sharp vertical track variations should be avoided.

The rock and roll motion due to variations in cross level results in an alternating transfer of weight from one rail to the other. If the vertical load on one rail is decreased in this manner while the lateral load is not then the L/V ratio increases. The variations in cross level may be such as to cause reduced vertical loads on diagonally opposite wheels of a truck or car.

A widened track gauge permits greater truck skew. The resulting greater angle of attack increases lateral forces and the tendency for the leading wheel to climb the rail. This is especially so in curves, where the leading wheel normally is already exerting a heavy outward load on the outer rail. Although the lateral force may be abnormally high, the L/V ratio will remain within safe limits if the vertical wheel load is sufficiently high. However, if it is significantly reduced for any reason and for a sufficient duration, the resulting large L/V ratio may cause the wheel to climb the rail. This tendency may be compounded by the combination of a heavily worn outer rail and wheel profile in nearly new condition.

Track alignment irregularities also can cause lateral force and L/V variations. Thus, derailment due to an unfortunate combination of various irregularities is always possible on a poorly maintained track.

2.3.3 Creep Forces and Moments

The shape of the contact area between wheel and rail is approximately elliptical. The length of the major and minor axes and the distribution of the normal load between the bodies across the contact area can be determined from Hertz' theory for two elastic bodies in contact. The contact area can consist of two regions: an adhesion region of no interfacial slip, and a region in which there is interfacial slip.

Tangential (creep) forces and moments result from the tangential motion of the wheel relative to the rail in the contact region. A manifestation of this tangential motion is the relative difference in forward velocity (called creepage or, sometimes, creep) between an ideally rolling wheel, having no slippage in the contact region, and a practical one. Specifically, creep (or creepage) is defined to be the difference in velocity divided by the ideal (pure rolling) forward velocity [3]. Spin creep is the wheel's angular velocity about an axis normal to the contact region divided by the ideal forward velocity. The relationship between creep force and creepage, shown in Fig. 3.9(b) is qualitatively similar to the relationship between spin creep moment and spin creepage.

In this figure, we see that creepage consists of creep and pure slip domains. The creep domain corresponds to the values of creepage for which adhesion and slippage regions exist simultaneously within the contact region. As seen in the illustration below the graph, the slippage region trails the adhesion region in the direction of motion. The limiting cases are: (a) pure (or, ideal) rolling, characterized by no slippage and zero creep force, and (b) pure sliding, where there is no adhesion and the creep force is given by the product of the coefficient of friction and the normal force. Between these two limiting values, the creep force varies almost linearly with creep.

The longitudinal and lateral creep force components and spin creep moment are illustrated in Fig. 3.9(a). The longitudinal creep force acts in the wheelset's forward direction of motion, the lateral creep force acts perpendicularly to it in the lateral creep direction, while the spin creep moment acts in the normal load direction, orthogonal to the creep force components and, hence, the plane of the contact region. It has been shown by Kalker [10] that, in the linear creep domain, the longitudinal creep force is proportional to longitudinal creep, whereas the lateral creep force and spin creep moment are both functions of lateral and spin creeps. An extensive treatment of the problem of creep can be found in the literature [7-11].

When the relative rates of strain in the contact area (or the creepages) becomes too large, gross sliding of the wheel over the rail occurs and the creep forces and moments are no longer proportional to the creepages but are determined by the level of adhesion.

Adhesion is defined as the coefficient of friction between the wheel and rail in the longitudinal direction, for acceleration or deceleration. It is a direct indicator of the amount of turning force the wheel can impart on the rail before wheel slip occurs.

A locomotive, with good rail conditions, can normally transfer between 20% and 25% of its weight into tractive effort. However, the parameter values in mathematical models of creep are functions of surface contamination. For example, the magnitudes of creep coefficients estimated in experiments by Pearce and May [18] have varied from Kalker's predicted values by as much as fifty percent for contaminated surfaces. Thus, environmental conditions resulting in surface contamination with water, oil, dirt, snow, ice, etc., can have a significant effect on wheel-rail interaction forces and, therefore, vehicle dynamics.

Wear is considered to include the removal of material, due to plastic flow, from the wheel and rail at the contact areas. Creepage produces wear also. Material properties that play an important role in wear are hardness, toughness and thermal conductivity, but they must be considered in conjunction with load magnitudes. Unfortunately, the axle loads of rail cars have been increasing significantly during the last couple of decades.

2.4 TRUCKS

Conventional railway vehicles are comprised of a carbody supported by trucks (or bogies) at each end. Basically, trucks are frames which hold one or more wheelsets. Most conventional railway vehicles use dual axle trucks. Truck configurations can be classified into three broad categories: passenger trucks, freight trucks, and locomotive trucks.

2.4.1 Passenger Trucks

The main components of a conventional passenger truck (Fig. 2.11) are wheelsets, primary suspension elements, truck frame and secondary suspension elements [5,6]. The wheelsets are connec-

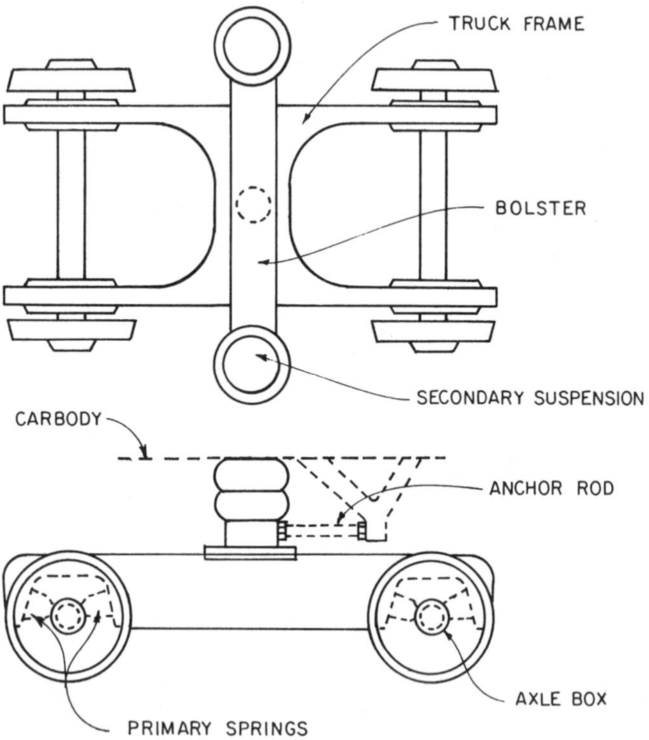

Fig. 2.11 Passenger Truck Configuration

ted to the truck frame (usually, an H-shape frame) through primary suspension elements. A truck is said to be equalized when the vertical wheel-rail interaction loads at its wheels are equal. To provide equalization, some trucks employ an equalizer beam which rigidly connects the wheelsets; then, the primary suspension elements are mounted between this beam and the truck frame which, in turn, supports the carbody through secondary suspension elements.

The primary suspension consists of coil springs with rubber nests and, in some designs, vertical dampers; or, coil springs with flexible links from axle boxes to the frame and vertical dampers; or, as in the Canadian LRC vehicle, Metalastic chevron-springs, which provide longitudinal, lateral and vertical stiffness, and vertical dampers; or, rubber rings between axles and sideframe, as in the Budd Pioneer III; or, an equalizer beam with coil springs mounted on it, as in the GSI Metroliner configurations. These elements allow wheelset motions relative to the truck frame and thereby assist in reducing vibration transmission to the carbody.

The secondary suspension, located between the truck frame and carbody, supports the latter either directly or through the bolster. It generally consists of springs and dampers between the truck bolster and carbody, friction which resists rotation at the centerplate and side bearings, and flexibility of side bearings and the bolster's anchor rod bushings. However, many configurations can be found in passenger trucks. For example, the Budd Pioneer III and GSI Metroliner use coil and air springs in parallel with hydraulic dampers for the vertical secondary suspension, whereas the Candian LRC, British BT 10 and Russian ER-200 use air springs and orifice, or hydraulic, dampers.

Lateral secondary suspension can be provided by swing links and hydraulic dampers, or lateral stiffnesses of vertical coil springs and lateral dampers. The Budd Pioneer III uses the lateral stiffnesses of vertical coil springs, lateral stabilizing rod bushings and lateral dampers. The lateral stiffnesses of coil and air springs and lateral dampers are used in the GSI Metroliner; or by rubber sandwiches and lateral dampers; combined shearing of traction and spring plank pads, in the Canadian LRC; lateral stiffnesses of air springs and lateral dampers, in the German ET-403, Japanese DT-200 and Russian DR-200.

The longitudinal secondary suspension generally consists of the longitudinal stiffnesses of coil, or air, springs and anchor rod rubber bushings. Other types of mechanisms can be found in secondary suspensions for other degrees of freedom; for example, anti-roll bars are included in the French Y-28, Italian Fiat-Eurofa, and German Minden Deutz. The French Y-32, Y-225 and Italian Fiat Eurofa includes yaw damping arrangements. The Budd Pioneer III has articulated sideframes, and the German ET-403 and Canadian LRC provide a tilting mechanism for the carbody.

2.4.2 Freight Trucks

Freight trucks [13] are less complex than passenger trucks. A typical North American freight truck is shown in Fig. 2.12(a). Conventional freight trucks are characterized by (1) truck frame flexibility, (2) an absence of primary suspension elements between

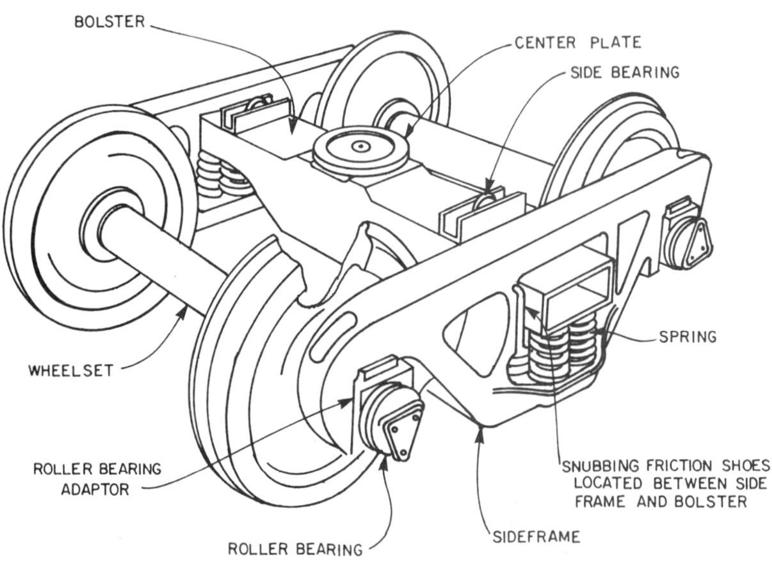

Fig. 2.12 (a) A Typical North American Freight Truck

the wheelsets and sideframes, and (3) the use of friction throughout the truck and in parallel with springs in the secondary suspension.

The frame of a typical freight truck consists of two sideframes and a bolster. The ends of the bolster rest on vertical spring groups which, in turn, rest in the sideframes. This three-piece frame can deform so that the wheelsets and sideframes can approximate the shape of a parallelogram. This type of deformation is known as "lozenging", "parallelogramming", or "warping". Warping is (a) resisted by friction between the bolster and sideframes, and the sideframes and wheelsets, (b) limited by solid contact between the bolster and sideframes, and (c) responsible for the yaw flexibility of wheelsets. The wheelsets of conventional freight tucks are connected to the sideframes through plain, or roller, bearings. Normally no resilient material exists between the bearings and sideframes. Thus, the soft primary suspension required for good curve negotiation, through wheelset yaw flexibility, does not exist in conventional freight trucks, but can be found in some European trucks.

Wheelset lateral motion is relatively unrestrained in plain, or journal, bearing boxes until the allowed clearance is taken up and the axle hits the stops. Thus, longitudinal and yaw motions of the wheelsets relative to the truck frame are virtually eliminated. On the other hand, roller bearings permit negligible lateral motion, but significant yaw motion, of the wheelsets relative to the truck frame.

As the two types of bearing constrain wheelset motions differently, the corresponding "truck hunting" motions are different. For roller bearing trucks, wheelset lateral and yaw motions are strongly coupled to truck frame motions, so that hunting usually involves large motions of the entire truck and carbody. In the case of plain bearing trucks, lateral motions of the wheelset can be large with respect to the truck frame; therefore, hunting usually produces large wheelset lateral motions and small motions of the truck frame and carbody.

The conventional secondary suspension consists of vertical spring groups in parallel with friction plates. The spring groups, which can "bottom", rest in the sideframes and support the bolster at its ends. The suspension forces tend to oppose lateral, vertical, and roll motions of the bolster relative to the sideframes.

The behaviour of the secondary suspension groups may be influenced by conditions at the bearing adaptor connections. Each sideframe is capable of a small amount of roll motion about those connections, which influences the behaviour of the secondary suspension. In the new condition, a point contact exists and allows the greatest amount of sideframe roll motion. With wear, the connections are flattened, thereby decreasing or eliminating sideframe roll motion entirely.

So-called "self-steering" trucks are obtained by interconnecting the wheelsets by a frame or yoke. Upon negotiating a curve, a yaw moment and lateral force are transmitted between the wheelsets of such a truck so that they tend to align themselves radially to the curve. Newland [17] showed that alignment of the wheelsets in this manner significantly reduces wheel slip and flange contact during curving. List [14] discussed the relative advantages and disadvantages of conventional and "self-steering" trucks on tangent and curved tracks. He concluded that self-steering trucks can improve both curving performance and stability. Interconnected wheelset designs were analyzed by Wickens [22], Scheffel [20], and List et al. [15]. Wickens analyzed the stability and curving performance of vehicles with wheelset interconnection. He characterized the interconnection by bending stiffness and shear stiffness, shown schematically in Fig. 2.12(b), and concluded that zero bending stiffness allows perfect curving but, without sufficient shear stiffness, the critical speed is zero. Scheffel analyzed the stability and curving performance of a pair of wheelsets interconnected by a lateral elastic restraint with negligible bending stiffness. He showed that this arrangement improves curving performance without degrading stability. List et al. [15] discussed the results of tests and analyses of several proposed

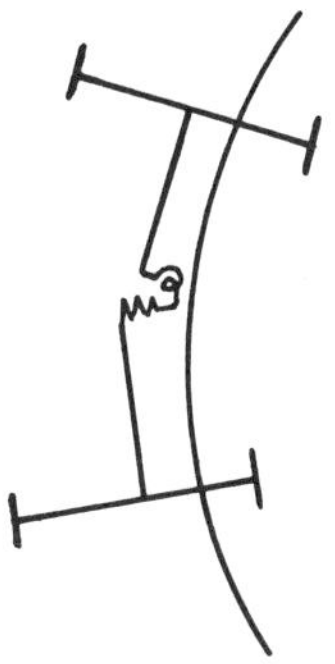

Fig. 2.12 (b) Interconnected Wheelset Curving Behaviour

truck designs with interconnected wheelsets. They claimed that wear, operating costs, derailments, truck hunting, and flange contact in curves could be reduced by these trucks. Several other designs have been proposed, including Scale's [19], and Scheffel's [21].

2.4.3 Locomotive Trucks

The locomotive truck supports the weight of the locomotive and provide a means for transmission of power to the rails. They are designed to withstand the stress resulting from road shock due to normal variations in the roadbed and other conditions encountered during operation. An important function of the truck assembly is to absorb and isolate these shocks so they will not be transmitted to the locomotive underframe and the equipment mounted on the underframe.

The General Motor's Electro Motive Division's (EMD) HT-C truck mainly used in the North American freight and Passenger train operations is shown in Fig. 2.13. The traction motors, located in the truck, convert electrical energy into locomotive tractive effort. The motors are geared to the driving axles which in turn apply force to the rail through the wheels. The driving force is transmitted to the truck frame by the axle journal boxes and from the truck frame to the bolster at the truck and bolster chafing plate interfaces. The bolster then transmits the force at the centre

Fig. 2.13 The General Motors' Electro Motive Division's (EMD) HT-C Truck

bowl to the carbody center bearing to provide the locomotive tractive effort.

The three traction motors are supported on their respective drive axles and at motor nose assemblies mounted on separate truck frame transoms. A main feature of the HT-C truck design is the orientation of the traction motors in one direction. This arrangement provides good motor accessibility and good adhesion characteristics.

The journal box (primary suspension) transmits the vertical load from the springs to the axle and provides a housing for the axle bearings. Each journal box is located between two pedestal jaws which are joined at the bottom by a pedestal tie bar. Renewable pedestal liners and journal box wear plates provide for control of clearances between the pedestals and journal boxes.

The locomotive carbody weight is tranferred to the truck frame through the "H" design bolster. Four truck mounted rubber spring pads (secondary suspension), located at the corners of the bolsters, provide the mating part between the truck and bolster. The primary suspension consists of twelve double coil journal springs; two located at each journal. A heavy duty shock absorber is mounted between the truck frame and each center journal box to provide damping of excessive carbody motions.

Lateral stops are provided at two locations on the bolster to limit the lateral movement between the truck and bolster. The bolster chafting plate mounting surfaces also function as stops for the vertical motion between the truck and bolster. Safety links, installed at three locations, provide a means of lifting the truck along with the carbody and prevent separation of the truck and bolster from the locomotive in case of derailment. Extreme truck rotation, in case of derailment, is prevented by stops welded to the underframe.

Air brake cylinders and brake rigging mounted on the trucks are used to apply retarding forces to the wheels to slow and stop the locomotive. Brake rigging systems are available which provide either one or two wheel tread brake shoes per wheel.

2.5 VEHICLES

The carbody of a freight vehicle is supported at the bolster centers through centerplate connections. These connections consist of a stub that is attached to the carbody and fits loosely in the centerplate, i.e., a shallow dish fixed to the center of the bolster. A "kingpin" is placed through holes in the centers of the stub and dish. The carbody may rock in the centerplate with little restraint from either the kingpin or the rim of the dish. Due to the small clearance allowed, the carbody may slide slightly in the dish until the stub contacts the rim.

In minor rocking situations, the carbody rolls in unison with the bolsters. But, as rocking increases, it begins to roll relative to the bolsters and eventually contacts the side bearings (Fig. 2.12). In extreme cases, the carbody can roll onto a bolster's side bearing with a sufficiently large moment to become separated from that bolster at its centerplate and fall off the truck. Carbody rocking is a nonlinear phenonemon due to friction at the centerplate connection, clearances between the centerplate stub and dish and clearance between the carbody and sidebearings.

Most dynamic analyses assume the carbody to be rigid. This approximation is usually valid for box cars and most passenger vehicles.

However, if a structural member of a rail vehicle is sufficiently flexible, coupling of its flexural modes with its rigid body modes can have a significant influence on vehicle dynamics. Thus, it is important to include the flexiblity of such members in dynamic analyses. The structural members of a conventional rail vehicle are the wheelsets, sideframes, bolsters and carbody. The flexural mode frequencies of bolsters and sideframes are usually large compared with their rigid body frequencies. On the other hand, the torsional modes of wheelsets and the flexural and torsional modes of certain carbodies may be strongly coupled with the rigid body modes.

There are about as many types of freight carbodies as there are types of rail cargoes (Fig. 2.14) and they have diverse flexural

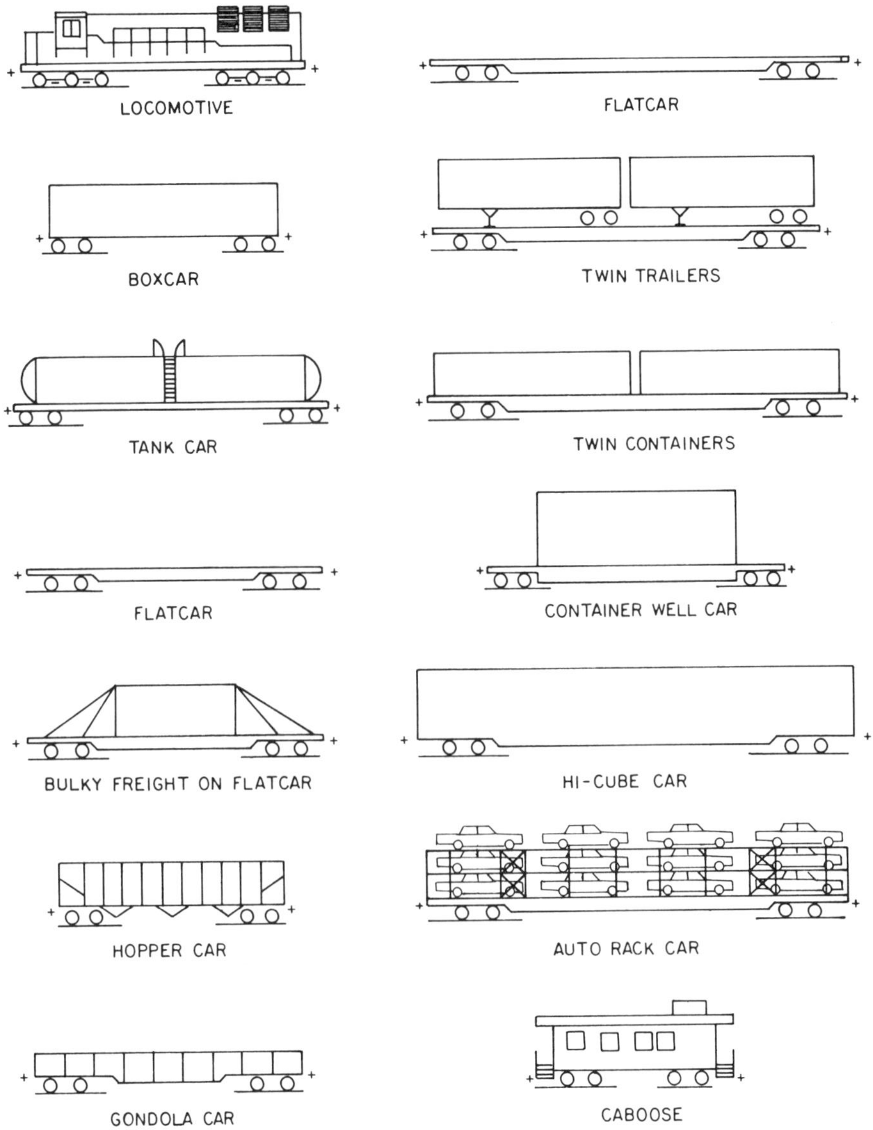

Fig. 2.14 Comparative Sizes and Shapes of Rolling Stock

properties [2]. For example, hopper cars (which carry loose bulk materials) and box cars (which carry solid bulk materials) have fundamental bending and torsional modes of lower frequencies than those of cylindrical tank cars (which are liquid carriers). Flat cars (which can carry large containers, automobiles, etc.) may be more flexible in torsion and stiffer in lateral bending than box, hopper or tank cars. The type and amount of loading of a carbody can also affect its flexural properties.

2.5.1 Harmonic Roll

Harmonic roll is the side-to-side rocking motion that shifts car weight alternately from one rail to the other. It can cause load shift, wear of truck components, or derailment due to wheel climb or lift. The following circumstances can contribute to harmonic roll [1]: (a) Truck center spacing approximately equal to the length of rail, laid with one-half stagger of joints, in combination with a series of low rail joints. Then a car encounters rail joints alternately with all wheels on one side and then on the other, as shown in Fig. 2.15. Each successive low joint excites

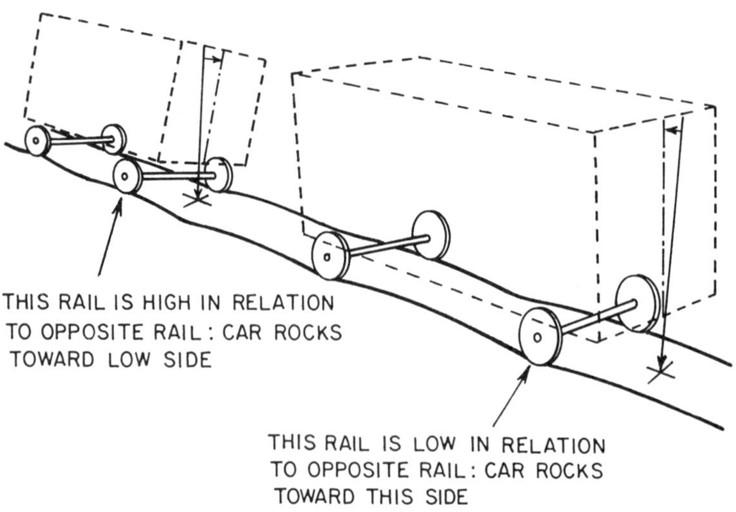

Fig. 2.15 Harmonic Roll

the carbody to rock through increasing roll angles, usually up to a finite limit; otherwise, there is derailment, (b) A rail car has a resonant frequency at which harmonic roll occurs readily. For example, on 39 foot rails, harmonic roll often develops at speeds of 19-29 km/h (11.81-18.02 mph) with loaded cars and 48.28-56.33 km/h (30-35 mph) with empty cars, (c) A car with high center of gravity [Fig. 2.16(a)] accumulates more energy in its rocking motion than a car with low center of gravity [Fig. 2.16(b)] and is, therefore, more likely to develop wheel lift, (d) A torsionally stiff carbody is more prone to wheel lift than a flexible one, (e) Worn, or broken, springs and harmonic roll damping devices can change the suspension characteristics of a car and, therefore, the severity of harmonic roll, (f) Harmonic roll derailments can occur when going from tangent to curved track, or vice versa. If the carbody rolls severaly enough to lift a wheel during such a transition, the track alignment can change sufficiently fast, due to the speed of the vehicle, that the wheel's flange can come down on top of the rail, (g) Travelling at less than equilibrium speed on a

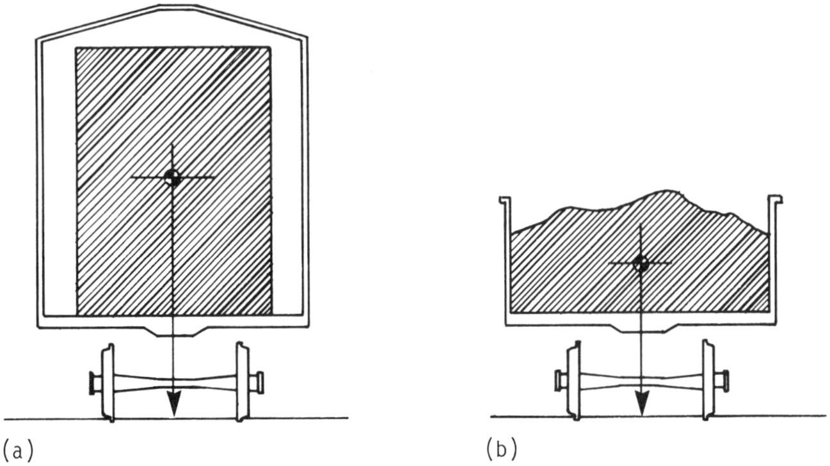

Fig. 2.16 (a) High Center of Gravity Car, (b) Low Center of Gravity Car

superelevated track unloads the wheels on the high rail and aggravates any wheel lift tendency due to harmonic roll, (h) Harmonic roll is more severe when a car is decelerating through its critical speed range than when it is accelerating through it.

2.5.2 Load Distribution

Eccentric loads create forces which can become hazardous when combined with the dynamic forces of a moving train. They affect the frame of the car in two ways: first, their weight acts vertically; second, a moment acts about the center axis of the car with a force equal to the weight of the load times the distance from the car's rotational axis.

Figure 2.17(a) illustrates a load concentrated on one side of a car and shows its tendency to rotate the car and cause wheel lift on the opposite side. A load of this nature seldom exerts sufficient force to cause wheel lift by itself but, when combined with

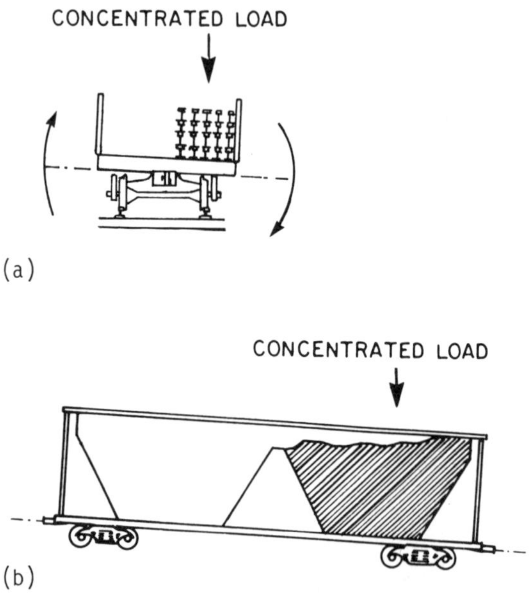

Fig. 2.17 (a) Load on One Side of Car, (b) Load on One End of Car

dynamic forces of train action, it may be sufficient to promote wheel climb or lift, particularly on superelevated track. Concentration of loads on one end of a car tends to unload the wheels on the opposite end of the car, as illustrated in Fig. 2.17(b) rendering them also susceptible to train action forces.

2.5.3 Centrifugal Forces

When a train goes around a curve it is subjected to horizontal centrifugal forces which act through the center of gravity of its cars and are directly proportional to the degree of curvature and the square of the speed. It was seen in Fig. 2.2 that, at equilibrium speed, the resultant of the weight and centrifugal force vectors is directed toward the center of the track, so that the load is distributed equally on the two rails as the car negotiates the curve.

Figure 2.18 shows the effects of centrifugal force when a car is travelling around a curve at overbalanced speeds, i.e., faster than the equilibrium speed. The resultant force acting through the center of gravity of the car is directed away from the center of the track toward the high rail. The L/V ratio on the high rail

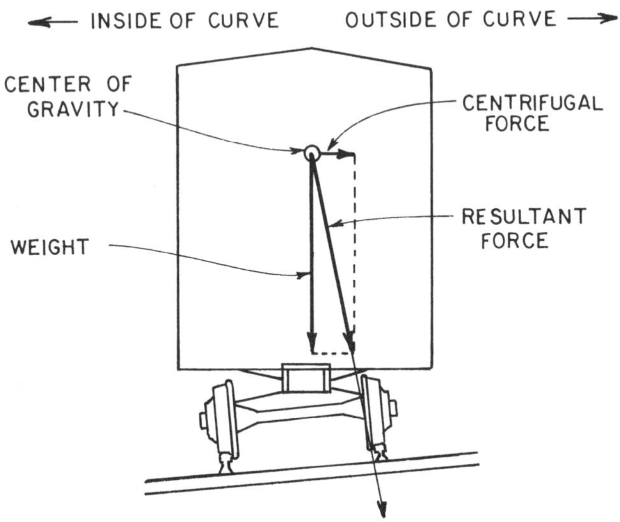

Fig. 2.18 Effects of Centrifugal Force

tends to increase with speed and may result in wheel climb or rail rollover.

When a car travels at underbalanced speeds, i.e., slower than the equilibrium speed, the resultant force is directed toward the low rail, as illustrated in Fig. 2.19, and there is an unloading of the high rail. A critical condition exists when the low rail carries the entire vertical load while the high rail is completely unloaded, signalling the onset of wheels lifting off the rail. At a given underbalanced speed, the resultant force falls farther from the track centerline as the height of the car's center of gravity is increased.

2.5.4 Stability and Guidance

Among the research efforts in the past decade are those which have centered on determining the details of the process by which guidance is achieved and on the stability problems which may arise in designing for good guidance. These stability problems are generally lumped under the term "hunting". Primary (or body) hunt-

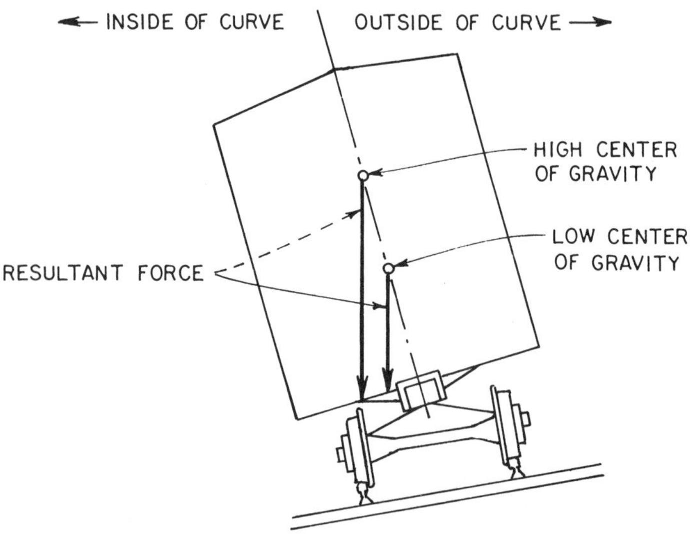

Fig. 2.19 Effects of Height of Center of Gravity at Low Speed on Elevated Curves

ing involves lateral, yaw, and rolling motions of the carbody. It often occurs at less than the critical speeds required for truck hunting. Secondary (or truck) hunting, which often occurs without primary hunting, is a yaw and lateral displacement oscillation of the wheelsets and truck frame. It is damped below a certain critical value of forward speed. Above the critical speed, the amplitude of oscillation increases until the truck is banging from rail to rail, limited only by the wheel flanges.

Truck hunting occurs on tangent track and it may occur with empty cars at speeds as low as 56 km/h (35 mph), in special cases, but normally does not begin until around 72-80 km/h (45-50 mph) with worn wheels, and 88-104 km/h (55-65 mph) with new wheels. Hard flange contact occurring at regular intervals of 9.14 to 15.24 metres (30 to 50 feet) has been observed. The critical speed is higher for locomotives and heavily loaded cars. Empty cars with roller bearings and worn wheels are very prone to hunting.

Truck hunting causes accelerated wheel and rail wear, fatigue and wear damage to car components such as centerplates, bolsters and sideframes, damage to lading of lightly loaded cars, gauge widening, loosening of track spikes, and disruption of ballast around tie ends. In extreme cases, it can cause derailments.

2.6 TRAINS

The movement of a train from its point of origin to its final destination involves several operations, including passage over various types of terrains, accelerations, deceleration and braking. All of these operations produce dynamic interactions between vehicles, and components within vehicles, as well as between vehicles and the track. The forces generated by these interactions are at the crux of the problem of safe, efficient and stable train operation. In this section, we introduce the basic train components, and types of equipment and procedures used in the operation of trains.

A train can be defined as a string of rail vehicles, including the propulsive unit(s), i.e., the locomotive(s), inter- connected by links, known as couplers (Fig. 2.20), through draft-gears (Fig. 3.28) or end-of-car cushioning devices (Fig. 3.29). The draft gears and end-of-car cushioning devices are the basic longitudinal energy-absorbing elements in a train. Thus, a train can be described as a multi-degree-of-freedom system, in which the vehicles are represented by masses, and couplers, draft-gears and end-of-car cushioning devices are represented by spring and damping elements.

Each individual car's brake system is activated by the locomotive engineer when he needs to stop the train. During braking, the kinetic energy of a train is dissipated as heat in the wheels and rails. In "dynamic braking", the locomotive's traction motors, which normally provide propulsion power, are used as electrical generators to dissipate the kinetic energy of the train as heat in on-board resistors.

Trains can be in "buff", or "draft", or a combination of these (Fig. 2.21). In the buff mode, vehicles are subjected to compressive forces; in the draft mode, they are subjected to tensile forces. Depending upon the terrain and the train's mass distribution, some parts may be in buff while others are in draft. Thus a train is a dynamic system of masses, springs, dampers and wheel-rail interactions. Its dynamic behaviour is a function of terrain and operating procedures as well as its constituent components,

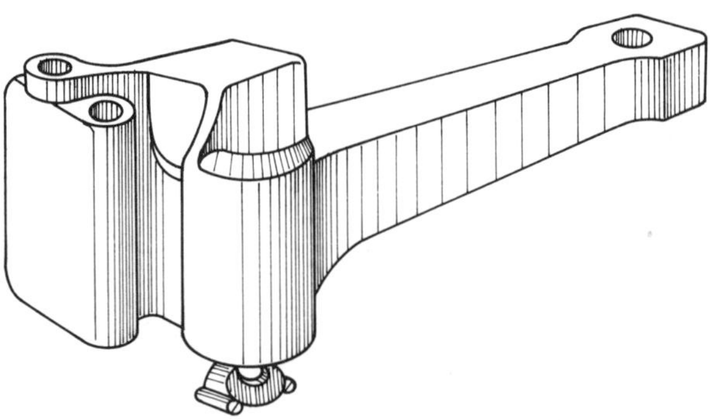

Fig. 2.20 A Typical Coupler

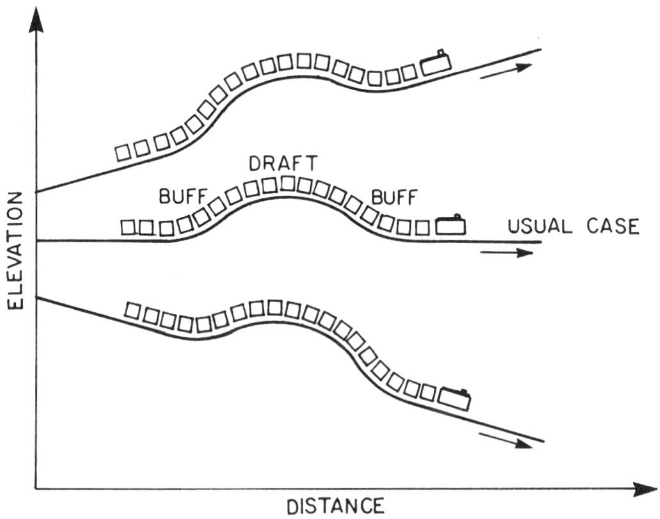

Fig. 2.21 A Train on Uphill, Level and Downhill Tracks Showing Draft (Tension) and Buff (Compression) Regions Along the Train Length

some of which are described below, e.g., couplers, draft gears, end-of-car cushioning devices, and braking and propulsion systems.

2.6.1 Train Components

Train length variations, which occur during normal operation, are accompanied by surges in coupler forces. When these variations are rapid, the resulting coupler forces may be large enough to break couplers and cause derailments. Coupler slack and the compliance of draft gears and end-of-car cushioning devices are the main causes of train length variations.

The coupler shown in Fig. 2.20 is one of several types [1,5,16]. Couplers are designed to allow for free motion or "slack" when they are connected. Coupler slack assists locomotives, in starting a train but may be dangerous, in the case of braking, accelerating or negotiating grades, if large run-in coupler forces are allowed to develop. It is useful in starting a train because, from rest, there is considerable journal resistance to overcome. Consequently, the engineman typically backs the engine, to bunch the cars, before starting forward. As the locomotive proceeds forward, only one car at a time is "broken away"; the cars are "picked-up" sequentially, starting with the one immediately behind the locomotive.

Draft gears are generally mounted in series with a car's underframe, so that the coupler forces are transmitted through them to the underframe. The maximum amount by which a draft gear can change its length is known as the draft gear travel. When this maximum travel is reached, a draft gear becomes a solid connection between the coupler and carbody. Various types of draft gear designs are available in the railroad industry, but they all use dry friction as the basic mechanism of energy dissipation; this is shown in the schematic diagram of Fig. 3.28.

Alternatives to draft gears are end-of-car cushioning devices. Similarly mounted, they also absorb longitudinal shocks. A typical schematic diagram is shown in Fig. 3.29. Energy dissipation is by viscous damping. Hydraulic shock absorber principles are used; mechanical energy is converted to heat when, under the action of coupler forces, liquid is forced through orifices in fluid-carrying cylinder-and-piston systems.

Sometimes a "sliding sill" is incorporated in the design of a vehicle for longitudinal shock absorption. It is a horizontal beam which underlays the vehicle body. The latter slides parallel to the former and is connected to it by spring and damper mechanisms. At each end of the sill are mounted an end-of-car cushioning device and coupler. Sliding sills do not contribute to changes in length of a train.

It is possible for a train to change violently from a state of extension to a state of contraction, and vice versa. For example, consider a train running up a hill; it is stretched by the force of gravity on each car while the locomotive pulls forward. Now, let the train run over the top of the hill and suppose the brakes are applied, in an emergency, when the rear car is on top of the hill. Assume, as is often true, that deceleration is greater for the locomotive than the cars; then, the change in state of the train, from extension to contraction, can be violent. Indeed, the coupler forces might be large enough to derail a light car.

Trains are usually equipped with an air brake system [1,5]. A schematic diagram of a typical system is given in Fig. 2.22. To apply the brakes, the locomotive engineer reduces the brakes pipe pressure by opening a brake pipe valve to atmosphere. A control valve on each car responds by permitting compressed air to flow from an auxiliary reservoir to the car's brake cylinder. The brake cylinder piston reacts to the compressed air by moving a linkage mechanism which, in turn, presses a brake shoe against each wheel tread. In order to release the brakes, (1) the brake pipe and auxilary reservoir on each car are again charged up, by locomotive power, and (2) the brake cylinder air is released to atmosphere, which has the effect of returning the brake cylinder piston back to its original position.

The actual braking effort exerted on each wheel depends upon the brake rigging, brake pipe pressure reduction and duration of

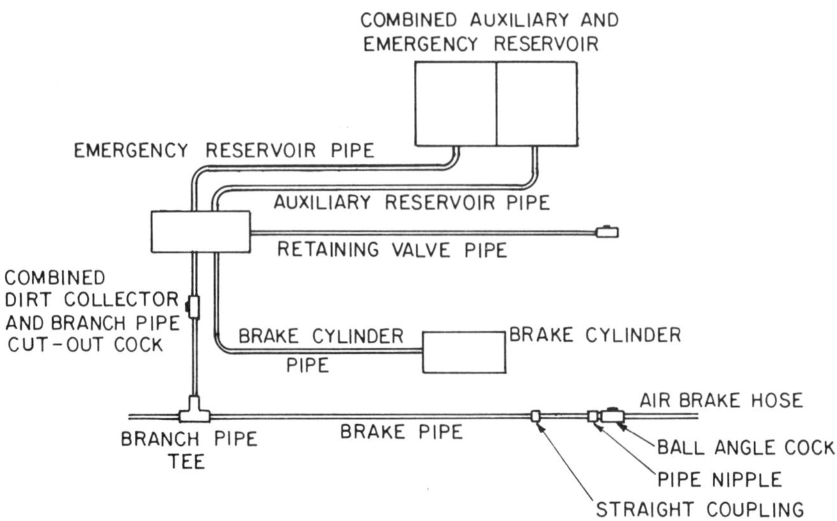

Fig. 2.22 Schematics of a Typical Air Brake System used in Trains

brake application. In an emergency, the maximum brake pipe pressure reduction is applied quickly, which produces the maximum brake cylinder pressures and braking efforts. The distance of a car from the location of application (usually the locomotive) of brake pipe pressure reduction can be important in any braking action. Since the pressure wave generated by venting the pressurized brake pipe to atmosphere travels at a finite speed, the cars have their brakes applied at different times, depending on their distances from the initial location of brake application. Thus, the analysis of braking systems and their operation is an essential part of train dynamics.

The tractive force in a train is provided by its locomotive(s). Although steam locomotives are still in use, diesel-electric and electric locomotives are becoming predominant. In diesel-electric locomotives, a diesel engine drives an electrical alternator, the output of which is full-wave rectified for use by the electrical traction motors geared to the wheelsets.

Although the basic function of a locomotive is to pull or push a train, it can also be used for dynamic braking. This is achieved by a commanded change in the electrical connections of the traction motors, to operate them as wheelset-driven generators connected to power resistors; thereby, some of the train's kinetic energy is dissipated as heat in the resistors. Typical tractive effort and dynamic braking characteristics are shown in Figs. 2.23 and 2.24, respectively. The higher throttle positions correspond to greater engine powers; therefore, the tractive effort increases with throttle position. The braking force capability is significantly inferior to the locomotive's ability to produce tractive effort, thereby asserting its basic function; nevertheless, it can assist the train's pneumatic braking system.

2.6.2 Dynamic Forces

Longitudinal forces, transmitted throughout a train, may be either tensile (draft) or compressive (buff). Ideally, when a train is in draft and on tangent track, all drawbar forces act vertically above track centerline and in a direction parallel to

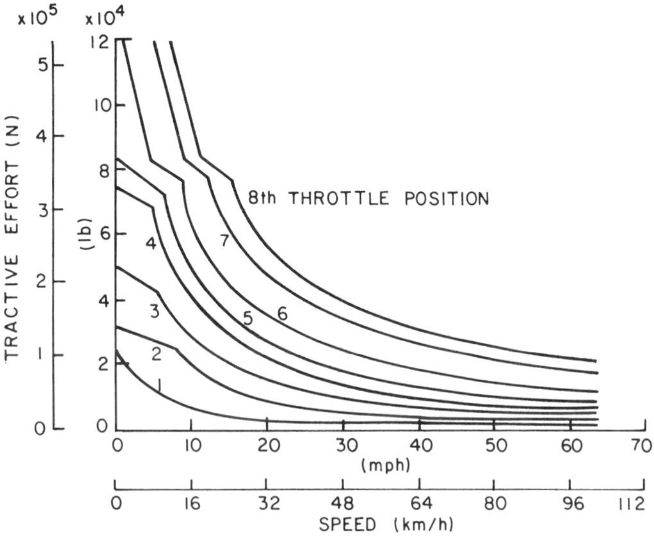

Fig. 2.23 Typical Tractive Effort Characteristics of a Locomotive

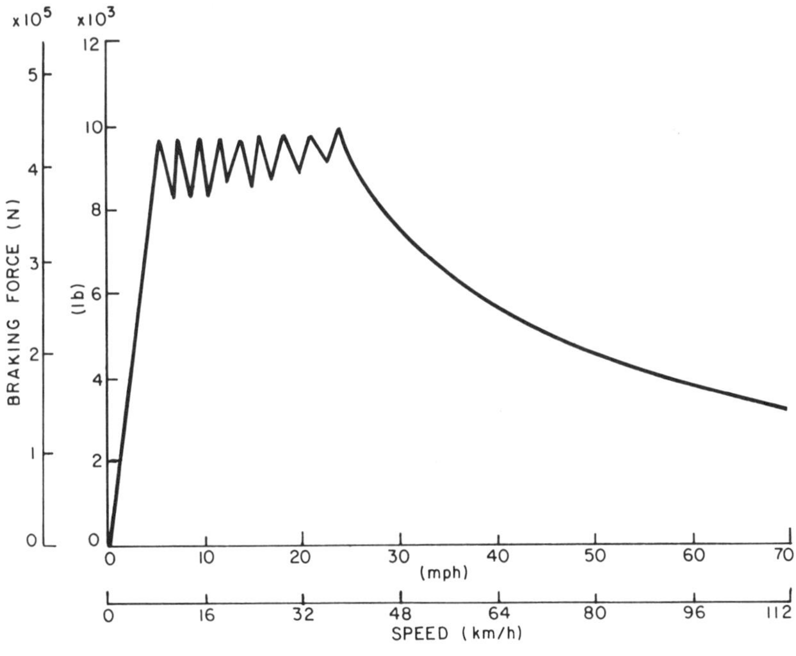

Fig. 2.24 Typical Dynamic Braking Characteristic of a Locomotive

it. Longitudinal forces are increased in the process of braking or accelerating. When there is a difference in velocity between cars (slack action) due to non-uniform braking or accelerating, longitudinal force variations are likely to increase substantially. Longitudinal forces are also influenced by grade; the forces due to gravity and, hence horsepower and braking requirements, increase with the magnitude of grades. The use, or misuse, of this installed tractive and braking power can result in large coupler forces.

The magnitude of lateral forces acting on a track is influenced by: irregularities in track surface, cross level, alignment and gauge, track superelevation, rail temperature, type of terrain, design of equipment, truck hunting, location of loads on car and center of gravity of equipment, train make-up, and train handling.

When a train is operating on curved track, the coupler forces, acting at related coupler angles, have lateral components which are transmitted to the rails. The magnitude of these lateral components increases with coupler force magnitude and angularity; the latter is directly related to the degree of curvature.

Long cars usually also have a long overhang beyond the truck bolster centers, which increases the coupler angle between cars. The coupler angle is often greatest when a long car is coupled to a short one. Draft forces act to stretch the train, thereby forcing the wheel flanges against the inside rail of the curve. Conversely, buff forces tend to cause the train to buckle outward and force the wheel flanges against the outside rail.

Slack action can produce large coupler forces and, therefore, lateral components, even on well-aligned tracks with good roadbeds. A locomotive engineer's ability to control a train is inversely related to the total amount of slack in the train. Under certain conditions, heavy slack action can occur in the middle of a train without being detected in either the engine or the caboose. A slack run-in can cause an increase in lateral force sufficient to turn a rail over. Even on tangent track, a heavy run-in can cause wheel climb or track shift.

The magnitude of dynamic lateral forces in track is influenced by abrupt changes in alignment, which are most likely to develop at turnouts, road crossings and when entering or leaving curves. Wide track gauges permit greater skewing of trucks, resulting in larger angles of attack between wheels and rails. This increases lateral forces and accelerates wheel and rail wear. Abrupt changes in gauge, such as may occur at joints, result in hard wheel flange contact and consequently large lateral forces of short duration.

For a given track with specified irregularities, the magnitude of wheel-rail force variations is a function of train speed. Vertical track irregularities cause vertical displacements of truck suspensions. As speeds increase, the maximum wheel load can reach levels in excess of twice that of the static wheel load. The resulting contact pressures between rail and wheel can sometimes stress the rail beyond its yield strength, causing corrugation, shelling, and/or progressive fractures. In curves, the force variations are also influenced by superelevation and its rate of change with distance, and by the location of the center of gravity of the vehicles. Under high buff or draft conditions weight transfer may occur due to the lateral component of drawbar forces, causing an increase in loading on the wheels of one side of a vehicle and a corresponding decrease on the other side.

2.7 Summary

Some physical aspects of railway tracks, vehicles and trains were examined. It was seen that the dynamic behaviour of a train, and the vehicles within it, depends on its speed and track geometry, among other factors. Superelevation, spirals, reverse curves, grade and irregularities are some of the relevant track parameters. The mechanics of guidance, involving wheel-rail kinematics, creep forces and spin creep moments, is fundamental to railway dynamics. Truck designs are different, in general, for passenger, freight and locomotive applications but, in any case, they have a strong influence on the guidance, stability and ride quality of vehicles. The eccentricity of a car's load can be detrimental to tracks, especially when the car is travelling at unbalanced speeds in curves. The vehicles of a train are interconnected by couplers; the forces transmitted through the latter may be damped by draft gears or end-of-car devices, possibly in combination with sliding sills. The magnitude of these forces depends on train operating procedures in conjunction with track geometries. In a curve, when changing states abruptly from "draft" to "buff", or vice versa, derailment is entirely possible.

REFERENCES

1. ANON. (1978): Track Train Dynamics - To Improve Freight Train Performance; *Association of American Railroads*, Report R-185, 2nd edn, Chicago, U.S.A.
2. ANON. (1980): The Car and Locomotive Cyclopedia; *Simmons-Boardman Publishing Corporation*, 4th edn, Omaha, U.S.A.
3. CARTER E.W. (1926): On the Action of a Locomotive Driving Wheel; *Proceedings of the Royal society of London*, Vol. 112, Series A, 151-157.
4. DUKKIPATI R.V., BOWLER E.H. and GARLICKI A.M. (1980): Dynamics of Independently Rotating Railway Wheel Systems - Past Experiences and Possible Future Exploitations; *Proceedings of the 5th Symposium on Engineering Applications of Mechanics*, Ottawa, Canada, June 1980, 259-263.
5. GARG V.K. and DUKKIPATI R.V. (1984): Dynamics of Railway Vehicle Systems; *Academic Press*, Toronto.
6. HEDRICK J.K., WORMLEY D.N., KAR A.K., MURRAY W. and BAUM W. (1979): Performance Limits of Rail Passenger Vehicles: Evaluation and Optimization; *U.S. Dept. of Transportation*, Report DOT/RSPA/DPB-50/79/ 32, Washington, D.C.
7. HOBBS A.E.W. (1967): A Survey of Creep; *British Railways Research Department*, Technical Note DYN/52, Derby, England.
8. JOHNSON K.L. (1958): The Effect of a Tangential Contact Force

Upon the Rolling Motion of an Elastic Sphere on a Plane; *Transactions of the ASME Journal of Applied Mechanics*, Vol. 25, 339-346.

9. JOHNSON K.L. (1964): Adhesion; *Proceedings of the Institution of Mechanical Engineers*, Vol. 178, Part 3E, 208-209.

10. KALKER J.J. (1973): On the Rolling Contact of Two Elastic Bodies in the Presence of Dry Friction; *Ph.D. Thesis*, Delft University of Technology, Delft, The Netherlands.

11. KUMAR S. (1980): Some Wheel-Rail Interaction Factors Influencing Vehicle Dynamics; *Rail International*, 599-610.

12. LAW E.H. and COOPERRIDER N.K. (1974): A Survey of Rail Vehicle Dynamics Research; *Transactions of the ASME, Journal of Dynamic Systems, Measurement, and Control*, Vol. 96, Series G, (2), 132-146.

13. LAW E.H., HADDEN J.A. and COOPERRIDER N.K. (1977): General Models for Lateral Stability Analyses of Railway Freight Vehicles; *U.S. Dept. of Transportation*, Report FRA-OR&D-77-36, Washington, D.C.

14. LIST H.A. (1971): An Evaluation of Recent Developments in Rail Car Truck Design; *IEEE-ASME Joint Railroad Conference*, Paper 71-RR-1, New York, U.S.A., 19-21 April 1971.

15. LIST H.A., CALDWELL W.N. and MARCOTTE P. (1975): Proposed Solutions to the Freight Car Truck Problems of Flange Wear and Truck Hunting; *ASME Winter Annual Meeting*, Paper 75-WA/RT-8, Dec. 1975.

16. MARTIN G.C. and TIDEMAN H. (1977): Technical Documentation-Detailed Longitudinal Train Action Model; *Association of American Railroads*, Report R-221, Chicago, U.S.A.

17. NEWLAND D.E. (1968): Steering a Flexible Railway Truck on Curved Track; *Transactions of the ASME, Journal of Engineering for Industry*, Vol. 91, Series B, (3), 908-918.

18. PEARCE T.G. and MAY B.J. (1969): A Study of the Stability and Dynamic Response of the Linear Induction Motor Test Vehicle; *U.S. Dept. of Transportation*, Office of High Speed Ground Transportation, Report FRA-RT-70-25, Washington, D.C.

19. SCALES B.T. (1972): Behaviour of Bogies on Curves; *Railway Engineering Journal*, Vol. 1, (4), 19-24.

20. SCHEFFEL H. (1974): The Hunting Stability and Curving Ability of Railway Vehicles; *Rail International*, February 1974, Vol. 5, (2), 154-176.

21. SCHEFFEL H. (1976): Self-Steering Wheelsets will Reduce Wear and Permit Higher Speeds; *Railway Gazette International*, Vol. 132, (12), 453-456.

22. WICKENS A.H. (1975): Steering and Dynamic Stability of Railway Vehicles; *Vehicle System Dynamics*, Vol. 5, (1-2), 15-46.

3

MATHEMATICAL FOUNDATION

3.1 INTRODUCTION

The purpose of this chapter is to present some pertinent aspects of railway vehicle dynamics as a guide in the development and use of analytical models and their computer implementations. Whereas only the form of models is stressed, a more complete treatment is made possible by numerous references to the literature.

The performance of rail vehicles is generally defined in terms of safety and productivity. Safety, as it relates to railway vehicles, is considered primarily in terms of derailment. Equipment failure, poor track conditions, train operation and train make-up as well as vehicle dynamics are the main sources of derailment. Derailment is concerned with those lateral vehicle motions which cause wheels to climb over the rail head or rails to roll over. The propensity for derailment is usually expressed in terms of the lateral (L) to vertical (V) wheel force ratio (L/V).

The productivity of a rail vehicle is a function of its revenue and operating, maintenance and capital costs. These factors are influenced by ride quality, vehicle speed, and wheel-rail forces which lead to wear and degradation of tracks and vehicles.

Safety and productivity, as they relate to vehicle dynamic performance, are expressed in terms of a set of indices dealing with specific aspects of dynamic performance. The most commonly-used performance indices for lateral stability, curve negotiation capability and ride quality are described in Section 3.2.

An all-inclusive dynamic model of a railway vehicle or train would be large and complex. Because size and complexity are factors which tend to reduce physical insight and increase the development and usage costs of computer programs, less general models are usually preferable. By addressing each performance index separately, instead of all of them together, it is possible to replace a large general model by a set of simpler specialized models.

Moreover, insight into the mechanisms underlying a whole vehicle's behaviour may be gained by investigating the individual behaviour of its trucks. In turn, the performance of a truck's wheelsets on various track layouts can help to understand its dynamics. Therefore, in this chapter, the mathematical aspects of modeling are treated in a building-block order, starting with the layout of tracks (Section 3.3); then, followed by the mechanics of wheel-rail interaction (Section 3.4), and the dynamics of trucks (Section 3.5), vehicles (Section 3.6) and entire trains (Section 3.7).

With the exception of the crucial wheel-rail contact forces, only the general form of analytical models is given. Furthermore, in addition to their pertinence as specific examples, several of the analytical models presented were chosen also because of their availability as computer programs for use in the case studies of Chapter 5. Included are parametric models of a passenger vehicle, a freight vehicle, a locomotive and a train. Although representative of such programs, they are not necessarily the best or the worst.

3.2 DYNAMIC PERFORMANCE INDICES

The investigation of the dynamic behaviour of vehicles and trains is generally classified into two groups as depicted in Fig. 3.1. They are:

(1) Response: the dynamic behaviour of the system due to external inputs such as track irregularities, and

(2) Stability: the stability of the system under various operating conditions.

In order to analyze rail vehicle dynamic performance effectively and to develop an analytical model appropriate for this analysis,

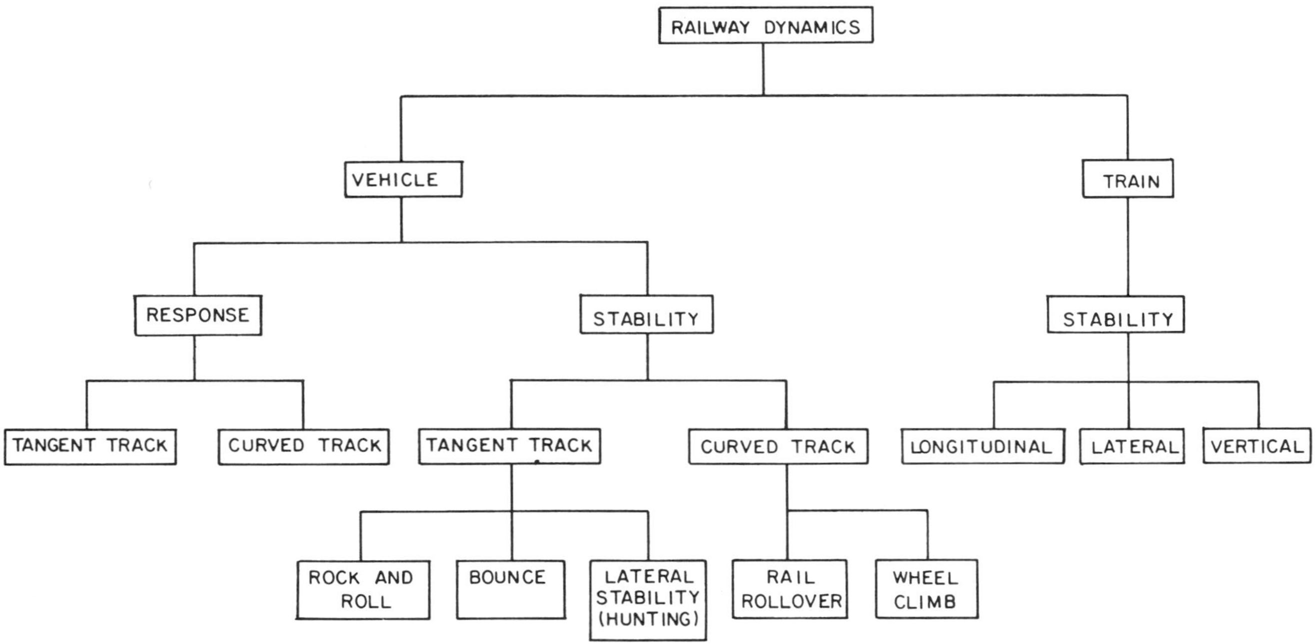

Fig. 3.1 Railway dynamics classification chart

it is important to define the performance indices quantitatively. Three major performance indices considered are: (a) vehicle lateral stability, (b) vehicle curve negotiation capability, and (c) vehicle ride quality. In addition, various design and dynamic constraints such as suspension stroke length, wayside clearance, wheel-load equalization and wheelset excursions etc., are also considered. The performance indices are briefly described as follows.

3.2.1 Vehicle Lateral Stability

Vehicle running stability or hunting is a phenomenon associated with conically shaped wheels. The conical wheels provide a lateral guidance or centering action which tends to force the wheels into the tread region away from the flanges. The friction or creep forces between the wheel and the rail are a function of the vehicle speed. As the speed increases, a value can be reached at which the vehicle-rail system has zero effective damping and a sustained periodic oscillation is set up. This state is called hunting and the speed at which it occurs is called the critical speed.

Rail vehicles exhibit several different modes of hunting, commonly classified as primary (carbody) hunting and secondary (truck) hunting. Carbody hunting usually occurs within a limited low-speed range and may involve relatively large lateral, roll and yaw motions of the carbody compared with the trucks. Truck hunting, which occurs at higher speeds, is strongly dependent on the lateral and yaw dynamics of wheelsets while being considerably less dependent on carbody dynamics.

Although a vehicle may be stable by operating well below critical speed, its response to a disturbance may nevertheless be underdamped. A dynamic system has as many natural frequencies and modes of vibration as the degrees of freedom. The general motion is the superposition of the modes. The differential equations of motion of a vehicle or a subsystem can be written in matrix form as

$$[M]\,\ddot{\underset{\sim}{y}} + [C]\,\dot{\underset{\sim}{y}} + [K]\,\underset{\sim}{y} = 0 \tag{3.1}$$

where [M], [C], and [K] are the mass and moment of inertia, damping and stiffness matrices respectively. $\underset{\sim}{y}$, $\dot{\underset{\sim}{y}}$ and $\ddot{\underset{\sim}{y}}$ are vectors representing the displacement, velocity and acceleration of each degree of freedom. These second-order differential equations are converted into a set of first-order differential equations and the solution of which will produce, in general, the complex eigenvalue of the form

$$\mu_i = \alpha_i + j\beta_i; \quad j = \sqrt{-1} \tag{3.2}$$

The natural frequency, ω_{ni}, damped frequency, ω_{di}, and damping ratio, ξ_i, are found from

$$\begin{aligned} \omega_{ni} &= [\alpha_i^2 + \beta_i^2]^{\frac{1}{2}} \\ \omega_{di} &= |\beta_i| \\ \text{and } \xi_i &= \frac{-\alpha_i}{\omega_{ni}} \end{aligned} \tag{3.3}$$

If the system is stable, ξ_i will be positive which implies α_i must be negative. The motion of an unstable vehicle is a divergent harmonic oscillation at frequency ω_{di}.

When the masses of a system are oscillating in such a manner that they reach maximum displacements simultaneously and pass their equilibrium points simultaneously, or all moving parts of the system are oscillating in phase with one frequency, such a state of motion is called *normal mode* or *principal mode of vibration.* To obtain a measure of the inherent damping in the principal modes of vibration, the damping ratios of the principal modes are computed for the operating speed range of the vehicle. The value of the least damped mode of vibration is used as a performance measure to reflect the need, not only to be stable, but also to have adequate damping in response to a disturbance. The higher the value of this performance measure, the more damped is the mode of vibration.

Nonlinearities such as flange contact introduce the possibility of limit cycle motion. Other nonlinearities such as Coulomb friction in the suspension, creep force saturation and suspension stops and clearances also produce amplitude-dependent vehicle dynamics. The limit cycle or hunting behaviour may be described in terms of a limit cycle amplitude vs speed plot such as shown in Fig. 3.2. Conceptually, a linear rail-vehicle system would have a unique critical speed above which the amplitudes of motion would diverge towards infinity, as shown by a vertical line A.

A system is stable about an equilibrium position if its motion, due to an arbitrary small initial disturbance, does not increase with time. In addition, if the resulting motion tends towards the equilibrium position with increasing time, the system is asymptotically stable.

The limit cycle, or self excited oscillation, is a nonlinear phenomenon. The solution of a differential equation of motion representing a vibrating system can be presented as a curve in a phase plane. Consider an autonomous system with single degree of freedom:

$$\ddot{x} + f(x,\dot{x}) = 0 \tag{3.4}$$

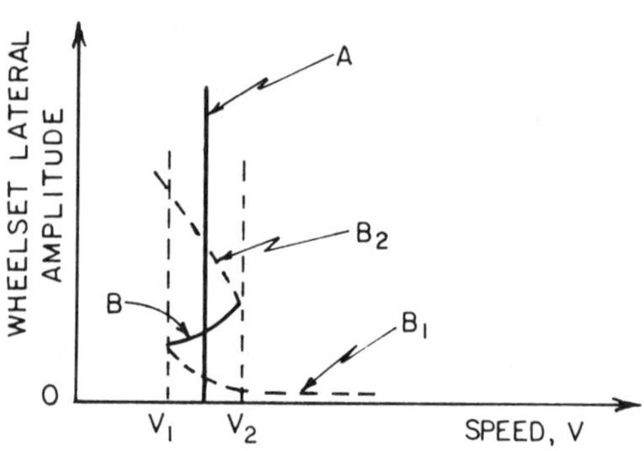

A : CONCEPTUAL LINEAR RAIL-VEHICLE SYSTEM

B : STABLE LIMIT CYCLE

B_1, B_2 : UNSTABLE LIMIT CYCLES

} PRACTICAL RAIL-VEHICLE SYSTEM

Fig. 3.2 Wheelset lateral limit cycle amplitude versus vehicle speed

This equation can be expressed as two simultaneous first-order differential equations:

$$\dot{x}_1 = x_2$$

$$\dot{x}_2 = -f(x_1, x_2) \tag{3.5}$$

or as a single equation in which time is implicit

$$\frac{dx_2}{dx_1} = \frac{-f(x_1, x_2)}{x_2} \tag{3.6}$$

A solution of the equation above can be portrayed in a *phase plane* with the coordinates $(x, \dot{x})$ or (x_1, x_2). The solution of the curve is called a phase trajectory. A family of trajectories in the phase plane gives a perspective of the general solution of the equation.

In practice, however, when the actual wheel-rail geometry and suspension nonlinearities are taken into account, limit cycle curves such as B, B_1 and B_2 are obtained. At speeds less than V_1, the system is asymptotically stable for all initial condition (IC) amplitudes. At speeds approaching V_1, the system is asymptotically stable for IC amplitudes less than unstable limit cycle curve B_2; otherwise, the response diverges to derailment. At speeds between V_1 and V_2 the system is asymptotically stable for IC amplitudes less than unstable limit cycle curve B_1. Initial conditions whose amplitudes fall between curves B_1 and B_2 result in oscillations that either grow or decay to the stable limit cycle amplitude defined by curve B, depending respectively on whether the IC amplitudes are below or above curve B. At speeds greater than V_2, the system is asymptotically stable provided IC amplitudes are below curve B_1; otherwise, the response diverges to derailment.

3.2.2 Curve Negotiation Capability

Most of the existing conventional rail vehicles achieve guidance on curved track through wheel-flange contact with the rail, resulting in poor ride quality and rapid wear of wheels and rails. In order to avoid flanging during curving, curve guidance must be achieved through creep forces. That is, the frictional forces generated at the wheel-rail interface must be sufficient to balance the net cant deficiency forces (centrifugal minus superelevation/ gravitational (see Section 3.4) forces). Thus to achieve curve negotiation using creep guidance, the following two conditions must be avoided: (1) wheel slippage, or the condition where the required friction force exceeds that available, and (2) flange contact due to excessive wheelset excursions. Curving performance is measured by calculating the maximum degree curve a particular vehicle design can negotiate without slipping or flanging.

Track damage can be caused by rail roll-over and lateral track shift. The ratio of the lateral-to-vertical wheel force (L/V) is usually used as an indicator of rail roll-over. Experimental data indicates that rail roll-over can occur with (L/V) ratios of 0.5 (for wood ties with old spikes) to 1.4 (for concrete ties) [see Garg and Dukkipati [15]].

Lading damage indices are often expressed by the shock spectra in the freight carbody. Graphs of force magnitude versus number of exceedances per mile are generally used to evaluate the relative vibration isolation provided by various freight trucks.

Vehicle and track maintenance costs are important economic indices that can be related to engineering indices such as hunting speeds and curving performance. If a vehicle's hunting speed is within its operational speed range, wheel and track wear will be accelerated. Wheel and track wear will also be accelerated if excessive flange contact occurs during curve negotiation.

Quantitative indices which are used for evaluating vehicle performance in curves include:

- truck-hunting critical speed
- carbody-hunting critical speed
- minimum radius of curvature negotiated without gross slippage
- maximum lateral-to-vertical wheel force ratio during curve entry and exit
- lading damage
- resistance to rail roll-over
- wheel load equilibrium

The dominant safety consideration in a rail vehicle is derailment. The tendency to derail is related to the "derailment quotient" or the ratio of the lateral-to-vertical track force (L/V) at the wheel/rail contact point. The different types of derailment criteria that are considered in practice include [1]:

- wheel climb criteria
- rail roll-over criteria
- track lateral shift criteria

Which derailment mechanisms are critical depends upon wheel/track conditions, for example, the type of ties used and the condition of the spikes [14,15,31,39].

3.2.3 Vehicle Ride Quality

The passenger's perception of ride comfort is a subjective evaluation of many factors including vibration, noise, temperature, and passenger compartment features. In most dynamics-based investigations on ride quality, the dynamic vibration environment is considered primarily, with emphasis on the vibration resulting from excitation of the vehicle due to track irregularities. Many quantitative descriptors have been used to describe rail passenger ride quality (see Hanes [16]). Two extensively-used measures are: (a) the root mean square (rms) acceleration levels, and (b) the distribution of the rms acceleration in one third octave frequency bands. They are described in the International Standards Organization (ISO) guidelines [5] for vertical and lateral motion and by Leatherwood and Dempsey [30] for roll motion.

Figure 3.3 illustrates the ISO recommended limits for the vertical and lateral 1 hour reduced comfort boundaries. When the vehicle vibration environment is a broad band, i.e., has a frequency content over a wide range, the recommended procedure is to compute the rms acceleration in 1/3 octave frequency bands and to compare it to Fig. 3.3 at the 1/3 octave center frequency.

Figure 3.4 shows roll equal discomfort curves where DISC = 1 corresponds to the threshold of discomfort and DISC = 2 provides twice the level of discomfort as DISC = 1. While these criteria are not exact, they are generally accepted useful indicators of

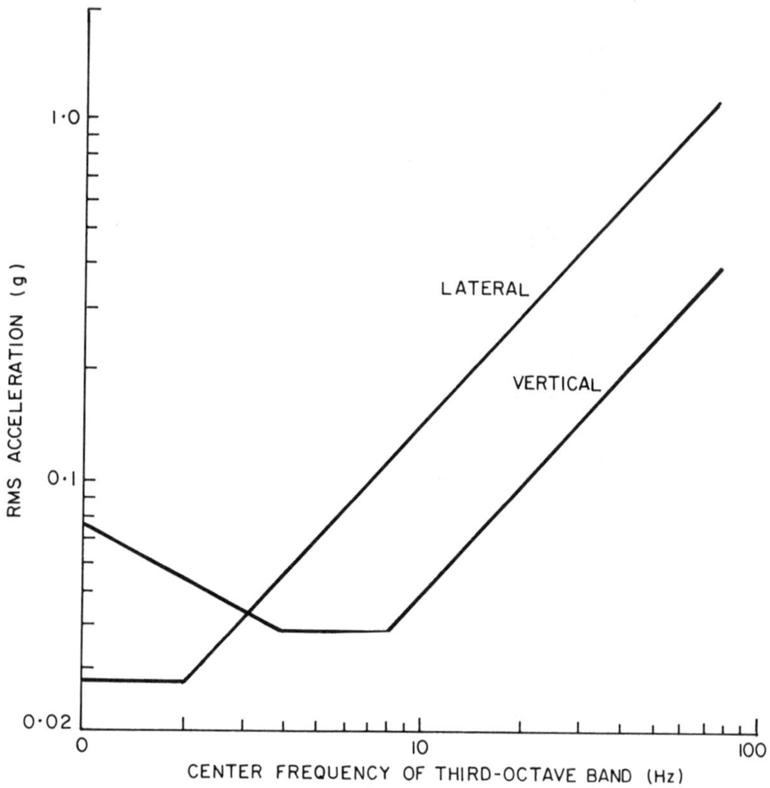

Fig. 3.3 Lateral and vertical acceleration limits as a function of frequency for 1 hour exposure time: "Reduced comfort boundary" (from Ref. [5])

general passenger acceptance levels. Although parametric study data are usually presented in terms of these criteria, they can also be expressed in terms of alternate criteria, such as the Sperling Ride Index [15] and the British Railways Ride Index [8].

In a freight vehicle, the ride vibration environment of primary importance is that of goods being transported. The diversity of goods carried in vehicles is so great, however, that it is not practical to characterize them individually and to develop explicit damage-avoidance ride vibration standards, analogous to the ISO passenger comfort standards. Nevertheless, the U.S. Federal

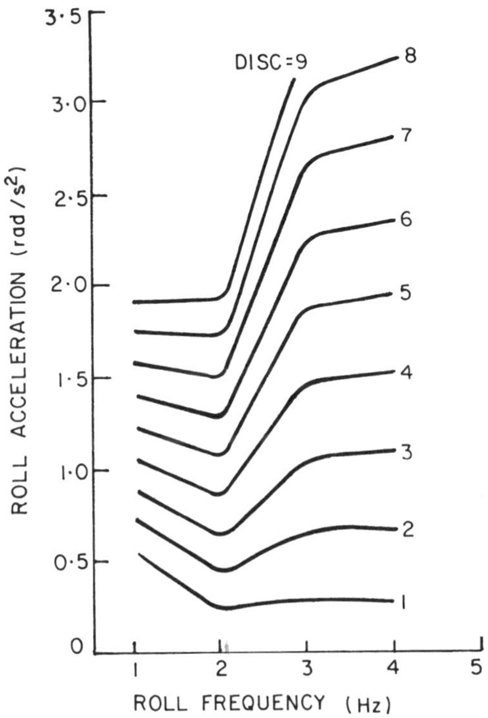

Fig. 3.4 Roll equal discomfort curves (from Ref. [30])

Railroad Administration (FRA) has proposed acceleration limits for trailers and containers mounted on rail cars. Specifically, the resultant acceleration from three orthogonal accelerometer measurements at any locations on the floor of a trailer or container is not to exceed 1.0 g during constant speed travel at any speed up to and including the maximum allowable speed for each class of track. Also pertaining to a trailer or container floor, the rms values of vertical and lateral acceleration, respectively are not to exceed 0.5 g and 0.1 g in the frequency range of 0 to 30 Hz under the same constant-speed operating conditions on all classes of track. Furthermore, resonances up to 30 Hz are required to have a damping ratio of not less than 0.2.

3.3 TRACK MODELS

Geometrical track irregularities, arising from initial installation tolerances as well as degradation through usage, are important external disturbances in the dynamics of rail vehicles. An accurate representation of these irregularities and track compliance is essential in predicting the dynamic behaviour of rail vehicles. Since track segments can be categorized conveniently as either tangent or curved, this breakdown is adopted in the following discussion of track modeling [4].

3.3.1 Tangent Track

The four geometrical parameters illustrated in Fig. 3.5 are commonly used to define tangent track irregularities. They are the alignment, gauge, cross-level and vertical profile. *Alignment* is the average of the two rail lateral positions. *Gauge* is the horizontal distance between the two rails. *Cross-level* is the difference between the elevations of the rails. *Vertical profile* (sometimes called 'surface') is the average of the two rail elevations. *Twist* defined as the variation in cross-level over a fixed distance, is a derived quantity which provides a measure of the rate of change of cross-level. The geometrical parameters are expressed as follows:

Gauge: $y_g = y_\ell - y_r$ (3.7a)

Cross-level: $z_c = z_\ell - z_r$ (3.7b)

Alignment: $y_a = \left(y_\ell + y_r\right)/2$ (3.7c)

Vertical profile: $z_v = \left(z_\ell + z_r\right)/2$ (3.7d)

where subscripts ℓ and r respectively denote the left and right rail, with y the lateral distance and z the vertical distance. Tangent track irregularities are defined as deviations of these parameters from nominal values.

Railroad track has not only static irregularities, but also dynamic irregularities caused by the vehicle passing over it. It has been shown that track flexibility can be modeled fairly accurately by a pure compliance at low frequencies and a one degree of

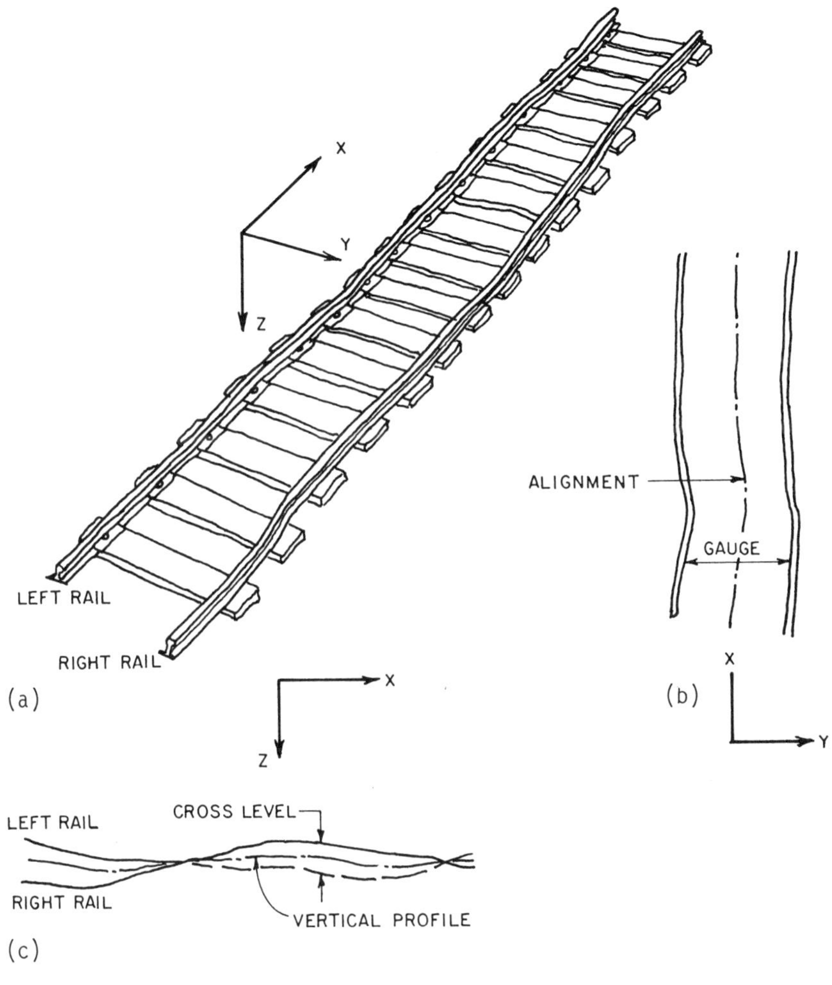

Fig. 3.5 Tangent track irregularities, (a) Track layout, (b) Gauge and alignment, (c) Cross-level and vertical profile

freedom spring-mass system for higher frequencies. On the other hand, each static irregularity may be classified as periodic, aperiodic or random.

(a) Periodic irregularities: The periodic components observed in measured vertical spectra are due to jointed rails which are subjected to rail end batter. The ends of each rail segment are flattened and driven down by impact forces occurring at the joints. For staggered joints shown in Fig. 3.6, the shape of each rail may be approximated roughly by a rectified cosine wave (RCW), i.e.,

$$z_r = B|\cos u| \tag{3.8a}$$

$$z_\ell = B|\cos(u - \tfrac{\pi}{2})| \tag{3.8b}$$

$$u = \Omega_p \, x/2 \tag{3.8c}$$

$$\Omega_p = 2\pi/L_R \tag{3.8d}$$

where B is the amplitude of the RCW (m), L_R is the rail length (m), x is along-track distance (m), and Ω_p is the wave number (rad/m). The Fourier series equivalent of Eq. (3.8a) is [40]:

$$z_r = \frac{2B}{\pi}\left[1 + \frac{2}{1\times3}\cos 2u - \frac{2}{3\times5}\cos 4u + \frac{2}{5\times7}\cos 6u - + \ldots\right] \tag{3.9a}$$

By substituting $\left(u - \frac{\pi}{2}\right)$ for u in Eq. (3.9a), we obtain

$$z_\ell = \frac{2B}{\pi}\left[1 - \frac{2}{1\times3}\cos 2u - \frac{2}{3\times5}\cos 4u - \frac{2}{5\times7}\cos 6u - - \ldots\right] \tag{3.9b}$$

Applying Eq. (3.8c) to Eq. (3.9) and performing coordinate transformations to eliminate the constant terms, we get

$$\begin{aligned} Z_r &= z_r - \frac{2B}{\pi} \\ &= \frac{4B}{\pi}\left[\frac{1}{1\times3}\cos \Omega_p x - \frac{1}{3\times5}\cos 2\,\Omega_p x + \frac{1}{5\times7}\cos 3\,\Omega_p x - + \ldots\right] \end{aligned} \tag{3.10a}$$

$$\begin{aligned} Z_\ell &= z_\ell - \frac{2B}{\pi} \\ &= \frac{4B}{\pi}\left[- \frac{1}{1\times3}\cos \Omega_p x - \frac{1}{3\times5}\cos 2\,\Omega_p x - \frac{1}{5\times7}\cos 3\,\Omega_p x - - \ldots\right] \end{aligned} \tag{3.10b}$$

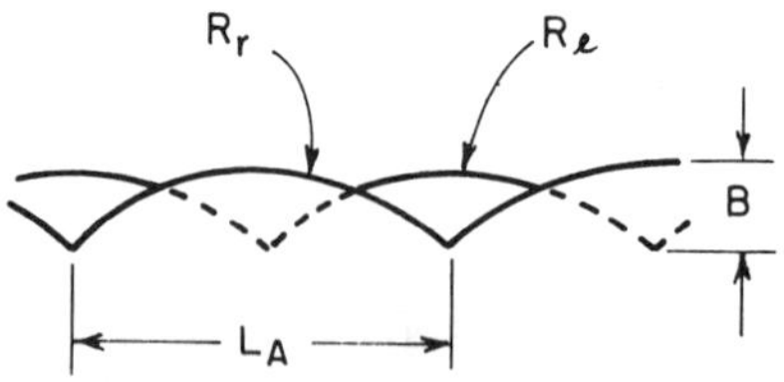

Fig. 3.6 Periodic vertical irregularity of left and right rails

Applying Eqs. (3.10) to Eq. (3.7d), the vertical profile variations are given by:

$$Z_v = (Z_r + Z_\ell)/2$$

$$= \sum_n A_n \cos n\,\Omega_p x, \; n = 2, 4, 6, \ldots \qquad (3.11a)$$

where

$$A_n = \frac{-4B}{\pi(4n^2-1)} \qquad (3.11b)$$

Likewise, using Eq. (3.7b), the cross-level variations are:

$$Z_c = Z_\ell - Z_r$$

$$= -2 \sum_n A_n \cos n\,\Omega_p x, \; n = 1, 3, 5, \ldots \qquad (3.12)$$

It can be shown [41] that the spectral density of a cosine function $v(x) = K \cos Mx$ is $S_v(\Omega) = (\pi/2)\, K^2\, \delta(\Omega - M)$, for $\Omega \geq 0$, where K is amplitude, Ω and M respectively are variable and constant wave numbers, and $\delta(\Omega - M)$ is the Dirac delta function at $\Omega = M$. The mean-square value of v(x) is $\overline{v^2} = K^2/2$. Furthermore, the spectral density of a sum of cosine functions is equal to the sum of their spectral densities, and a similar statement holds for the mean-square value.

Therefore, the spectral densities of the vertical profile variations [Eqs. (3.11)] and the cross-level variations [Eq. (3.12)] are:

$$S_v(\Omega) = \frac{\pi}{2} \sum_n A_n^2\, \delta(\Omega - n\Omega_p), \; n = 2, 4, 6, \ldots \qquad (3.13)$$

and

$$S_c(\Omega) = 2\pi \sum_n A_n^2\, \delta(\Omega - n\Omega_p), \; n = 1, 3, 5, \ldots \qquad (3.14)$$

If we assume that the alignment is a pure RCW, then the alignment variations are:

$$Y_a = y_a - \frac{2C}{\pi}$$

$$= \frac{4C}{\pi} \left[\frac{1}{1\times 3} \cos \Omega_p x - \frac{1}{3\times 5} \cos 2\,\Omega_p x + \frac{1}{5\times 7} \cos 3\,\Omega_p x - + \ldots\right] \qquad (3.15)$$

where C is the amplitude of the RCW. The corresponding spectral density is:

$$S_a(\Omega) = \frac{\pi}{2} \sum_n B_n^2 \, \delta(\Omega - n\Omega_p), \; n = 1, 2, 3, \ldots \tag{3.16a}$$

where

$$B_n = \frac{4C}{\pi(4n^2-1)} \tag{3.16b}$$

The mean-square values of vertical profile, cross-level and alignment variations are:

$$\overline{Z_v^2} = \frac{1}{2} \sum_n A_n^2, \; n = 2, 4, 6, \ldots \tag{3.17}$$

$$\overline{Z_c^2} = 2\sum_n A_n^2, \; n = 1, 3, 5, \ldots \tag{3.18}$$

and

$$\overline{Y_a^2} = \frac{1}{2} \sum_n B_n^2, \; n = 1, 2, 3, \ldots \tag{3.19}$$

For parallel rail-joints, it follows that $Z_\ell = Z_r$, where Z_r is given by Eq. (3.10a); therefore, we have:

$$Z_v = Z_r \tag{3.20}$$

$$Z_c = S_c(\Omega) = \overline{Z_c^2} = 0 \tag{3.21}$$

$$S_v(\Omega) = \frac{\pi}{2} \sum_n A_n^2 \, \delta(\Omega - n\Omega_p), \; n = 1, 2, 3, \ldots \tag{3.22}$$

and

$$\overline{Z_v^2} = \frac{1}{2} \sum_n A_n^2, \; n = 1, 2, 3, \ldots \tag{3.23}$$

(b) Aperiodic irregularities: Such occasional special track structures as switches and frogs can be simulated by using measured track geometry as the input. In addition, several aperiodic irregularities have been identified as being amenable to representation by analytical expressions [15]. The cusp, bump, jog, plateau, trough, sine cycle and damped cosine, described in Table 3.1, are among them.

Table 3.1. Aperiodic Track Irregularities (Adapted from Garg and Dukkipati [15]).

NAME AND FORMULA	SHAPE

CUSP

$y = Ae^{-k|x|}$

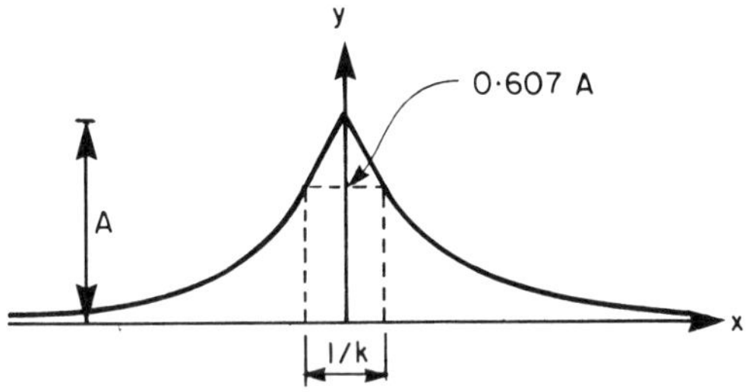

BUMP

$y = Ae^{(-1/2)(kx)^2}$

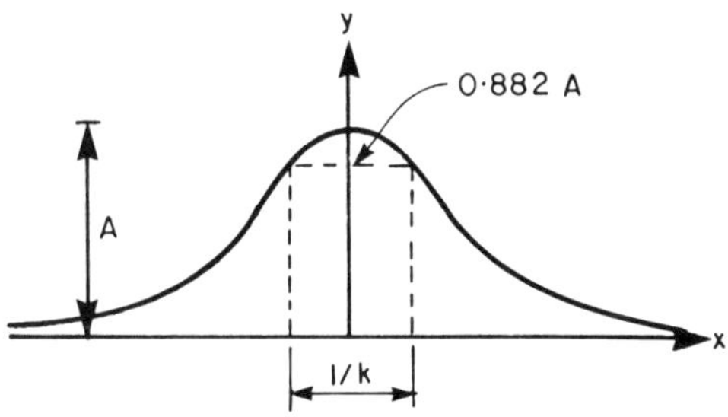

JOG

$$y = \frac{Akx}{(1+4k^2x^2)^{1/2}}$$

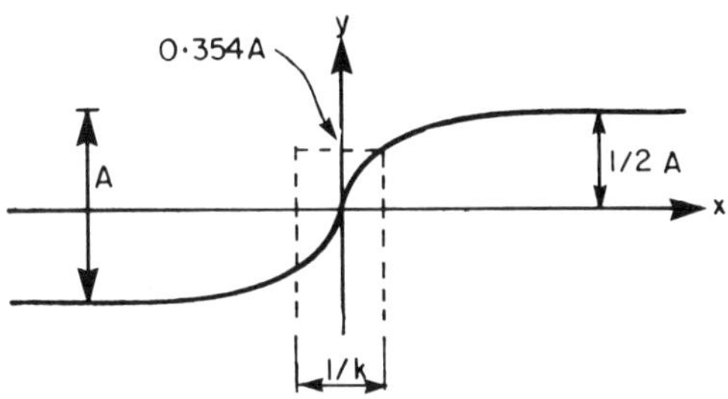

PLATEAU

$$y = \left(\frac{A^2}{1+(kx)^8}\right)^{1/2}$$

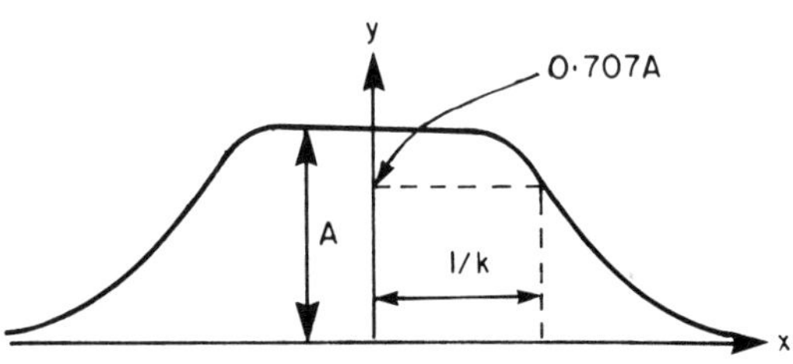

Table 3.1. (continued).

NAME AND FORMULA	SHAPE
TROUGH $y = Ak\left[(1/k)^2 - x^2\right]^{1/2}$	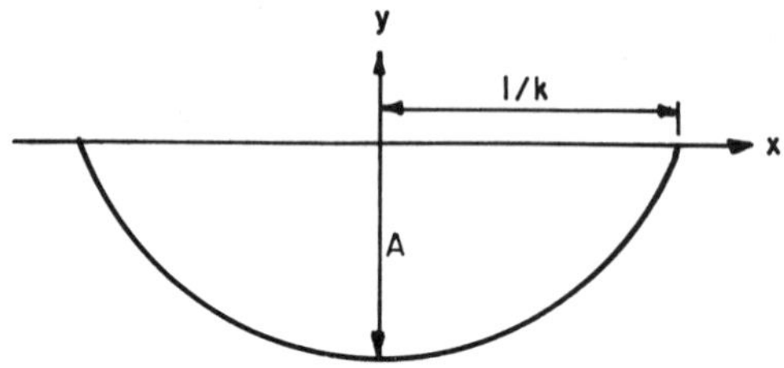
SINE CYCLE $y = A \sin \pi kx$	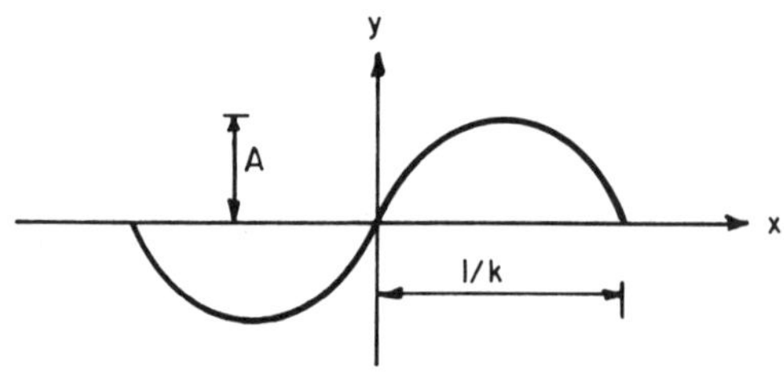
DAMPED COSINE $y = Ae^{-kx} \cos \pi kx$	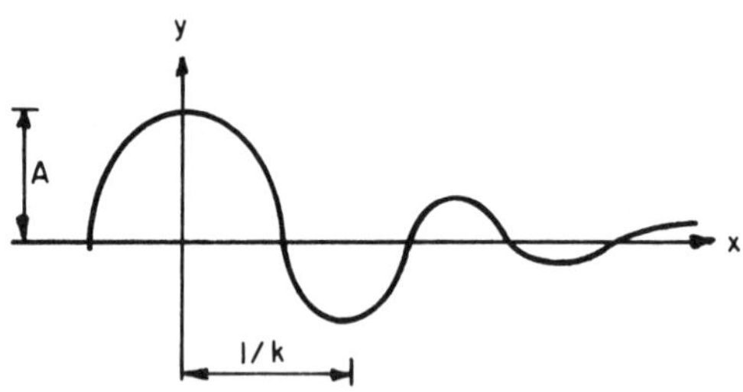

(c) Random irregularities: Stationary random track irregularities are well described by their power spectral densities (PSDs), also called 'spectral densities'. Spectral density is a distributed mean-square value of a given variable as a function of frequency or wave number, if in the time or space domain, respectively. Early spectral density measurements for various types of road and track surfaces were approximated by:

$$S(\Omega) = A/\Omega^2 \tag{3.24}$$

where A depends on the type of surface measured (see Houbolt [23]).

A description of rail irregularities by the above formula, however, yields values which are higher than those obtained by measurement for large values of wave number Ω. In order to resolve this difference, modifications to the above equation have been proposed in which either the exponent of Ω is changed [27] or piecewise continuous functions over various wave number ranges are used [13]. An extension of the above two modifications was suggested by Anis and Hedrick [3] for the alignment power spectral density $\{cm^2\ m/rad\ (in^2\ ft/rad)\}$:

$$S_a(\Omega) = \frac{A_a \Omega_c^2}{\Omega^2(\Omega^2 + \Omega_c^2)} \tag{3.25}$$

where A_a is a roughness parameter, Ω is the wave number $\{rad/m\ (rad/ft)\}$ and Ω_c is a break wave number.

Since wavelengths shorter than 3.048 m (10 ft) were found to have little effect on carbody lateral acceleration at moderately high speeds, no adjustment for low wavelength discrepancies was made in this formula. A similar formula was found to hold for vertical profile irregularities:

$$S_v(\Omega) = \frac{A_v \Omega_c^2}{\Omega^2(\Omega^2 + \Omega_c^2)} \tag{3.26}$$

where A_v is a roughness parameter. However, cross-level irregularities required a further adjustment to better fit the measured data for small wave numbers. It has been shown that to a good approximation, the cross-level spectral density S_c and gauge spectral density S_g are statistically independent of both the alignment and vertical profile spectral densities, S_a and S_v:

$$S_g(\Omega) = S_c(\Omega) = \frac{4A_v\Omega_c^2}{(\Omega^2 + \Omega_c^2)(\Omega^2 + \Omega_s^2)} \tag{3.27}$$

where Ω_s is a break wave number.

Alignment, vertical profile and cross-level irregularities may be approximated as being statistically independent of each other, so that, if linear models are used, the total response is the sum of the responses due to the three inputs. This independence is justified by the low values of coherence and cross-spectral density functions of any two of them.

Track PSDs are directly applicable to the frequency domain analysis of rail vehicles. They can be used to calculate the mean square values of rail deviations, rail curvatures, vehicle vibration levels, forces at the wheel-rail interface, and the relative displacements between various vehicle components. However, they have limitations. Without a detailed knowledge of the probability distributions governing each input and response mode, mean square values cannot predict peak values. Also, because of their averaging property, identical PSDs can result from a wide variety of time histories; therefore, isolated geometric variations are obscured.

3.3.2 Curved Track

The simplest form of curved track has a constant radius, R, a constant superelevation and no irregularities. The superelevation, ϕ_{SE}, defined in Fig. 3.7, is the angle by which one rail is elevated with respect to the other. The degree of curve, D, shown in Fig. 3.8(a), is the angle subtended by a 30.48 m (100 ft) chord at the center of a circular alignment. The curvature, R^{-1}, superelevation, ϕ_{SE}, and vehicle speed, V, are sufficient to determine the cant deficiency angle, ϕ_d:

$$\phi_d = \frac{V^2}{gR} - \phi_{SE} \tag{3.28}$$

where g is the acceleration due to gravity. Balanced running or zero cant deficiency occurs when the speed of the vehicle is such

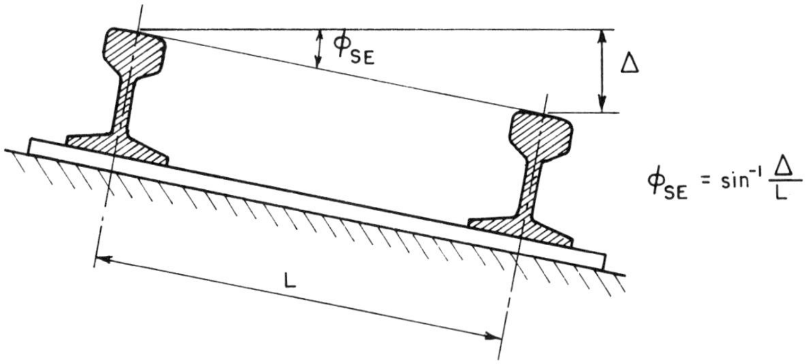

Fig. 3.7 Definition of superelevation

that the lateral component of weight cancels the centrifugal force during curve negotiation. A nonzero cant deficiency angle is the angle by which the superelevation would have to be increased in order to achieve balanced running. Plots of cant deficiency versus degree of curve are often used as measures of curving performance.

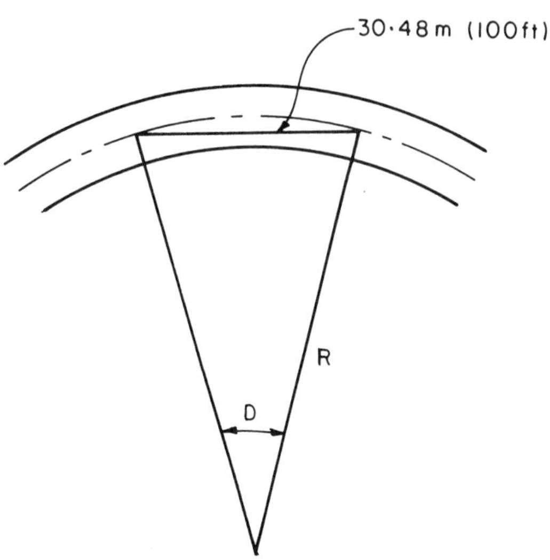

Fig. 3.8 (a) Definition of track curvature

In a dynamic analysis, one may also be interested in a vehicle's behaviour during curve entry and exit as well as in steady curving. The track geometry may then involve the following four types of track segments:

(1) Tangent: The curvature and superelevation are zero.

(2) Transition: A transition segment is often used between tangent and spiral, or spiral and steady curve segments. Such a segment may not be required with certain types of spiral segments which provide a smooth enough transition by themselves.

(3) Spiral: Spiral segments are used between tangent and steady curve segments. Their curvature and superelevation vary according to specified functions of along-track position. Linear, sine or arc tangent functions are commonly used. Figure 3.8(b) demonstrates the use of linear spiral segments with quadratic transition segments. A given spiral, together with its associated transition segments, may be described further as an entry or exit

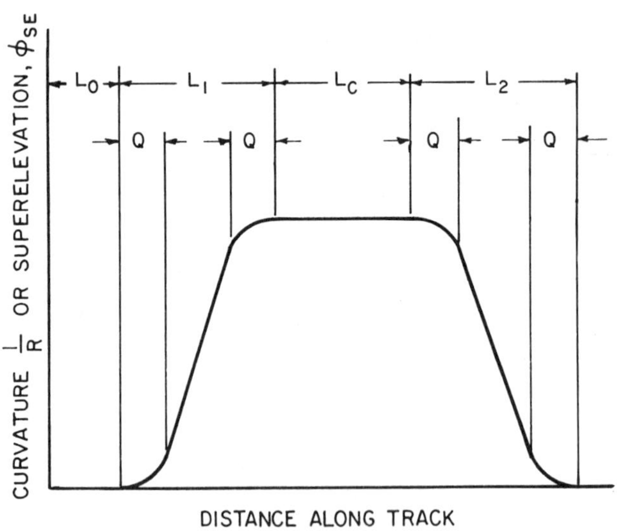

Fig. 3.8 (b) Curve geometry

curve (i.e., to or from a steady curve), depending on the intended direction of travel.

(4) Steady curve: The curvature and superelevation are constant.

The curve geometry can be completely specified by the following parameters once the functions defining the transition and spiral segments are specified:

L_0 - length of tangent before curve
L_1 - length of entry curve
L_2 - length of exit curve
L_c - length of steady curve
R_s - radius of steady curve
ϕ_{SES} - superelevation angle of steady curve

The choice of a function for the spiral entry is not critical and linear spiral segments with quadratic spiral transition segments have been used extensively in practice.

Radii of curvature can vary from a minimum near 30 m(98 ft) in low speed streetcar systems to large values on high speed track such as 3048 m (10,000 ft) on the test track at the U.S. Department of Transportation Test Center in Pueblo, Colorado.

3.4 WHEEL–RAIL MODELS

The basic component of rail vehicle steering and support systems is the wheelset. The forces and moments which occur in the wheel-rail contact regions have a dominant effect on vehicle dynamics. Central to wheel-rail interaction modeling is the *creep* phenomenon. Creep may be described as a part-elastic and part-frictional behaviour in which an elastic body rolling over another elastic body share an area of contact (approximately elliptical in the case of spheres) where both slip and adhesion occur simultaneously. A creep region of contact, therefore, may be regarded as a transition state between pure rolling and pure sliding.

When a wheelset rolls over a rail, both the wheel and the rail deform elastically at the contact area and deviations from pure rolling occur. These deviations, called creepages, were discovered by Carter [9] and are defined as follows:

Longitudinal creepage: $\xi_x = (v_x - v_x^*)/v_x^*$ (3.29a)

Lateral creepage: $\xi_y = (v_y - v_y^*)/v_x^*$ (3.29b)

Spin creepage: $\xi_{sp} = (\omega_u - \omega_\ell)/v_x^*$ (3.29c)

where v_x and v_x^* respectively are actual and pure rolling forward velocities, v_y and v_y^* are actual and pure rolling lateral velocities, ω_u and ω_ℓ are angular velocities of the upper and lower elastic bodies about the normal to the contact surface (Fig. 3.9(a)).

Creep forces and moments are generated by these creepages. The relationship between a creep force and its corresponding creepage is essentially nonlinear (Fig. 3.9(b)). However, a creep force varies linearly over most of the creepage range, starting from a value of zero at zero creepage, to a limiting frictional force value of μN at large creepages, where μ is the coefficient of friction and N is the normal force. For small values of creepage, approximate linear relationships exist between creep forces and creepages. The constants of the proportionality, called creep coefficients, were determined by Kalker [25].

Three commonly used creep models are presented: (1) a linear creep force and spin moment model, (2) a nonlinear creep force model assuming negligible spin, and (3) a nonlinear creep force model including the effect of spin. Finally the dynamics and guidance of wheelsets are discussed.

3.4.1 Linear Creep Force and Spin Moment

Kalker [26] developed a linear theory of rolling contact which suggested that, for very small creepages ξ_x, ξ_y and ξ_{sp}, the area of slip is small and the adhesion zone can be assumed to cover the entire elliptical contact area. His widely-used linear model for longitudinal creep force, F_x, lateral creep force, F_y, and spin-creep moment, M_z, consists of the following three equations:

$$F_x = f_{33}\xi_x \qquad (3.30a)$$

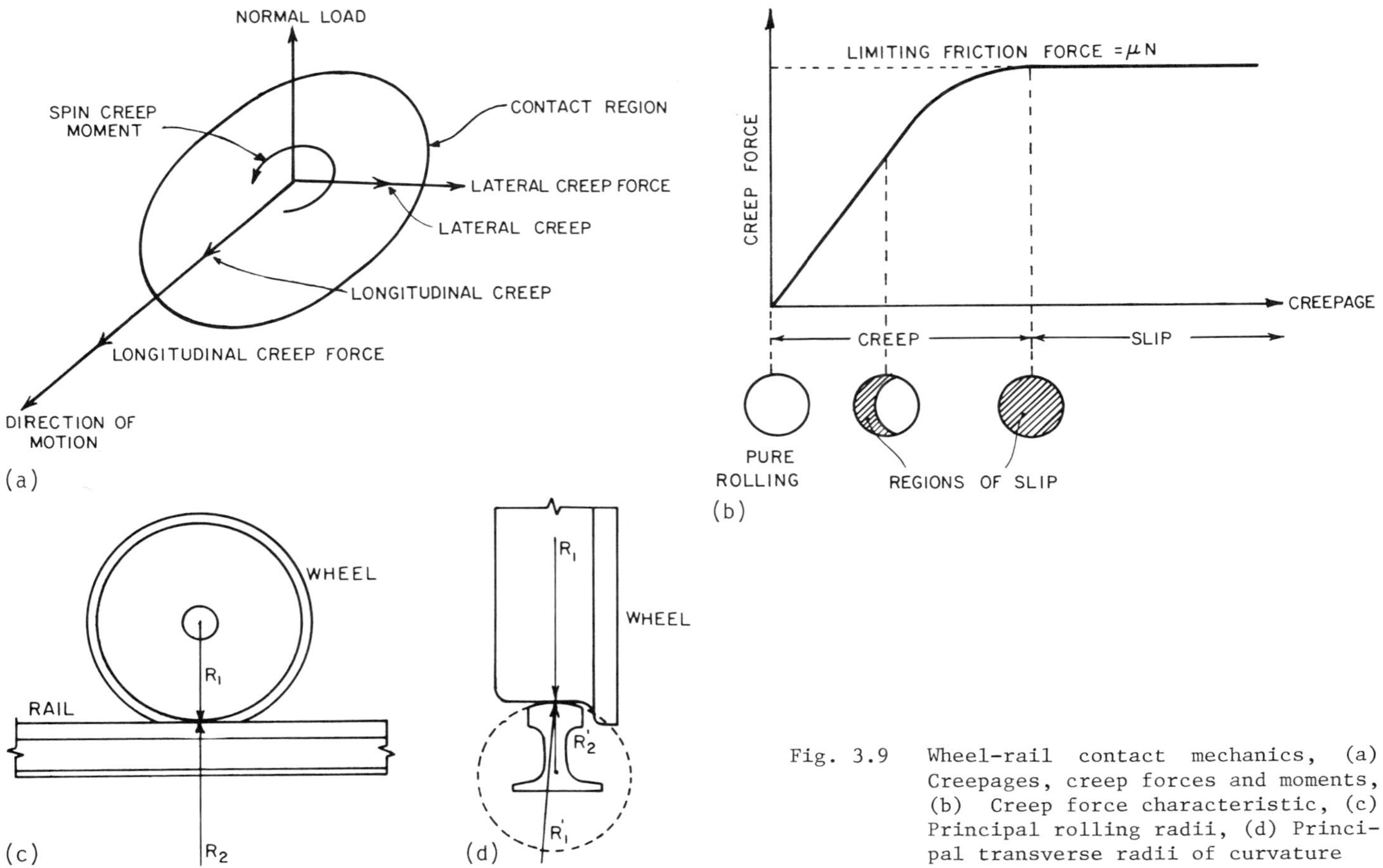

Fig. 3.9 Wheel-rail contact mechanics, (a) Creepages, creep forces and moments, (b) Creep force characteristic, (c) Principal rolling radii, (d) Principal transverse radii of curvature

$$F_y = f_{11}\xi_y + f_{12}\xi_{sp} \tag{3.30b}$$

$$M_z = -f_{12}\xi_y + f_{22}\xi_{sp} \tag{3.30c}$$

where f_{ij} are the creep coefficients, functions of the wheel-rail geometry, material properties and the normal force acting at the centre of the contact surface.

The longitudinal and lateral semiaxes, a and b, of the contact ellipse are given by (Hertz [15]):

$$a = m[3\pi\ N(K_1 + K_2)/4K_3]^{1/3} \tag{3.31}$$

$$b = n[3\pi\ N(K_1 + K_2)/4K_3]^{1/3} \tag{3.32}$$

where

$$K_1 = \frac{1 - \sigma_w^2}{\pi E_w} \tag{3.33a}$$

$$K_2 = \frac{1 - \sigma_R^2}{\pi E_R} \tag{3.33b}$$

$$K_3 = \frac{1}{2}\left[\frac{1}{R_1} + \frac{1}{R_1'} + \frac{1}{R_2} + \frac{1}{R_2'}\right] \tag{3.33c}$$

$$G = (2G_wG_R)/(G_w + G_R) \tag{3.33d}$$

$$\sigma = [G(G_w\sigma_R + G_R\sigma_w)]/(2G_wG_R). \tag{3.33e}$$

and

N = total normal force

R_1, R_2 = principal rolling radius of the wheel and rail respectively,

R_1', R_2' = Principal transverse radius of curvature of the wheel and rail profiles respectively,

σ_w, σ_R = Poisson's ratio for the wheel and rail materials respectively,

E_w, E_R = Young's modulus of elasticity of the wheel and rail materials respectively, and

G_w, G_R = Shear modulus of rigidity of the wheel and rail materials respectively.

Table 3.2. Coefficients m and n (Adapted from Hertz [42]).

θ (deg)	m	n	θ (deg)	m	n	θ (deg)	m	n
0.5	61.40	0.102	10	6.604	0.311	60	1.486	0.717
1	36.89	0.131	20	3.813	0.412	65	1.378	0.759
1.5	27.48	0.152	30	2.731	0.493	70	1.284	0.802
2	22.26	0.169	35	2.397	0.530	75	1.202	0.846
3	16.50	0.196	40	2.136	0.567	80	1.128	0.893
4	13.31	0.219	45	1.926	0.604	85	1.061	0.944
6	9.79	0.255	50	1.754	0.641	90	1.000	1.000
8	7.86	0.285	55	1.611	0.678			

The radius of curvature of a body is considered to be positive if the corresponding center of curvature is within the body. In Fig. 3.9(c) and (d), for example, only R_1' is shown to be negative.

Moreover, the coefficients m and n in Eqs. (3.31) and (3.32) are functions of the ratio K_4/K_3, where

$$K_4 = \frac{1}{2}\left[\left(\frac{1}{R_1} + \frac{1}{R_1'}\right)^2 + \left(\frac{1}{R_2} + \frac{1}{R_2'}\right)^2 + 2\left(\frac{1}{R_1} - \frac{1}{R_1'}\right)\left(\frac{1}{R_2} - \frac{1}{R_2'}\right) \cos 2\psi\right], \quad (3.34)$$

and ψ is the angle between the normal planes that contain the curvatures $1/R_1$ and $1/R_2$. Values of m and n are given in Table 3.2 as functions of θ, where

$$\theta = \cos^{-1}(K_4/K_3). \quad (3.35)$$

Values of the creepage and spin coefficients c_{ij}, which depend only on σ and the ratio a/b [25], are listed in Table 3.3. The linear creepage and spin coefficients are then computed by:

$$f_{11} = ab\ G\ C_{22} \quad (3.36a)$$

$$f_{12} = (ab)^{3/2}\, G\, C_{23} \tag{3.36b}$$

$$f_{22} = \left(\frac{a}{b}\right)^2 G\, C_{33} \tag{3.36c}$$

$$f_{33} = ab\, G\, C_{11} \tag{3.36d}$$

3.4.2 Nonlinear Creep Force Without Spin

The nonlinear creep model of Johnson and Vermeulen [24] is limited to the case of longitudinal and lateral creepage without spin. Johnson and Vermeulen showed that the error in calculated creep forces using their mathematical model was less than 25% compared with experimental results.

The resulting tangential force can be formulated as follows. Let 'a' be the semiaxis in the rolling direction of unit vector 'i' and 'b' the semiaxis in the lateral direction of unit vector 'j' of the contact ellipse in the plane of contact. Then, given modulus of rigidity G, Poisson's ratio σ, coefficient of friction μ, longitudinal creepage ξ_x, lateral creepage ξ_y, and total normal force N, the tangential (creep) force F is given by

$$F/fN = \begin{cases} (1/\tau)[(1-\tau/3)^3 - 1](\xi i + \eta j) & \text{for } |\tau| \le 3, \\ -(1/\tau)(\xi i + \eta j) & \text{for } |\tau| \ge 3. \end{cases} \tag{3.37}$$

where the normalized longitudinal and lateral creepages, ξ and η, and the normalized resultant creepage τ are calculated by

$$\xi = \pi abG\xi_x/(\mu N\phi) \tag{3.38a}$$

$$\eta = \pi abG\xi_y/(\mu N\psi) \tag{3.38b}$$

and

$$\tau = (\xi^2+\eta^2)^{\frac{1}{2}} \tag{3.38c}$$

in which

$$\left.\begin{aligned} \phi &= B-\sigma(D-C) \\ \psi &= B-\sigma(a^2/b^2)C \end{aligned}\right\} \quad \text{for } a \le b,\ k = (1-a^2/b^2)^{\frac{1}{2}} \tag{3.39a}$$

$$\left.\begin{aligned} \phi &= [D-\sigma(D-C)](b/a) \\ \psi &= [D-\sigma C](b/a) \end{aligned}\right\} \quad \text{for } a > b,\ k = (1-b^2/a^2)^{\frac{1}{2}} \tag{3.39b}$$

wherein

Table 3.3. Creepage and Spin Coefficients (Adapted from Kalker [25]).

	C_{11}			C_{22}			$C_{23} = -C_{32}$			C_{33}		
g^{+}	$\sigma=0$	$\frac{1}{4}$	$\frac{1}{2}$	$\sigma=0$	$\frac{1}{4}$	$\frac{1}{2}$	$\sigma=0$	$\frac{1}{4}$	$\frac{1}{2}$	$\sigma=0$	$\frac{1}{4}$	$\frac{1}{2}$
(a/b)												
0.1	2.51	3.31	4.85	2.51	2.52	2.53	0.33	0.47	0.73	6.42	8.28	11.7
0.2	2.59	3.37	4.81	2.59	2.63	2.66	0.48	0.60	0.81	3.46	4.27	5.66
0.3	2.68	3.44	4.80	2.68	2.75	2.81	0.61	0.72	0.89	2.49	2.96	3.72
0.4	2.78	3.53	4.82	2.78	2.88	2.98	0.72	0.82	0.98	2.02	2.32	2.77
0.5	2.88	3.62	4.83	2.88	3.01	3.14	0.83	0.93	1.07	1.74	1.93	2.22
0.6	2.98	3.72	4.91	2.98	3.14	3.31	0.93	1.03	1.18	1.56	1.68	1.86
0.7	3.09	3.81	4.97	3.09	3.28	3.48	1.03	1.14	1.29	1.43	1.50	1.60
0.8	3.19	3.91	5.05	3.19	3.41	3.65	1.13	1.25	1.40	1.34	1.37	1.42
0.9	3.29	4.01	5.12	3.29	3.54	3.82	1.23	1.36	1.51	1.27	1.27	1.27
(b/a)												
1.0	3.40	4.12	5.20	3.40	3.67	3.98	1.33	1.47	1.63	1.21	1.19	1.16
0.9	3.51	4.22	5.30	3.51	3.81	4.16	1.44	1.59	1.77	1.16	1.11	1.06
0.8	3.65	4.36	5.42	3.65	3.99	4.39	1.58	1.75	1.94	1.10	1.04	0.95
0.7	3.82	4.54	5.58	3.82	4.21	4.67	1.76	1.95	2.18	1.05	0.96	0.85
0.6	4.06	4.78	5.80	4.06	4.50	5.04	2.01	2.23	2.50	1.01	0.89	0.75
0.5	4.37	5.10	6.11	4.37	4.90	5.56	2.35	2.62	2.96	0.96	0.82	0.65
0.4	4.84	5.57	6.57	4.84	5.48	6.31	2.88	3.24	3.70	0.91	0.75	0.55
0.3	5.57	6.34	7.34	5.57	6.40	7.51	3.79	4.32	5.01	0.87	0.67	0.45
0.2	6.96	7.78	8.82	6.96	8.14	9.79	5.72	6.63	7.89	0.83	0.60	0.34
0.1	10.70	11.70	12.90	10.70	12.80	16.00	12.20	14.60	18.00	0.79	0.52	0.23

+For $g = 0$; $c_{11} = \pi^2/4(1-\sigma)$; $c_{22} = \pi^2/4$; $c_{23} = -c_{32} = \pi\sqrt{g/3}$; and $c_{33} = \pi^2/16(1-\sigma)g$.

$$B = \int_0^{\pi/2} (\cos^2 \theta)(1-k^2 \sin^2 \theta)^{-\frac{1}{2}} d\theta \tag{3.40a}$$

$$C = \int_0^{\pi/2} (\cos^2 \theta \sin^2 \theta)(1-k^2 \sin^2 \theta)^{-\frac{1}{2}} d\theta \tag{3.40b}$$

$$D = \int_0^{\pi/2} (\sin^2 \theta)(1-k^2 \sin^2 \theta)^{-\frac{1}{2}} d\theta \tag{3.40c}$$

Since function ϕ is rarely equal to ψ, the creep force is usually not exactly in the direction of the creepage.

3.4.3 Nonlinear Creep Force With Spin

A heuristic modification to the Johnson-Vermeulen formulation was made by White et al. [38] to include the effect of spin creepage on creep forces. In this model, the creep forces and moments are initially computed by means of Kalker's linear theory (Eqs. (3.30)). The creep coefficients which are functions of the instantaneous normal load N, are calculated in the following manner:

$$f_{11} = \left(\frac{N}{N_0}\right)^{2/3} f_{110} \tag{3.41a}$$

$$f_{12} = \left(\frac{N}{N_0}\right) f_{120} \tag{3.41b}$$

$$f_{22} = \left(\frac{N}{N_0}\right)^{4/3} f_{220} \tag{3.41c}$$

$$f_{33} = \left(\frac{N}{N_0}\right)^{2/3} f_{330} \tag{3.41d}$$

where f_{ij0} are the nominal values computed for the nominal load N_0 and f_{ij} are the values for the instantaneous load N. The unlimited resultant creep force obtained from the linear model is:

$$F_R = \left(F_x^2 + F_y^2\right)^{\frac{1}{2}} \tag{3.42}$$

However, the magnitude of the resultant creep force cannot exceed the pure slip value, μN (Fig. 3.9(b)). Therefore, a creep force saturation coefficient, ε, is computed with the following modified Johnson-Vermeulen formulation:

$$\epsilon = \begin{cases} \dfrac{\mu N}{F_R}\left[\left(\dfrac{F_R}{\mu N}\right) - \dfrac{1}{3}\left(\dfrac{F_R}{\mu N}\right)^2 + \dfrac{1}{27}\left(\dfrac{F_R}{\mu N}\right)^3\right] & \text{for } F_R \le 3\mu N \\ \dfrac{\mu N}{F_R} & \text{for } F_R > 3\mu N \end{cases} \tag{3.43}$$

Finally, the heuristic nonlinear creep force model is given by:

$$F'_x = \epsilon F_x \tag{3.44a}$$

$$F'_y = \epsilon F_y \tag{3.44b}$$

where F'_x and F'_y are the longitudinal and lateral components of creep force F', incorporating saturation of the force-creepage characteristic shown in Fig. 3.9(b). In the general case of arbitrary creepage without spin, the results obtained from the heuristic model are in close agreement with both the Johnson-Vermeulen model and Kalker's simplified theory [26].

3.4.4 Wheelset Dynamics and Guidance

Represented in Fig. 3.10(a) is a wheelset rolling on a straight track directed into the paper, so that the left and right wheel are as shown from the point of view of a vehicle-mounted observer. Neither rail is superelevated with respect to the other and the left and right rail cant-angles, β_L and β_R, are equal. The track gauge, $2a_r$, is the horizontal distance between rail head centerlines. The axial distance between wheel-tread tape lines is $2a_w$. The wheelset, shown displaced to the right by a distance y, has a nonzero negative roll angle ϕ with respect to the rail plane, which is parallel to the horizontal plane in this case. Distances of the left and right contacts above the rail plane are z_L and z_R, respectively. Whereas the rolling radius r_o, at the tape lines, is normally the same on both sides of a centered wheelset, the left rolling radius r_L is smaller than the right rolling radius r_R when the wheelset is displaced laterally to the right, as shown. Likewise, defining the "contact angle" as the angle between a contact plane and the axle centerline, the left contact angle δ_L is seen to be smaller than the right contact angle δ_R.

Figure 3.10(b), a free-body diagram of the wheelset in Fig. 3.10(a) using right-handed Cartesian coordinates with x pointing in the intended direction of travel and z pointing downwards, shows the lateral suspension force F_{SY}, the vertical force F_Z (consisting of a constant axle load L_A and a time-varying component of vertical suspension force, F_{SZ}) and the roll suspension moment M_{SX}, all translated to the center of mass. The remaining suspension force and moment components, F_{SX}, M_{SY} and M_{SZ}, are omitted. The axle load L_A is the net effective weight of the vehicle at the axle's center of mass. At the left wheel-rail contact, we see the lateral and vertical creep force components, F_{LY} and F_{LZ}, and the normal force N_L; however, the longitidunal creep force component

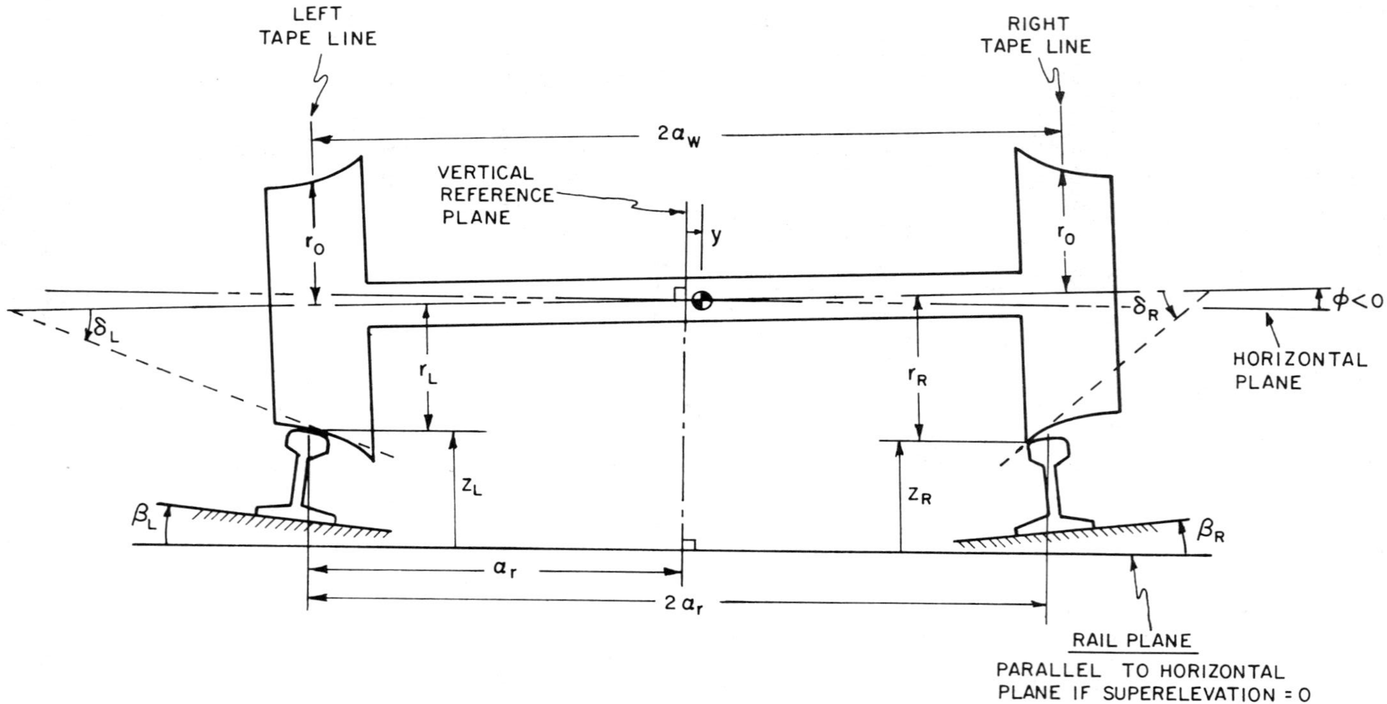

Fig. 3.10 (a) Wheel-rail geometry

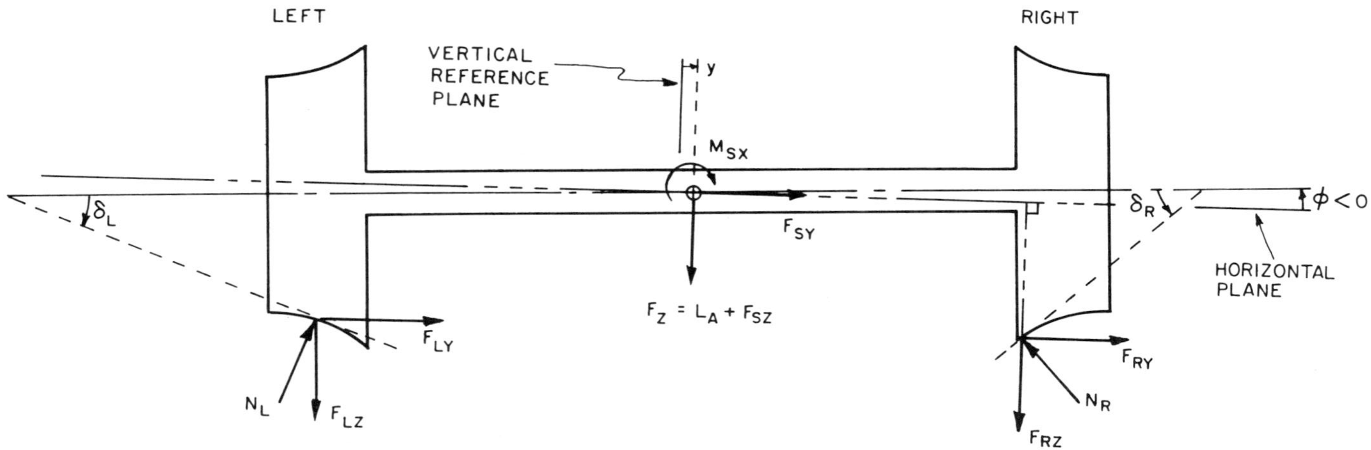

Fig. 3.10 (b) Free-body diagram of wheelset

F_{LX} and spin-creep moment components, M_{LX}, M_{LY} and M_{LZ}, are not shown. A similar description is applicable to the right contact.

The equations of motion of a wheelset may be expressed in the form

$$[M] \ddot{\underset{\sim}{x}} + [C] \dot{\underset{\sim}{x}} + [K] \underset{\sim}{x} = \underset{\sim}{F} \tag{3.45}$$

where $\underset{\sim}{x}$, $\dot{\underset{\sim}{x}}$ and $\ddot{\underset{\sim}{x}}$ are vectors representing the displacement, velocity and acceleration of each degree of freedom, respectively, and $\underset{\sim}{F}$ is a vector representing the generalized force due to wheel-rail interactions. Matrices [M], [C] and [K] represent respectively the mass (and inertia), damping and stiffness of the wheelset. For a complete derivation of the equations of motion of a wheelset with nonlinear wheel/rail profiles the reader is referred to Hedrick et al. [18].

As it rolls along a track, a wheelset is constrained to move laterally and vertically in a prescribed manner determined by the geometry of the wheels, rails and track structure. The characteristics of these geometric constraints determine, to a large extent, the nature of the lateral motions of the wheelsets. The wheelset position may be described by two independent variables: the lateral position of its geometric center relative to the track centerline, y, and the angular rotation, or "yaw angle", of the wheelset about a vertical axis, ψ. The remaining motions of the wheelset such as roll or vertical movement, are determined by the geometric constraints.

The equations of motion of a wheelset require that the constraint variables r_L, r_R, z_L, z_R, δ_L, δ_R and ϕ be known as functions of the independent variables, y and ψ. However, their dependence on the yaw angle may be neglected for most applications.

The wheelset equations of motion utilize the difference between the rolling radii r_L and r_R. When the wheelset is displaced from a centered position between the rails, the difference between r_L and r_R requires that the longitudinal velocity of the wheel at the contact point on the wheel with a larger rolling radius be greater than the velocity at the contact point on the other wheel. The result is partial slip, or creepage, at the wheel/rail interfaces giving rise to a moment due to the difference between the left and right creep forces. This moment tends, in most cases, to steer the wheelset towards the centered position on the rails.

The relationship between the difference in rolling radii and wheelset lateral position varies widely. For a new wheel with a conical taper, the difference in rolling radii depends linearly on the wheelset lateral displacement, y, until flange contact is made. The constant of proportionality, called the wheelset conicity, is the wheel profile taper (usually 1/20 or 0.05). Although essential for guidance, conicity is also a destabilizing factor in the sense that the critical speed for onset of hunting is lower with large values of conicity. As the wheels wear, the change in rolling radii difference with wheelset lateral displacement becomes more nonlinear, and, over certain ranges, is much greater than the change with new wheels.

The contact angle constraints enter the rail vehicle equations of motion in the description of the magnitude and direction of the wheel-rail contact forces. Of particular importance in stabilizing a vehicle is the gravitational stiffness force F_g, defined as the net lateral component of the normal contact forces. From Fig. 3.10(b), we have

$$F_g = N_L \sin(\delta_L+\phi) - N_R \sin(\delta_R-\phi) \tag{3.46a}$$

where, neglecting dynamical loads,

$$N_L = \frac{L_A}{2} \cos(\delta_L+\phi) \tag{3.46b}$$

and

$$N_R = \frac{L_A}{2} \cos(\delta_R-\phi) \tag{3.46c}$$

For small contact and roll angles Eqs. (3.46) reduce to

$$F_g = L_A \left(\frac{\delta_L-\delta_R}{2} + \phi\right) \tag{3.47}$$

In Fig. 3.10 it is seen that δ_R is greater than δ_L and ϕ is negative; therefore, F_g is negative, acting to the left in a direction tending to return the wheelset to a centered position on the track. In general, the gravitational stiffness force acts in a direction to move the wheelset toward the centered position and, as such, is a stabilizing factor associated with differences in conicity as evidenced by differences in contact angles.

When centered, the contact angles of new wheels are determined by the wheel tread taper. They remain constant until the wheelset is shifted laterally far enough for the contact point to move to the flange. For either new or worn wheels, the contribution of the roll angle to the gravitational stiffness effect is about the same. However, as long as the contacts remain in the tread region, the contribution of the difference in contact angles to the gravitational stiffness is always zero for new wheels but potentially very large for worn wheels.

This gravitational stiffness effect can provide more restoring force than the vehicle suspension for a heavily loaded wheelset. The stabilizing benefit of this effect, however, is far less important for light vehicles. In general, as the wheels wear, both the wheelset conicity and the contact angle difference increase. Because the former is destabilizing whereas the latter is stabilizing, the resultant effect on vehicle stability depends on the relative magnitudes of the terms. The dependence of the gravitational stiffness effect on wheel load means that wheel wear will be more destabilizing for light vehicles than for loaded or heavy vehicles.

The third geometric constraint to enter the equations of motion describes the roll of the wheelset about a longitudinal axis as the wheelset moves laterally. Although the roll inertia of the wheelset enters only as a second order effect, the roll angle appears in terms associated with "gyroscopic" effects, creep forces, and the gravitational stiffness discussed previously. The gyroscopic effects are normally small but can become important at high speeds.

The roll constraint relationship may vary considerably for different wheel profiles. When the contact between wheel and rail jumps laterally across the wheel tread, the rolling radii and contact angles also change discontinuously, and only the wheelset roll angle remains a continuous function of lateral position. The most obvious of these jump discontinuities occurs when the flange is contacted. Worn wheels often have several lateral positions where the contact jumps laterally across the wheel tread.

The relationships which express the rolling radii, contact angles and wheelset roll angle as functions of the wheelset lateral position are determined by the wheel and rail profiles. For actual profiles, these functions may be computed numerically using computer programs such as WHRAILA [22]. Furthermore, it has been shown by Cooperrider and Law [12] that offsets from nominal wheel and rail profiles affect vehicle dynamics in a manner similar to track irregularities.

3.5 TRUCK MODELS

The primary functions of a rail vehicle truck are: (a) to safely support the carbody under the full range of operating loads; (b) to provide guidance with stable operation on tangent track and flange free operation on curved track; (c) to provide acceptable vibration isolation of the carbody from track irregularities; (d) to provide acceptable adhesion during powering and braking operations.

The required complexity and accuracy of a rail-vehicle model depends upon the simulation objectives. Therefore it is pertinent to have a grasp of the relative importance of the various truck suspension elements. In this section, we discuss qualitatively that aspect of passenger, freight and locomotive trucks.

3.5.1 Passenger Trucks

A typical passenger truck is characterized by a rigid frame with primary and secondary suspensions designed for ride comfort. The wheelsets are connected to the frame by means of restoring and dissipative elements such as elastomer pads, coil springs or steel/rubber chevrons. These elements reduce vibration transmission to the carbody by allowing the wheelsets to move relative to the truck frame. The secondary suspension supports the carbody either directly or through a bolster. For example, vertical airspring groups in parallel with hydraulic dampers may act between the bolster and truck frame, and the carbody is supported by the bolster at a centerplate connection. In most cases, passenger truck suspension components may be represented by linear characteristics. A generic model for most two-wheelset passenger, freight or locomotive trucks was formulated by Hedrick et al. [19] and Murray et al. [32]. A schematic diagram of this model, including bending and shear stiffness between the wheelsets, is drawn in Fig. 3.11. Since it has been shown by Wickens [37] that the stiffness and kinematic properties of all possible interwheelset connections can be characterized by shear and bending stiffnesses between the wheelsets, this generic model is applicable also to radial trucks.

The degrees of freedom required for modeling passenger trucks are wheelset lateral and yaw displacements, and truck frame lateral, yaw and roll displacements. The bolster may have lateral, yaw and roll displacements relative to the truck frame.

3.5.2 Freight Trucks

A conventional three-piece freight truck, consisting of two sideframes and a bolster, differs from a passenger truck by the lack of primary suspension elements between its wheelsets and truck frame, and the use of friction in parallel with springs in its secondary suspension. The wheelset to sideframe connection, consisting of only a bearing and bearing adapter with associated friction, can be modeled as a ball joint with friction, by representing the wheelsets and side frames as a four bar linkage. When elastomeric pads are placed between the bearing adapters and sideframes, the inclusion of a primary suspension with individual stiffnesses in the longitudinal, lateral and yaw directions may be required. The spring group between the bolster and sideframe usually has both vertical and lateral flexibility in parallel with dry friction from the friction wedges. Lateral motion is limited

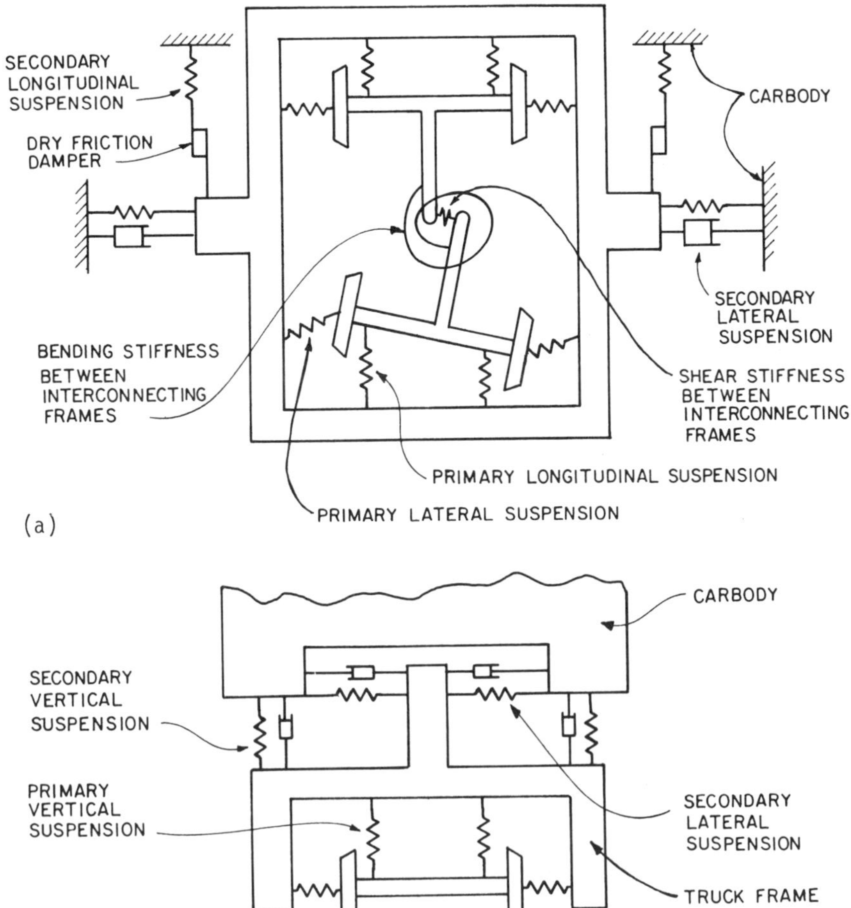

Fig. 3.11 Passenger truck generic model, (a) Plan view (b) Elevation

by the bolster gibs. Truck rotation relative to the centerplate is resisted by dry friction at the centerplate as well as stiffness and friction of constant-contact side bearings, when fitted. Nonlinear expressions are required in the modeling of large rolling motions to describe carbody contact with sideframes.

Nonlinearities such as Coulomb or dry friction, gaps, and stops in the suspension can be modeled with a generic suspension element

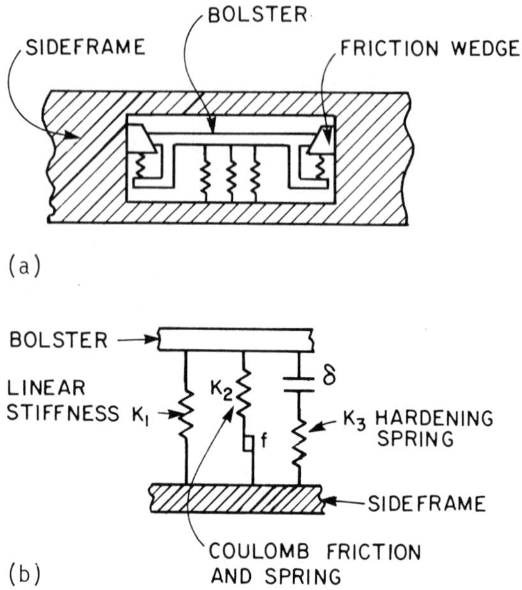

Fig. 3.12 Freight-car truck suspension model, (a) Physical model (b) Mathematical model (adapted from Cooperrider and Law [12])

as shown, for example, in Fig. 3.12(b) to represent the vertical secondary suspension of the freight truck [Fig. 3.12(a)]. This generic element is composed of a parallel combination of a linear spring K_1, dry friction f in series with a linear spring K_2, linear spring K_3 with clearance δ, and viscous friction (when present). The prevalence of nonlinearities in freight suspensions tends to make the analysis more difficult for freight vehicles than passenger vehicles.

Sometimes, in addition to the regular truck frame components direct interconnections are made between wheelsets. Such components may be modeled by the addition of bending and shearing suspension elements as shown in Fig. 3.13.

3.5.3 Locomotive Trucks

Figure 3.14 illustrates the lateral and vertical suspension elements of a linear locomotive truck model useful for lateral

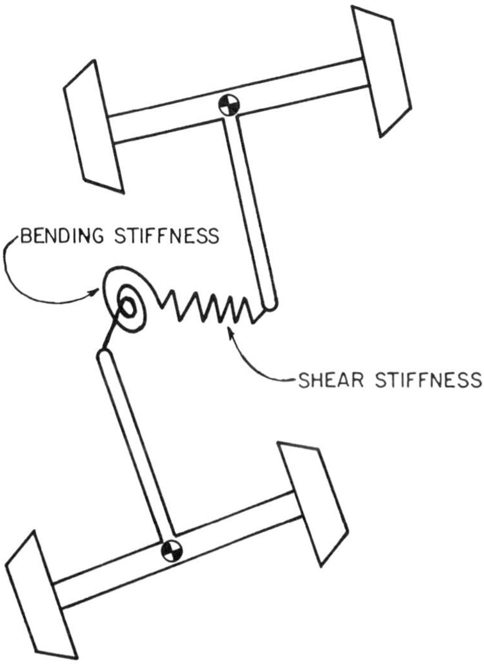

Fig. 3.13 Wheelset interconnection model

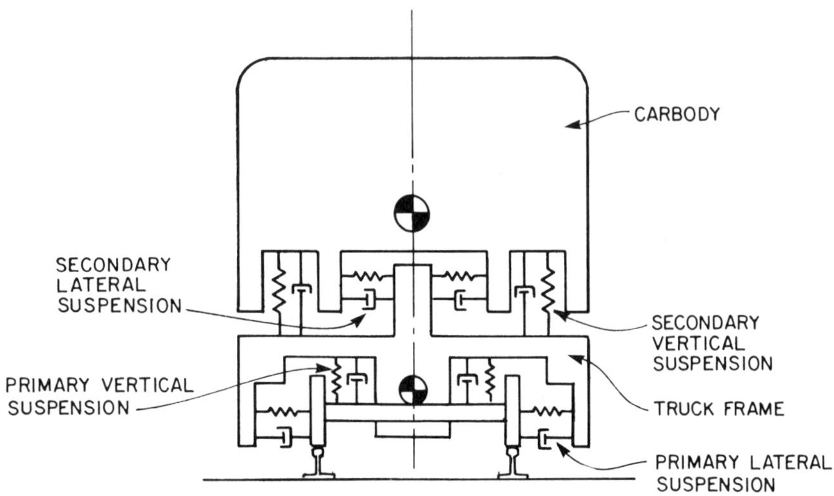

Fig. 3.14 Locomotive truck model

stability analysis. The wheelset is considered to be isolated from the truck frame by a primary suspension of parallel springs and dampers in the longitudinal, lateral and vertical directions. The truck frame is attached to the carbody through a secondary suspension of parallel springs and acting in the same three directions as the primary dampers. The carbody has three degrees of freedom - lateral, roll and yaw motions. The wheelset has lateral and yaw motions only.

The following approximations are made for the use of this truck model in the lateral stability analysis of locomotives: (1) the truck frame and carbody stiffnesses are lumped in the suspension elements; (2) the axles are assumed to run freely without friction in the journal bearings; (3) any lateral clearance between the wheelset and truck frame is neglected; (4) all displacements are considered small; (5) all springs and viscous dampers are assumed to be linear; (6) the vehicle is considered to be symmetrical about a vertical plane. This symmetry results in a set of equations of motion in which the vertical motions (bounce and pitch) are uncoupled from the lateral motions (roll, yaw and lateral displacement) when the equations are linearized. Coupling of the vertical and lateral motions through the nonlinear characteristics of certain suspension elements and wheel-rail interaction is negligible. Equations for longitudinal motion are also uncoupled due to the assumptions of symmetry and small displacement; (7) nonlinearities arising from suspension stops, wheel flange contact, dry friction in suspension elements, and adhesion limits between wheel and rail are neglected.

The inclusion of suspension nonlinearities in a locomotive model is discussed in Section 3.6.3.

3.6 VEHICLE MODELS

An all-inclusive dynamical model of a railway vehicle would be large and complex. Because size and complexity are factors which tend to reduce physical insight and increase the development and usage costs of computer programs, less general models are usually preferable. Specialized models may be classified as follows:

(a) Vertical/Lateral Dynamics Models: to study the dynamic response of a vehicle to track irregularities.

(b) Lateral Stability Models: to predict critical (hunting) speed, wheel/rail forces, and suspension and carbody displacements and forces.

(c) Curving Models: to calculate the dynamic or quasi-static forces of a vehicle during curve negotiation.

Three special-purpose models with existing computer programs have been selected as examples. Their analytical models are discussed in the following sections in preparation for the computer simulation studies of Chapter 5. The selected vehicle models are: (1) a linear model of passenger-vehicle vertical dynamics in which carbody heave, pitch and roll motions are excited by vertical and cross-level track disturbances acting through the wheelsets and trucks (Section 3.6.1). The determination of vertical ride quality is the objective of this model; (2) a model for the lateral dynamics or curving performance of a typical North American freight car on tangent, spiral or constant curvature track (Section 3.6.2); (3) a nonlinear 29 degree of freedom analytical model for the vertical and lateral response of a six-axle locomotive to tangent track irregularities (Section 3.6.3).

3.6.1 Passenger-Vehicle Vertical Dynamics Model

A linear model of the vertical dynamics of a passenger vehicle for the analysis of ride comfort, developed by Murray et al. [32], is presented here. The schematic diagram of Fig. 3.15 represents a passenger vehicle with a conventional secondary suspension, whose airspring stiffness and hydraulic damping are modeled by a linear spring and damper. Assumed to be travelling at a constant velocity on a nominally level, rigid track, with no vertical/lateral coupler or aerodynamic forces, the powered or drawn vehicle is assumed also to be longitudinally and laterally symmetrical, with its wheels always in contact with the rails. Based on the latter constraint,

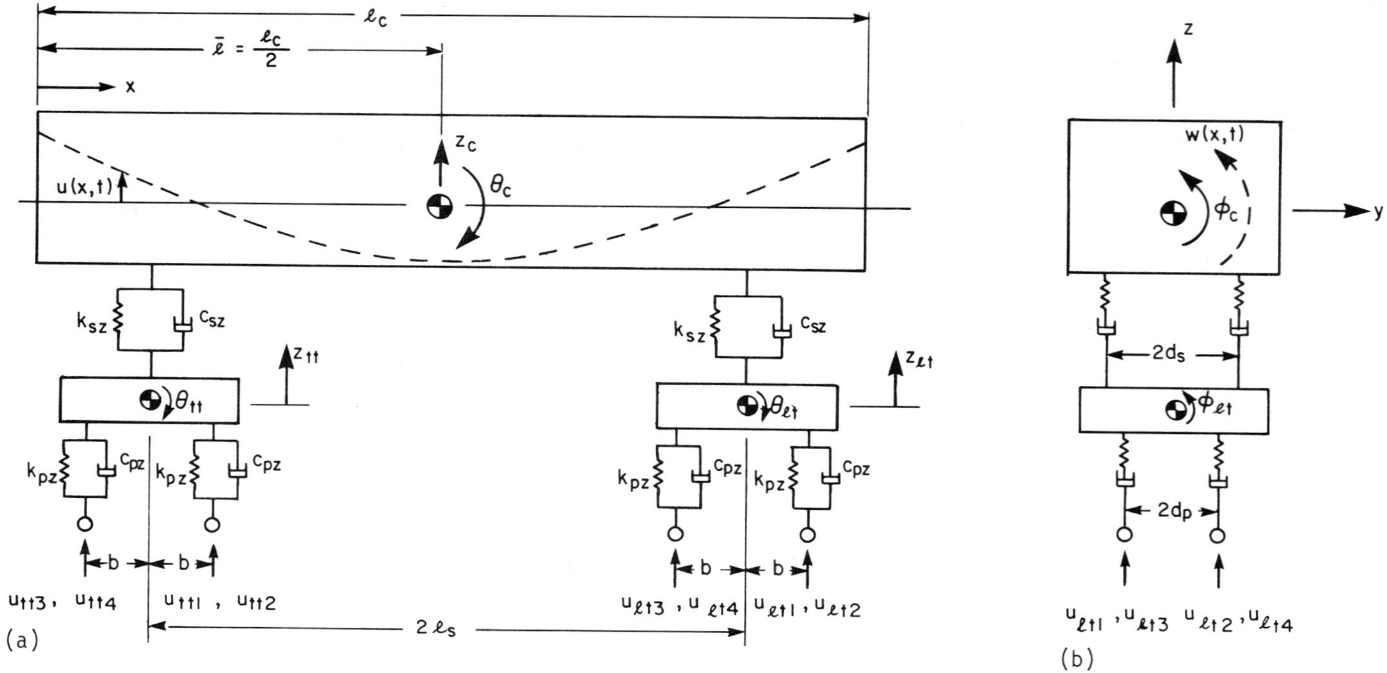

Fig. 3.15 Passenger vehicle vertical dynamics model, (a) Side view, (b) Front view (from Murray et al. [32])

the wheels are modeled as massless-point followers. This model uses a z-upwards orientation of Cartesian coordinates.

3.6.1.1 Nomenclature

The nomenclature used in the passenger-vehicle vertical dynamics model, which will also be referred to in a Chapter 5 case study, is as follows:

a half of wheelset contact lateral distance

b half of wheelbase

c_{pz} primary suspension vertical damping (4 per truck)

c_{sz} secondary suspension vertical damping (2 per truck)

d_p half of lateral distance between primary suspension springs

d_s half of lateral distance between secondary suspension springs

I_{cx} roll moment of inertia of carbody

I_{cy} pitch moment of inertia of carbody

I_{tx} roll moment of inertia of truck

I_{ty} pitch moment of inertia of truck

k_{pz} primary suspension vertical stiffness (4 per truck)

k_{sz} secondary suspension vertical stiffness (2 per truck)

$\bar{\ell}$ half of carbody length in the longitudinal direction

ℓ_c carbody length in the longitudinal direction

ℓ_s half of truck center pin spacing

M_c carbody mass

M_t truck mass

T_b first modal amplitude in bending of the uniform unconstrained beam model of the carbody

T_{t_1} first modal amplitude in torsion of the uniform unconstrained beam model of the carbody

T_{t_2} second modal amplitude in torsion of the uniform unconstrained beam model of the carbody

V vehicle forward speed

x_d longitudinal distance from the vehicle cg to the passenger point

y_d lateral distance from the vehicle cg to the passenger point

z_f front truck secondary suspension vertical stroke

z_r rear truck secondary suspension vertical stroke

ω frequency

3.6.1.2 Equations of motion

The degrees of freedom represented include the heave, roll, and pitch of both trucks and the carbody. In addition to these rigid-body modes, carbody flexibility has been incorporated in the model by including the first bending mode and first two torsional modes of the carbody.

The degrees of freedom for the leading truck are:

$z_{\ell t}$ the heave or vertical position of the leading truck center of mass,

$\phi_{\ell t}$ the roll or rotation of the leading truck about the x-axis, and

$\theta_{\ell t}$ the pitch or rotation of the leading truck about the y-axis.

The corresponding degrees of freedom for the trailing truck are z_{tt}, ϕ_{tt}, and θ_{tt} respectively.

The degrees of freedom of the carbody are:

z_c the heave or vertical position of the carbody center of mass,

ϕ_c the roll or rotation of the rigid carbody about the x-axis,

θ_c the pitch or rotation of the rigid carbody about the y-axis,

$u(x,t)$ the first bending mode of the carbody,

$w_1(x,t)$ the first torsion mode of the carbody, and

$w_2(x,t)$ the second torsion mode of the carbody.

Denoting u_v and u_c as the vertical profile and cross-level irregularities, respectively, the displacement inputs transmitted by the track to the leading wheelsets of the leading truck at time t are:

$$u_{\ell tj} = u_v(t) \mp \frac{d_p}{2a} u_c(t) \qquad j=1,2 \tag{3.48a}$$

$$u_{\ell tj} = u_v(t-t_1) \mp \frac{d_p}{2a} u_c(t-t_1) \qquad j=3,4 \tag{3.48b}$$

$$u_{ttj} = u_v(t-t_2) \mp \frac{d_p}{2a} u_c(t-t_2) \qquad j=1,2 \tag{3.48c}$$

$$u_{ttj} = u_v(t-t_3) \mp \frac{d_p}{2a} u_c(t-t_3) \qquad j=3,4 \tag{3.48d}$$

where

$$t_1 = 2b/V \tag{3.48e}$$

$$t_2 = 2\ell_s/V \tag{3.48f}$$

$$t_3 = 2(b+\ell_s)/V \tag{3.48g}$$

and V is the vehicle's forward velocity.

The vertical dynamics model is used primarily to calculate the vertical accelerations at a specified passenger location in the vehicle due to vertical and cross-level disturbances of the track. The vertical stroke of the secondary suspension elements is also calculated. The vertical force between a wheel and the track is estimated as the force in the primary suspension elements associated with that wheel. A detailed derivation of the 12 second-order differential equations describing the vertical dynamics is give by Murray et al. [32] and the model can be expressed in the following vector-matrix form:

$$[M]\ \ddot{\underset{\sim}{y}}+[C]\ \dot{\underset{\sim}{y}}+[K]\ \underset{\sim}{y} = [B]\ \underset{\sim}{u} \tag{3.49}$$

where [M], [C], and [K] are the inertia, damping, and stiffness matrices, respectively, [B] is the input distribution matrix, $\underset{\sim}{y}$ is the vector of displacements and $\underset{\sim}{u}$ is the vector of track inputs. Once the time domain solution of Eq. (3.49) has been obtained, the vertical acceleration a_v at an arbitrary passenger position, $x = \bar{\ell} + x_d$, within the carbody can be calculated as

$$\begin{aligned} a_v = -\omega^2[&z_c-x_d\theta_c+y_d\phi_c+u(\bar{\ell}+x_d)T_b+y_dw_1(\bar{\ell}+x_d)T_{t1} \\ &+ y_dw_2(\bar{\ell}+x_d)T_{t2}] \end{aligned} \tag{3.50}$$

and the secondary suspension vertical stroke lengths may be computed by

$$\begin{aligned} z_f = {}&z_c-\ell_s\theta_c+d_s\phi_c+u(\bar{\ell}+\ell_s)T_b+d_sw_1(\bar{\ell}+\ell_s)T_{t1} \\ &+ d_sw_2(\bar{\ell}+\ell_s)T_{t2}-z_{\ell t}-d_s\theta_{\ell t} \end{aligned} \tag{3.51}$$

$$z_r = z_c+\ell_s\theta_c+d_s\phi_c+u(\bar{\ell}-\ell_s)T_b+d_s w_1(\bar{\ell}-\ell_s)T_{t1} + d_s w_2(\bar{\ell}-\ell_s)T_{t2}-z_{tt}-d_s\phi_{tt} \qquad (3.52)$$

3.6.1.3 Ride quality

This section describes the frequency domain computation of spectral densities of output variables of interest, such as passenger accelerations and suspension stroke lengths given the track spectral densities and the linear, constant coefficient dynamic models. Then the ride quality can be evaluated by integrating the passenger acceleration spectral density over 1/3 octave bands to obtain the rms acceleration in 1/3 octave frequency bands and compared with ISO ride quality criteria. The total rms value of any variable can be computed by integrating its spectral density over the entire frequency band [36].

The Fourier Transform of Eq. (3.49) yields

$$[[K]-\omega^2[M]+j\omega[C]]\{y(j\omega)\} = [\underset{\sim}{B}_1(j\omega)]\{u_1(j\omega)\} + [\underset{\sim}{B}_2(j\omega)]\{u_2(j\omega)\} \qquad (3.53)$$

where u_1 and u_2 respectively are the vertical and cross-level scalar input terms, and $\underset{\sim}{B}_1$ and $\underset{\sim}{B}_2$ are complex, frequency dependent vectors that include the effects of derivatives of the inputs and the time delays associated with the input to each wheelset. (For the explicit form of these terms, see Murray et al. [32]). Equation (3.53) can be used to compute the transfer functions $[H_1]$ and $[H_2]$ relating the input and output variables, i.e.,

$$\{y(j\omega)\} = [H_1(j\omega)]\{u_1(j\omega)\} + [H_2(j\omega)]\{u_2(j\omega)\} \qquad (3.54)$$

The inputs u_1 and u_2 are assumed to be statistically independent. Therefore the cross-spectral densities between them are zero and the spectral density of any element of the $\underset{\sim}{y}$ vector is the sum of the spectral densities due to u_1 and u_2, e.g.,

$$S_{y_i}(\omega) = |H_{1i}(j\omega)|^2 S_{u_1}(\omega) + |H_{2i}(j\omega)|^2 S_{u_2}(\omega) \qquad (3.55)$$

The mean square value $E(y_i^2)$ of y_i is computed by integrating the spectral density S_{y_i} over all positive frequencies, i.e.,

$$E(y_i^2) = \int_0^\infty S_{y_i}(\omega)d\omega \tag{3.56}$$

Alternatively, the mean square value of y_i in a given 1/3 octave band can be found by evaluating Eq. (3.56) over the 1/3 octave band of interest.

A case study example of the application of this model to evaluate the influence of parametric variations on vehicle ride quality is presented in Chapter 5.

3.6.2 Freight Car Dynamic Curving Model

In this section, a digital simulation program developed by Hedrick and Krolewski [20] that models the lateral dynamics of a typical North American freight car on tangent, spiral or constant curvature track is presented. The model can be used for the simulation of conventional three-piece freight car trucks, including radial trucks, with a primary suspension. Vertical, pitch and longitudinal dynamics are assumed to be decoupled and are excluded from the model. However, rail lateral flexibility, nonlinear wheel/rail geometry, creep force saturation and nonlinear suspension characteristics are included. A typical North American freight car, shown schematically in Fig. 3.16 consists of a carbody supported by a pair of dual-axle three-piece trucks. Wheelset motion is described in terms of lateral and yaw degrees of freedom. An extra state describing the spin rate perturbation about the axle centerline is used to describe the change in nominal spin rate during curving. The wheelsets are assumed to never lose contact with the rails and the roll angle is dependent on the wheelset lateral displacement. Lateral, yaw and warp are the three degrees of freedom used to represent truck motion. The truck's roll angle is assumed to be the average of the roll angles of the two wheelsets. The carbody is modeled by the lateral, roll and yaw degrees of freedom. Lateral flexibility and torsion of the carbody are

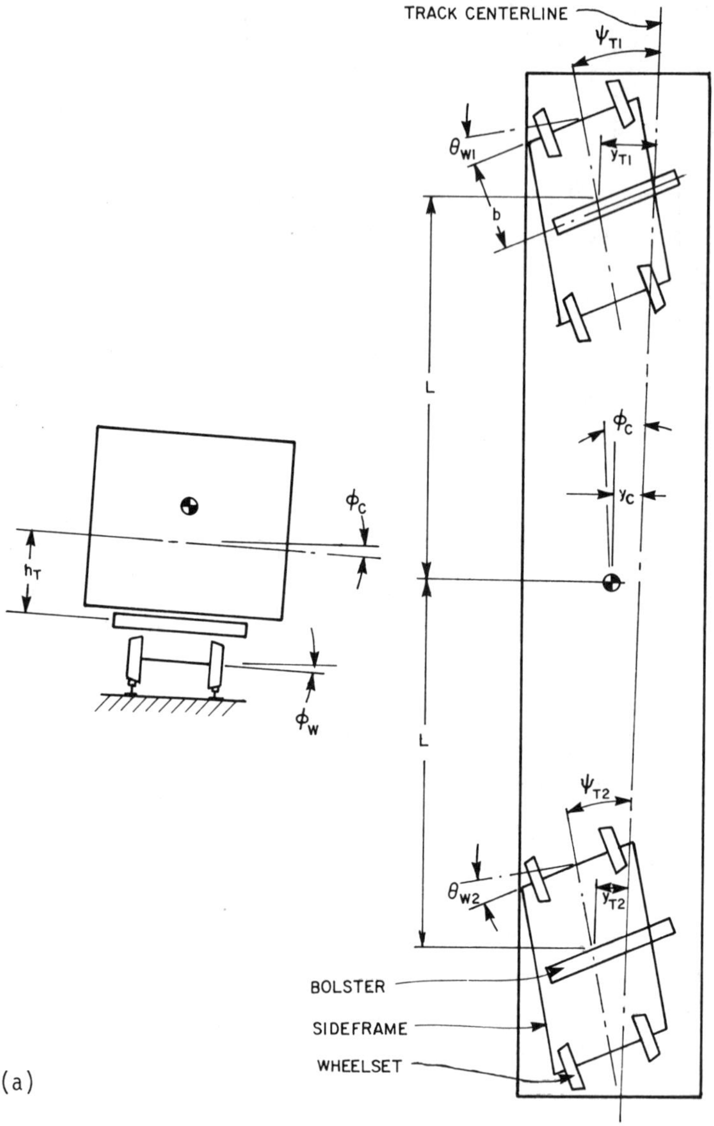

TRACK CENTERLINE
ψ_{T1}
θ_{W1}
y_{T1}
b
L
ϕ_C
y_C
h_T
ϕ_W
L
ψ_{T2}
θ_{W2}
y_{T2}
BOLSTER
SIDEFRAME
WHEELSET

(a)

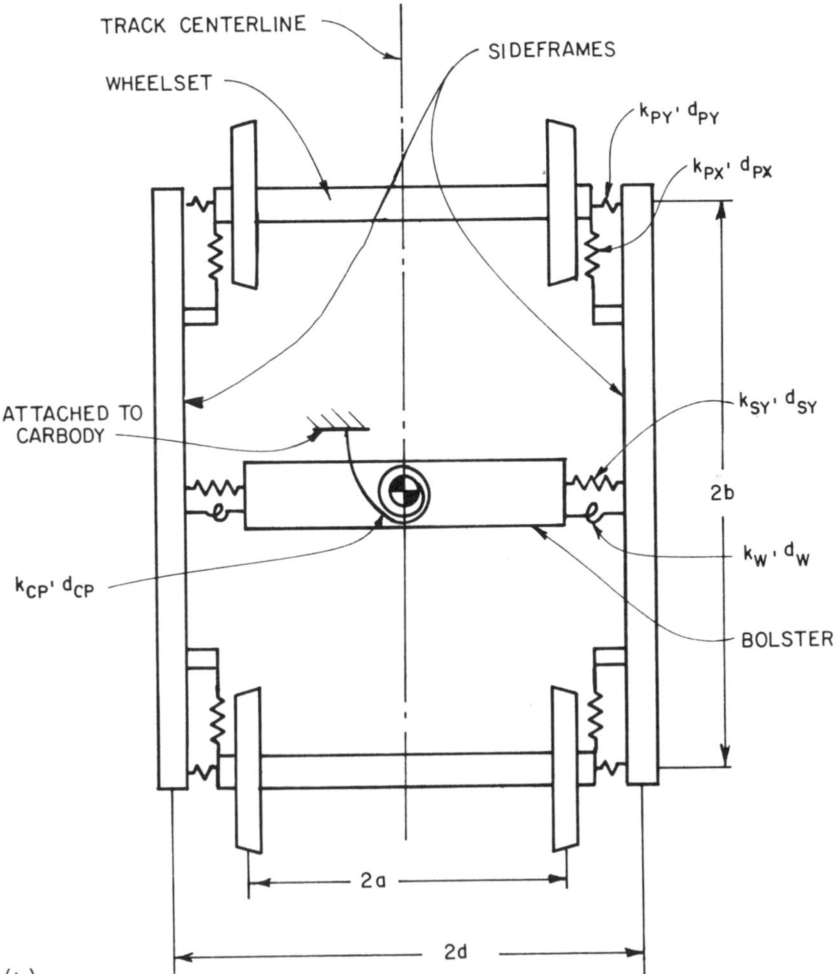

(b)

Fig. 3.16 Freight car dynamic curving model, (a) Typical north american freight car, (b) Truck equilibrium configuration plan view, (c) Truck disturbed configuration plan view, (d) Truck equilibrium configuration elevation, (e) Truck disturbed configuration elevation

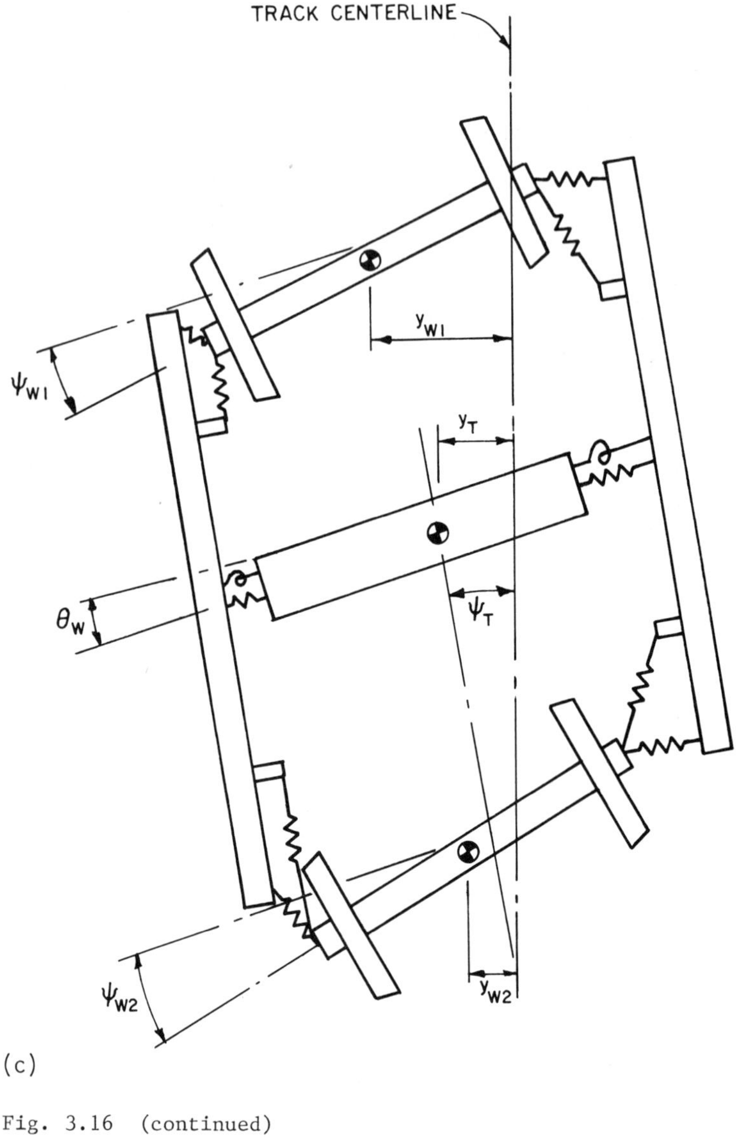

(c)

Fig. 3.16 (continued)

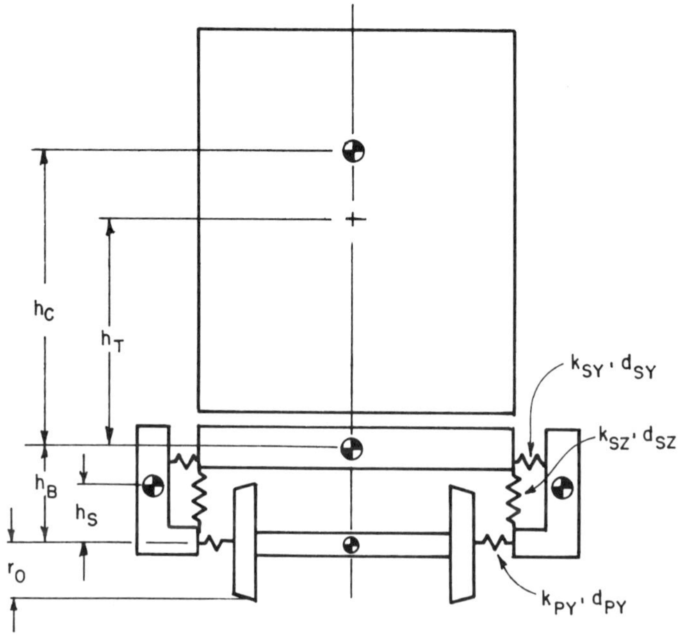

(d)

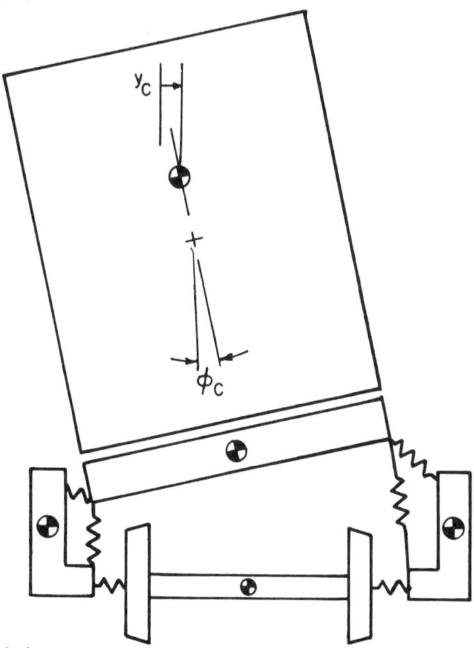

(e)

neglected. The carbody is assumed to roll and translate in unison with the bolster whereas its yaw rotation relative to the bolster is assumed to be resisted by Coulomb friction. The resulting seventeen independent degrees of freedom are listed in Table 3.4. All rigid bodies are represented by lumped masses and inertias and the connections between rigid bodies are modeled by linear springs in parallel with either Coulomb friction elements or linear viscous dampers. The wheelsets are assumed to run freely in the bearings with neither friction nor applied tractive or braking torques. Aerodynamic, coupler, and other external forces are not included in the model.

Table 3.4. Freight Car Dynamic Curving Model Degrees of Freedom.

	Type of motion				
Component	Lateral	Roll	Yaw	Warp	Spin perturbation rate*
Front truck-leading wheelset	y_{W1}	-	ψ_{W1}	-	$\dot{\beta}_1$
Front truck-trailing wheelset	y_{W2}	-	ψ_{W2}	-	$\dot{\beta}_2$
Rear truck-leading wheelset	y_{W3}	-	ψ_{W3}	-	$\dot{\beta}_3$
Rear truck-trailing wheelset	y_{W4}	-	ψ_{W4}	-	$\dot{\beta}_4$
Front truck	y_{T1}	-	ψ_{T1}	θ_{T1}	
Rear truck	y_{T2}	-	ψ_{T2}	θ_{T2}	
Carbody	y_C	ϕ_C	ψ_C	-	

* These are not independent generalized coordinates.

3.6.2.1 Nomenclature

The nomenclature used in the development of the equations of motion for the freight car dynamic curving model, which will also be referred to in a Chapter 5 case study, is as follows:

a	half the wheelset contact distance
b	half the truck wheelbase
b_T	rail damping
d	half horizontal spacing between side frames
d_{CP}	effective centerplate damping
d_{PX}	primary suspension longitudinal damping
d_{PY}	primary suspension lateral damping
d_{SY}	secondary suspension lateral damping
d_{SZ}	secondary suspension vertical damping
d_W	effective warp damping
f_{11}	lateral creep coefficient
f_{12}	lateral/spin creep coefficient
f_{22}	spin creep coefficient
f_{33}	longitudinal creep coefficient
F_{Y0}	lateral friction breakout force between bolster and side frames
F_{Z0}	vertical friction breakout force between bolster and side frames
h_B	height of the bolster center of gravity above the axle cg
h_C	height of the carbody center of gravity above the bolster cg
h_S	height of the side frame centre of gravity above the axle cg
h_T	height of the carbody hinge point above the bolster cg
I_{BX}	bolster roll moment of inertia
I_{BZ}	bolster yaw moment of inertia
I_{CX}	carbody roll moment of inertia
I_{CZ}	carbody yaw moment of inertia
I_{WY}	wheelset pitch moment of inertia
I_{WZ}	wheelset yaw moment of inertia

k_B	interaxle bending stiffness
k_{CP}	effective centerplate stiffness
k_{PX}	primary suspension longitudinal stiffness
k_{PY}	primary suspension lateral stiffness
k_{SY}	secondary suspension lateral stiffness
k_{SZ}	secondary suspension vertical stiffness
k_T	rail stiffness
k_S	interaxle shear stiffness
k_W	effective warp stiffness
L	half the distance between truck centerplates
L_A	axle load
m_B	bolster mass
m_C	carbody mass
m_S	sideframe mass
m_W	wheelset mass
N_0	nominal normal force
r_L	wheelset left rolling radius
r_0	wheelset nominal rolling radius
r_R	wheelset right rolling radius
R	track radius of curvature
T_{CPO}	centerplate friction breakaway torque
T_{WO}	warp friction breakaway torque
V	vehicle forward speed
Y_L	lateral displacement of the left rail
Y_R	lateral displacement of the right rail
μ	coefficient of friction

3.6.2.2 Model nonlinearities

The model nonlinearities, namely, the wheel/rail characteristics, suspension characteristics, curve geometry and track flexibility included in the model are briefly described below.

(a) Wheel/Rail Characteristics: The wheel/rail geometry shown in Fig. 3.10 is described by the wheelset roll angle, the rolling radii of the wheels and the contact angles between the wheels and the rails. Within the tread region, the contact parameters can be expressed as linear functions of the wheelset lateral excursion. Rapid changes and sudden discontinuities in contact geometry, however, are associated with flange contact. Since flanging is common during curving, tabulated wheel/rail profiles developed by Cooperrider and Law [11] are used to determine rolling radii, roll angle and contact angles. The first and second derivatives of roll angle are approximated using first order differences of data points and the time derivatives of wheelset excursions. The vertical velocity of the wheelset is neglected. The nonlinear creep forces with spin model described in Section 3.4.3 is used to calculate the wheel/rail forces and moments.

(b) Nonlinear Suspension Characteristics: The effects of freight car suspension nonlinearities are often more pronounced in curve negotiation than in tangent track operation. The only intentional suspension components are the coil springs and friction wedges between the bolster and sideframes. The lateral, vertical and warp motions of the truck are resisted by dry friction at each sideframe/bolster connection. Each of these connections is modeled by linear springs in parallel with a Coulomb friction element. The characteristics are shown in Figs. 3.17(a) through (c). The frictional resistance to rotation of the bolster centerplate relative to the carbody is modeled by a breakout torque.

A mathematical representation of an ideal Coulomb friction element [21] is illustrated in Fig. 3.18(a). For computational reasons, the discontinuity at zero velocity is avoided by the piecewise linear approximation of Fig. 3.18(b), where the selection of the width of the linear viscous band in the small velocity region is important. Too wide a band could introduce viscous damping effects, whereas too narrow a band could cause the discrete integration method to miss the stopped condition.

An alternative method of overcoming the difficulty of simulating Coulomb friction digitally has been investigated in which the connections between bolster and sideframe, and bolster and carbody, were modeled by linear springs in parallel with linear viscous damper equivalents of Coulomb friction. A drawback of this

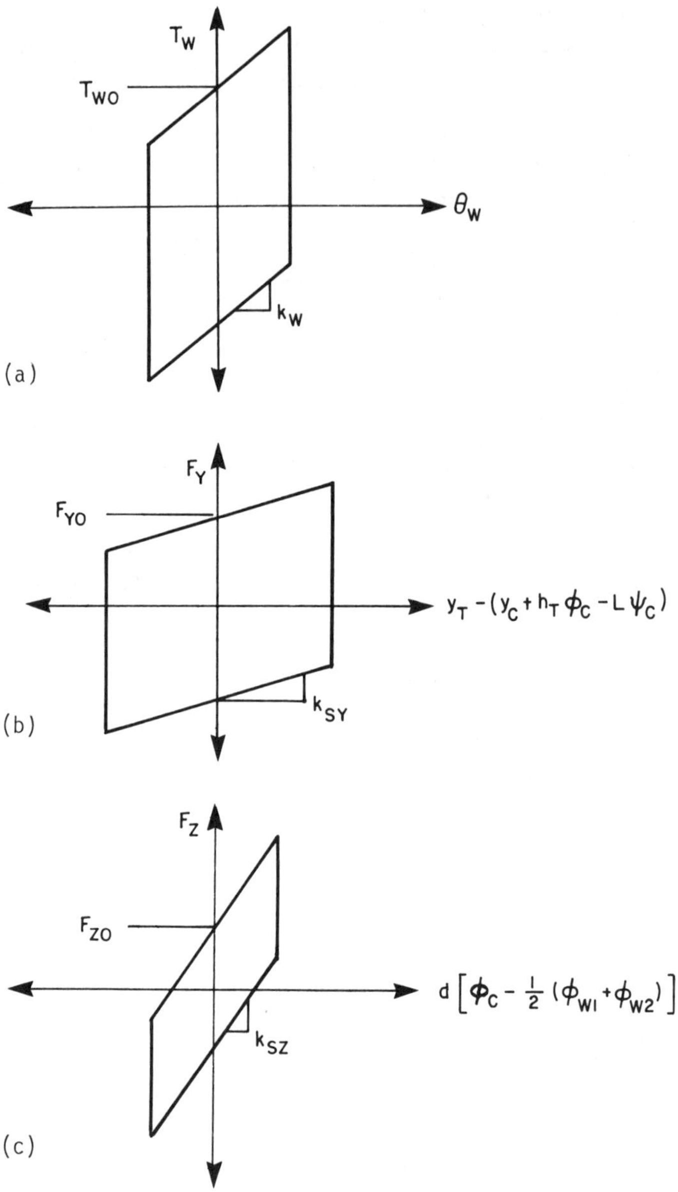

Fig. 3.17 Freight car suspension characteristics (a) Warp suspension characteristic (b) Lateral suspension characteristic (c) Vertical suspension characteristic (from Hedrick et al. [20])

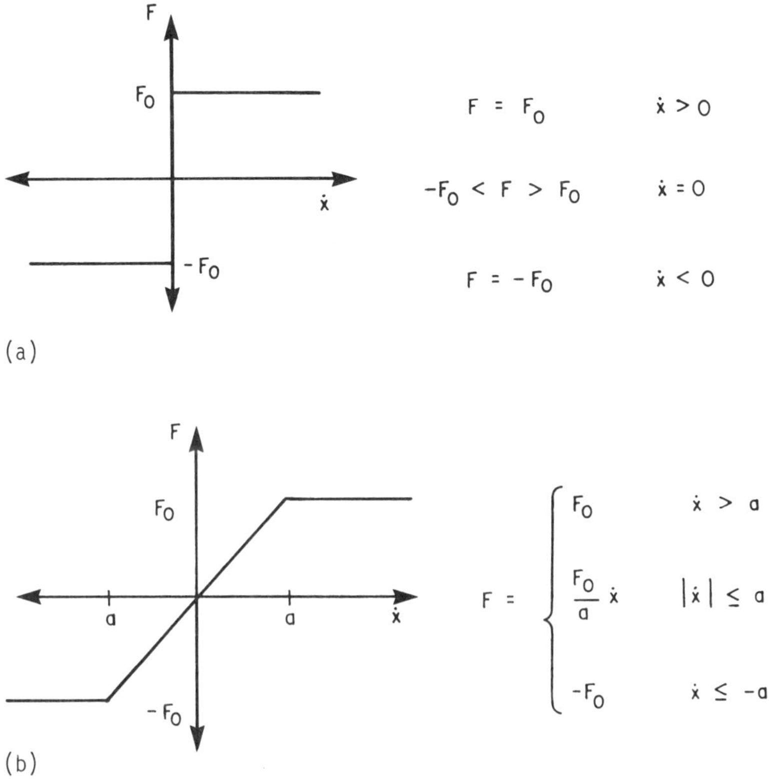

Fig. 3.18 Coulomb friction element, (a) Ideal Coulomb friction element, (b) Piecewise linear approximation of Coulomb friction (from Hedrick et al.

approach is that the phenomenon of gross slipping cannot be modeled by a linear viscous damper.

(c) Curve Geometry: The superelevation angle, track curvature and vehicle speed, all combined into the cant deficiency (Section 3.3.2), and track irregularities constitute the inputs to the vehicle model. The track, shown in Fig. 3.19, is composed of a tangent segment followed by a transition curve and constant radius curve. Along the transition curve, whose purpose is to reduce the magnitude of the peak flange force during the transition, the superelevation angle and radius of curvature vary quadratically.

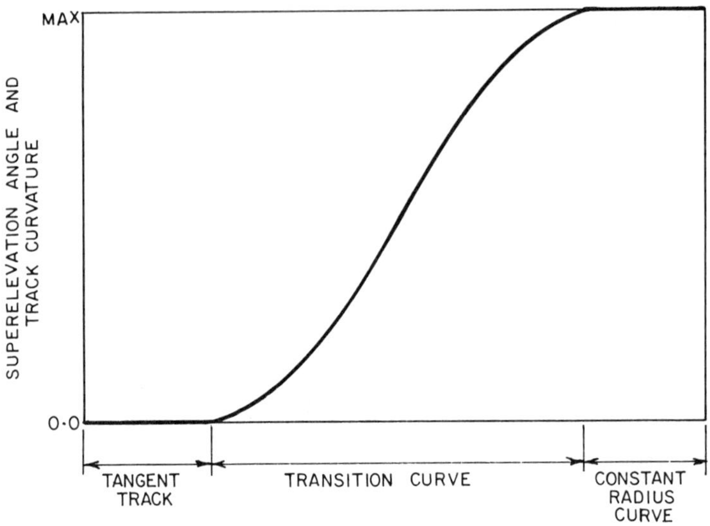

Fig. 3.19 Superelevation and track curvature as a function of distance

Whereas track irregularities have not been included directly, the implementation of cross level and alignment irregularities is possible by adding a random disturbance to the superelevation angle and the lateral excursion term in the wheel/rail geometry, respectively.

(d) Track Flexibility: Allowances are made for the rail's lateral displacement by including a finite lateral track stiffness to avoid predictions of excessive lateral forces during flanging [6,10].

Rail movement at each of the four wheels is assumed to be resisted by a linear spring in parallel with a viscous damper model as shown in Fig. 3.20. Neglecting the mass of the rails, the motion of the left and right rails is determined by:

$$b_T\dot{y}_L + k_T y_L = F_{LY} + N_{LY} \tag{3.57}$$

and

$$b_T\dot{y}_R + k_T y_R = F_{RY} + N_{RY} \tag{3.58}$$

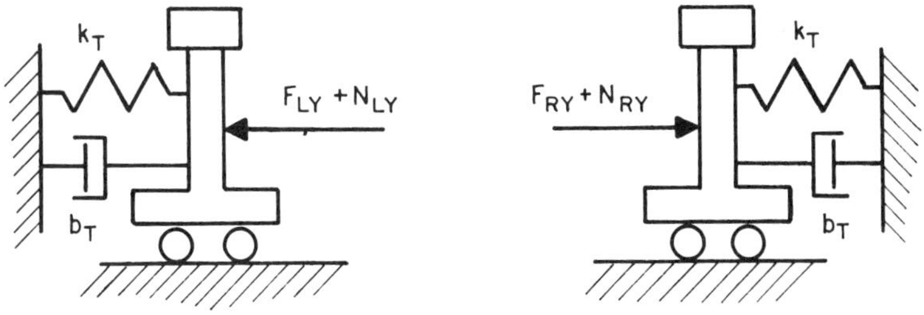

Fig. 3.20 Lateral rail flexibility model

An effective lateral excursion, found by subtracting the rail deflection from the wheelset lateral excursion, is used to determine the wheel/rail contact geometry. This method neglects the inertia of the rails and the influence of rail velocity on lateral creepage on the basis that the lateral creep force is usually in saturation during flange contact.

Values of track stiffness and damping can vary significantly between different types of track and loadings. Track lateral force/deflection data presented by Ahlbeck [2] illustrates the highly nonlinear dependence of lateral stiffness on both the lateral and vertical load, as well as the type and condition of rails, ties, fasteners and roadbed.

3.6.2.3 Equations of motion

A detailed derivation of the forty-six first order nonlinear differential equations of motion for the complete freight vehicle are presented by Krolewski [28]. The computerized model [20] uses a standard fourth order Runge-Kutta numerical integration algorithm and its outputs include wheel/rail forces, L/V ratios, wheel wear indices, axle angle-of-attacks, and carbody accelerations. A case study is presented in Section 5.4.

3.6.3 Six–Axle Locomotive Nonlinear Response Model

A nonlinear 29 degree of freedom mathematical model of a six-axle locomotive on tangent track, developed by Arslan [7], is presented in this section. The nonlinearities in the model include: (a) wheel/rail geometrical nonlinearities, (b) suspension nonlinearities, (c) nonlinear creepages and creep force saturation, and (d) rail flexibility. The degrees of freedom were chosen such as to include all the rigid body modes. A nonlinear model is appropriate since it has been shown by Law and Cooperrider [29] that a locomotive's nonlinearities have a significant influence on its dynamic behavior.

A schematic diagram of the locomotive is shown in Fig. 3.21. The model consists of a carbody, two bolsters, two trucks, six axles and the associated suspension elements (not shown), and the analysis is based on the following assumptions: (1) the vehicle operates at constant forward speed on tangent track; (2) all elements are rigid, and their stiffnesses are lumped in the suspension elements; (3) the wheels are always in contact with the rails, i.e., there is no wheel lift. The degrees of freedom pertaining to each mass are listed in Table 3.5. The carbody and each truck have five degrees of freedom: lateral, vertical, yaw, roll and pitch. It is assumed that there is no relative motion, except in yaw,

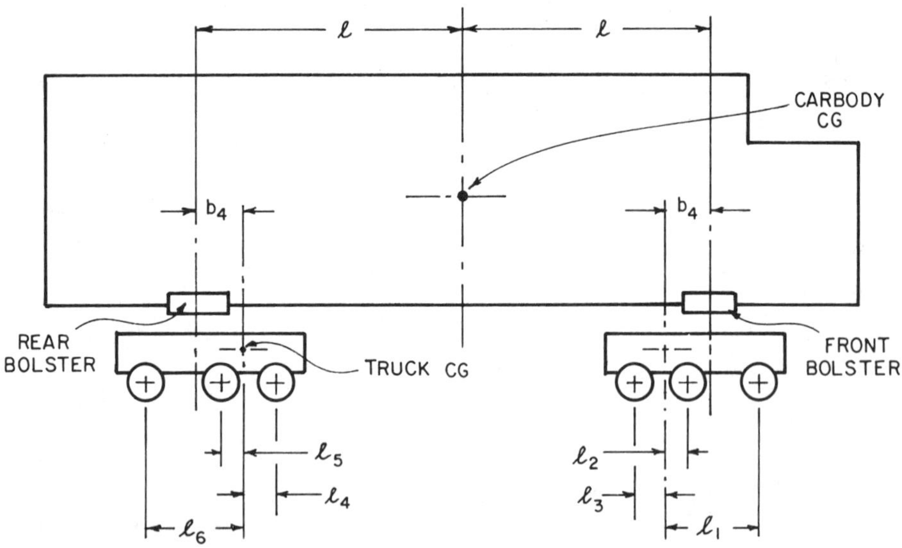

Fig. 3.21 Six-axle locomotive model (from Arslan [7])

Table 3.5. Six-axle Locomotive Nonlinear Response Model Degrees of Freedom.

	Type of motion				
Component	Lateral	Vertical	Roll	Pitch	Yaw
Axle 1	y_1	-	-	-	ψ_1
Axle 2	y_2	-	-	-	ψ_2
Axle 3	y_3	-	-	-	ψ_3
Axle 4	y_4	-	-	-	ψ_4
Axle 5	y_5	-	-	-	ψ_5
Axle 6	y_6	-	-	-	ψ_6
Front truck	y_{t1}	z_{t1}	ϕ_{t1}	θ_{t1}	ψ_{t1}
Rear truck	y_{t2}	z_{t2}	ϕ_{t2}	θ_{t2}	ψ_{t2}
Front bolster	-	-	-	-	ψ_{b1}
Rear bolster	-	-	-	-	ψ_{b2}
Carbody	y_b	z_b	ϕ_b	θ_b	ψ_b

between the bolster and carbody. Each axle has lateral and yaw degrees of freedom and its vertical and roll motions can be obtained from the wheel/rail geometrical constraints and various track irregularities.

3.6.3.1 Nomenclature

The nomenclature used in the development of the equations of motion for the six-axle locomotive nonlinear response model, which will also be referred to in a Chapter 5 case study, is as follows:

a half length of wheel base

a_{1i} roll coefficient for axle i

b_1 lateral distance between secondary suspensions

b_2 longitudinal distance between secondary suspensions

b_3 lateral distance between primary suspensions

b_4 longitudinal distance between bolster and truck cg

c_{pxi} longitudinal damping in primary suspension at location i

c_{pyi} lateral damping in primary suspension at location i

c_{pzi} vertical damping in primary suspension at location i

c_{sxi} longitudinal damping in secondary suspension at location i

c_{syi} lateral damping in secondary suspension at location i

c_{szi} vertical damping in secondary suspension at location i

c_{cp1} viscous damping between front bolster and carbody

c_{cp2} viscous damping between rear bolster and carbody

c_{RL} lateral damping of the rail d_{pxi} clearance in longitudinal primary suspension at location i

d_{pyij} jth clearance in lateral primary suspension at location i

d_{sxi} longitudinal clearance in secondary suspension at location i

d_{syi} lateral clearance in secondary suspension at location i

f_{11} lateral creep coefficient

f_{12} lateral/spin coefficient

f_{22} spin creep coefficient

f_{33} longitudinal creep coefficient

h_{ap} height of primary spring center above axle cg

h_{cs} height of carbody cg above bolster spring center

h_{tp} height of truck frame cg above axle center

h_{ts} height of bolster spring center above truck cg

I_{bz} yaw moment of inertia of the bolster

I_{cx} roll moment of inertia of the carbody

I_{cy} pitch moment of inertia of the carbody

I_{cz} yaw moment of inertia of the carbody

I_{tx} roll moment of inertia of the truck

I_{ty} pitch moment of inertia of the truck

I_{tz} yaw moment of inertia of the truck

I_{wx} roll and yaw moment of inertia of the wheelset

I_{wy} spin moment of inertia of the wheelset

k_{gi} k_g for axle i

k_{pxij} jth stiffness in longitudinal primary suspension at location i

k_{pyij} jth stiffness in lateral primary suspension at location i

k_{pzi} vertical stiffness in primary suspension at location i
k_{RL} lateral stiffness of the rail
k_{RV} vertical stiffness of the rail
k_{sxij} jth longitudinal stiffness in secondary suspension at location i
k_{syij} jth lateral stiffness in secondary suspension at location i
k_{szi} vertical stiffness in secondary suspension at location i
$k_{\delta i}$ k_{δ} for axle i
ℓ distance between truck centre of gravities
ℓ_1 distance between front truck cg and lead axle
ℓ_2 distance between front truck cg and middle axle
ℓ_3 distance between front truck cg and trailing axle
ℓ_4 distance between rear truck cg and lead axle
ℓ_5 distance between rear truck cg and middle axle
ℓ_6 distance between rear truck cg and trailing axle
m_b bolster mass
m_c carbody mass
m_t truck mass
m_w wheelset mass
r_0 wheel tread radius
T_{cp1} front truck Coulomb breakaway torque
T_{cp2} trailing truck Coulomb breakaway torque
W nominal axle load
δ_{0i} centered wheelset contact angle for axle i
λ_i wheelset conicity for axle i
μ coefficient of friction

3.6.3.2 Model nonlinearities

The model nonlinearities mentioned previously are briefly discussed here.

(a) Wheel/Rail Geometrical Nonlinearities: It has been established by several investigators that the wheel/rail geometrical parameters of significance to locomotive response include rolling radii, contact angles, and roll angle and vertical displacement of each axle. These parameters appear in the wheelset's equations of motion [29] as nonlinear functions of its lateral excursion.

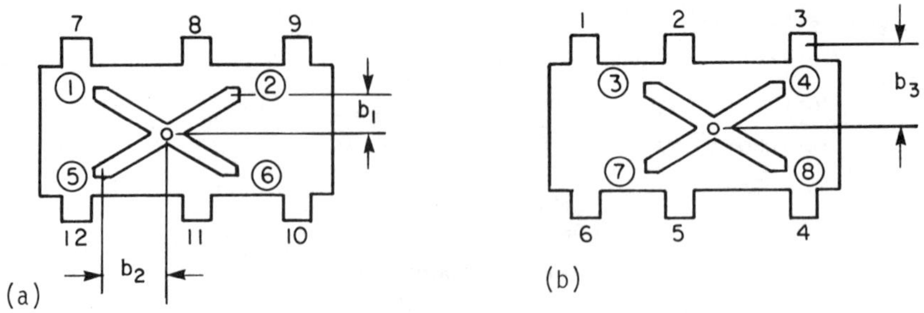

Fig. 3.22 Location of locomotive primary and secondary suspension elements, (a) Rear truck, (b) Front truck

(b) Suspension Nonlinearities: In the locomotive model, there are nonlinear elements in both the primary and secondary suspensions. The locations of these various suspension elements are shown in Fig. 3.22.

Consider the primary suspension. In the longitudinal direction, it is modeled by a piecewise linear hardening spring in parallel with a linear viscous damper, at the ends of each axle, for a total of twelve suspension groups. The force-displacement characteristic of the springs is shown in Fig. 3.23. In the

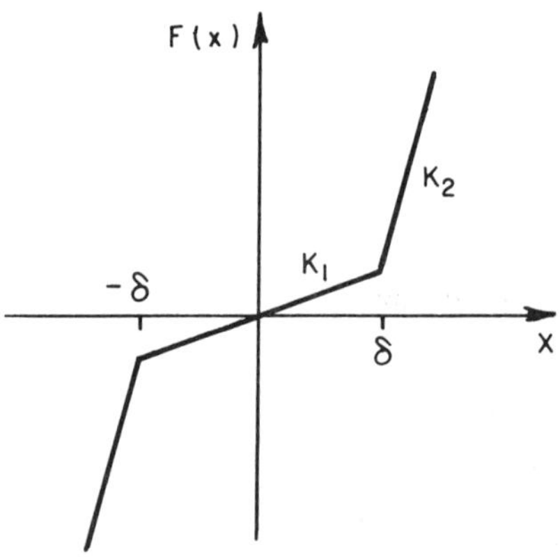

Fig. 3.23 Force-displacement characteristic for primary or secondary suspension longitudinal hardening springs

lateral direction, one parallel combination of a piecewise linear hardening spring and linear viscous damper is used, per axle, for a total of six suspension groups. Figure 3.24 shows the force-displacement characteristic of the lateral primary suspension springs. In the vertical direction, a parallel combination of linear spring, and linear viscous damper is used at both ends of each axle, for a total of twelve suspension groups.

Let us now examine the secondary suspension. Due to the rubber pads between the bolsters and truck frames, a bolster can move in any direction with respect to its truck frame. In the longitudinal and lateral directions, the secondary suspension is modeled by linear springs with stops (i.e., hardening springs) in parallel with linear dampers, for a total of two sets per truck. The hardening spring characteristics are shown in Figs. 3.23 and 3.25. In the vertical direction, the suspension is modeled by a linear spring in parallel with a linear damper at four locations, for a total of four sets, per truck. Figure 3.26 shows the relationship used to model the yaw moment between the carbody and each bolster as a function of the relative yaw rate.

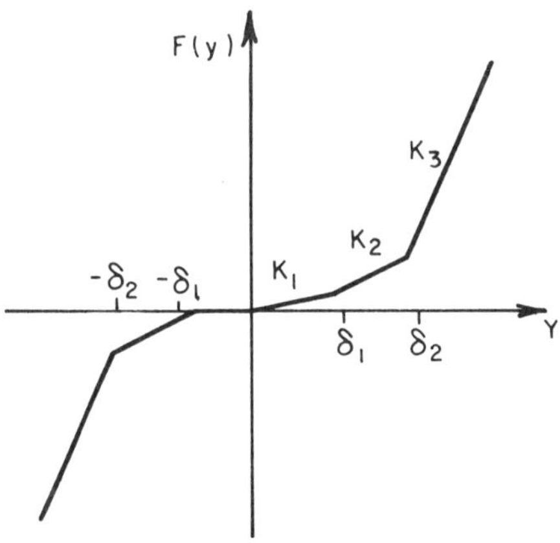

Fig. 3.24 Force-displacement characteristic for primary suspension lateral hardening springs

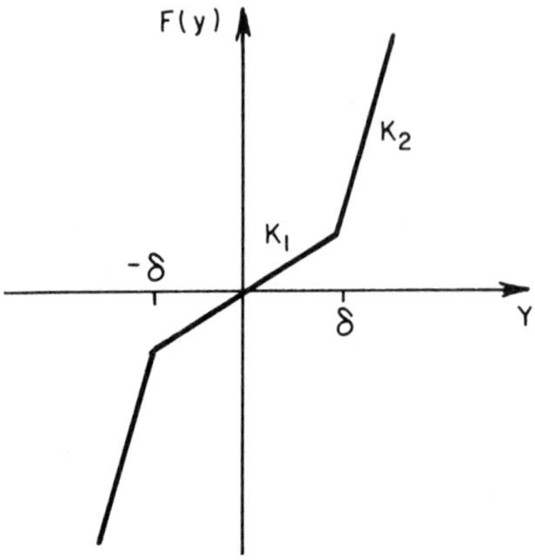

Fig. 3.25 Force-displacement characteristic for secondary suspension springs

(c) Nonlinear Creepages and Creep Force Saturation: The locomotive model presented here makes use of the nonlinear creep force with spin model of the interaction between wheels and rails discussed in Section 3.4.3.

(d) Rail Flexibility: Lateral loads cause the rails to deflect laterally, especially during flanging. The rails move vertically, also, under vertical loads. Thus, the stiffness of each rail is

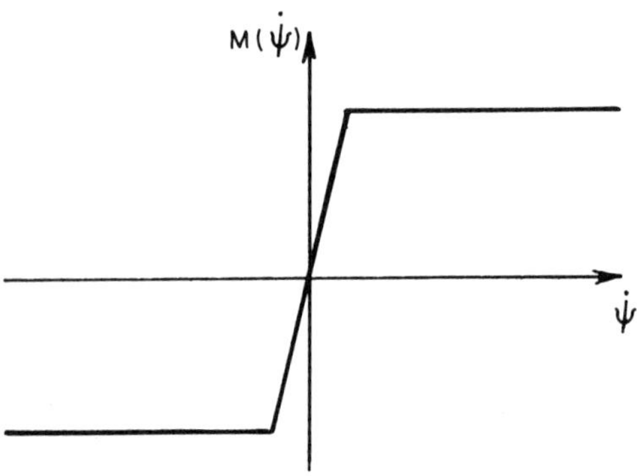

Fig. 3.26 Yaw moment characteristic for the suspension between a carbody and bolster

modeled by a lateral spring of stiffness k_{RL} and a vertical spring of stiffness k_{RV}. A lateral viscous damper of coefficient c_{RL} is included. The lateral motion of the rails is accounted for in the wheel/rail contact geometry equations and the vertical displacement of each wheelset is modified by the vertical deflection of the rails.

3.6.3.3 Equations of motion

For detailed derivation of the equations of motion, the reader is referred to the Reference [6]. The resulting 29 second-order, nonlinear, ordinary differential equations are written in matrix form as

$$[M]\ddot{\underset{\sim}{x}} = \{f(\underset{\sim}{x},\dot{\underset{\sim}{x}},\underset{\sim}{u},\dot{\underset{\sim}{u}})\} \tag{3.59}$$

where [M] is a 29 × 29 mass matrix, $\underset{\sim}{x}$ is a displacement vector and $\underset{\sim}{u}$ is an input vector. A dot or double dot over a variable denotes the first or second derivative with respect to time. By defining

$$\underset{\sim}{y}^T = [\underset{\sim}{x}^T,\dot{\underset{\sim}{x}}^T], \tag{3.60}$$

Eq. (3.59) can be transformed into 58 first-order, nonlinear, ordinary differential equations, i.e.

$$\dot{\underset{\sim}{y}} = \{g(\underset{\sim}{y},\dot{\underset{\sim}{y}},\underset{\sim}{u},\dot{\underset{\sim}{u}})\}. \tag{3.61}$$

Arslan's [7] digital computer implementation uses a fourth-order Runge-Kutta integration algorithm to solve these equations. A case study, in Sec. 5.5, will compare computed and measured wheel/rail loads and carbody accelerations.

3.7 TRAIN MODELS

There are at least three categories of train dynamics models:

(i) Longitudinal train dynamics models: These models [34] are used to predict coupler forces and train speed, distance and time relationships, and are useful in the study of train make-up, and handling, and the design of braking systems and cushioning devices.

(ii) Lateral train dynamics models: These models [35] are used primarily to investigate the lateral stability of trains under various operating conditions. They are used to study the effects of coupler length and alignment, and vehicle geometry on lateral stability. They may also provide information on gauge widening, rail rollover and wheel climbing.

(iii) Vertical train dynamics models: These models [33] serve as tools to study the heave action of trains under various operating conditions. The tendency of certain operating and impact conditions to produce coupler disengagements and carbody separations from their trucks can be investigated.

A longitudinal train dynamics model known as DTAM (Detailed Longitudinal Train Action Model), developed by Martin and Tideman [34], is presented here as a specific example whose computer implementation will be used for a case study in Section 5.6. It analyzes the longitudinal behaviour of a whole train, or

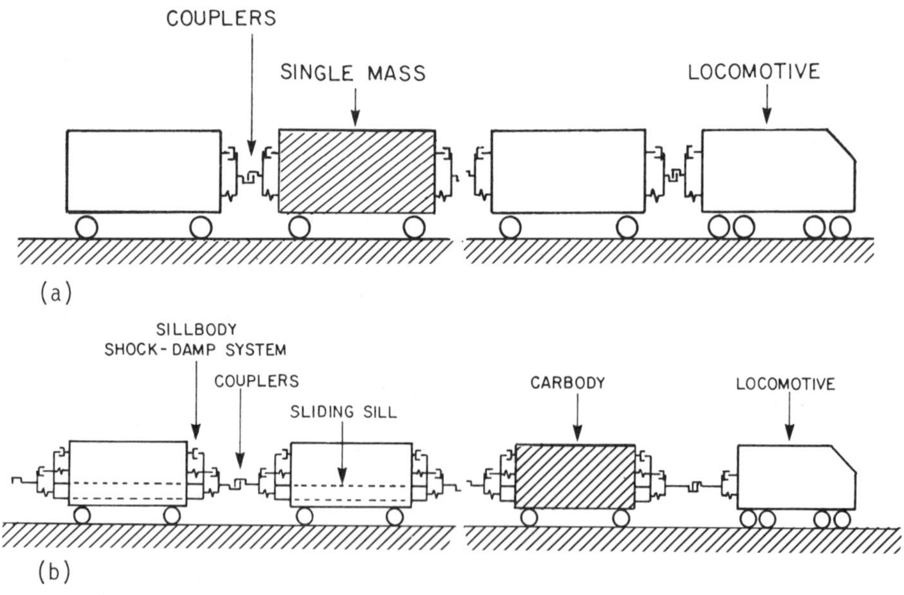

Fig. 3.27 Longitudinal train dynamics model for vehicles, (a) Without sliding sill, (b) With sliding sill

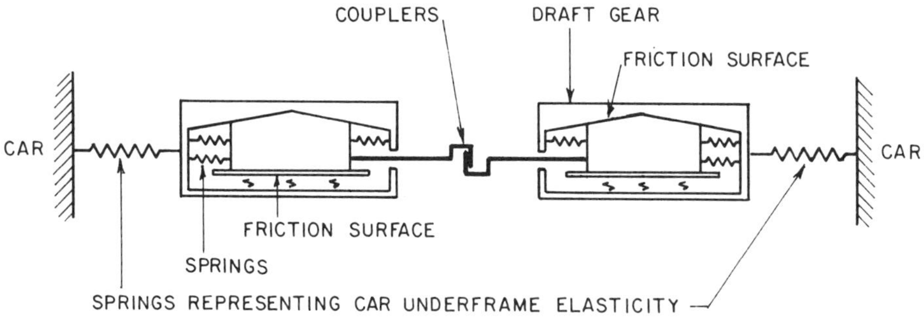

Fig. 3.28 Schematic of draft gear and coupler assemblage

the vehicles within it. A train is modeled as a string of masses interconnected by spring-dashpot-friction assemblages representing draft gears and cushioning devices. Each mass is that of an entire vehicle, as shown in Fig. 3.27(a), unless the vehicle is equipped with a sliding sill; then, it is considered to be composed of two masses, its sill and its body, as shown schematically in Fig. 3.27(b).

A draft gear and coupler model is drawn in Fig. 3.28. Figure 3.29 shows the representation of two end-of-car devices with inherent coupler slack.

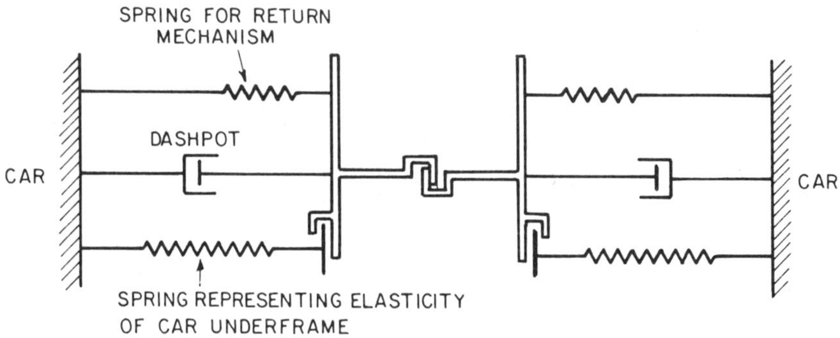

Fig. 3.29 Schematic of an end-of-car cushioning device and coupler assemblage

The equations of motion in program DTAM are based on the following approximations: (1) only one longitudinal degree of freedom is considered for each vehicle; (2) the trucks are rigid; hence, suspension effects are neglected; (3) the effects of track irregularities, worn wheels and stuck brakes are not considered; (4) the car underframe is assumed to be elastic, even under impact conditions; (5) the lading does not slip and is integral, therefore, with its carbody; (6) the effects of coupler misalignments on longitudinal inter-car forces are neglected; (7) the longitudinal dynamics are independent of the vertical position of a vehicle's center of mass.

Let m_i and x_i be the mass and longitudinal position of the i'th vehicle, x_{i-1} and x_{i+1} being the longitudinal positions of the fore and aft vehicles, respectively. Also, let θ be the track grade angle shown in Fig. 3.30. The functional relationships of the forces acting on the i'th vehicle are:

(a) Force due to brake application

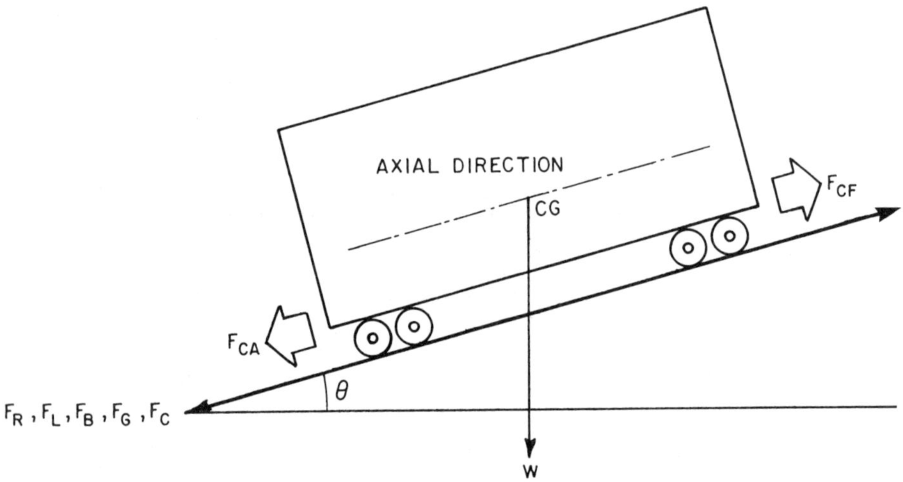

Fig. 3.30 Idealized vehicle and the forces acting on it.

$$F_B = F_B(\dot{x}_i) \tag{3.62}$$

(b) Curve resistance force

$$F_C = F_C(\dot{x}_i) \tag{3.63}$$

(c) Rear coupler force

$$F_{CA} = F_{CA}(x_i, x_{i+1}, |\dot{x}_i - \dot{x}_{i+1}|) \tag{3.64}$$

(d) Front coupler force

$$F_{CF} = F_{CF}(x_i, x_{i-1}, |\dot{x}_{i-1} - \dot{x}_i|) \tag{3.65}$$

(e) Dynamic braking force

$$F_D = F_D(\dot{x}_i) \tag{3.66}$$

(f) Grade resistance force

$$F_G = F_G(\theta) \tag{3.67}$$

(g) Tractive effort

$$F_L = F_L(\dot{x}_i) \tag{3.68}$$

(h) Rolling and wind resistance

$$F_R = F_R(\dot{x}_i) \tag{3.69}$$

Then, the equation of motion of the i'th vehicle is

$$m_i \ddot{x}_i = F_{CF} \pm F_L - F_{CA} - F_R - F_G - F_C - F_D \tag{3.70}$$

and, for an entire train of n vehicles, we have

$$[M]\ddot{\underset{\sim}{x}} = \{\underset{\sim}{g}(\underset{\sim}{x}, \dot{\underset{\sim}{x}}, \theta)\} \tag{3.71}$$

where [M] is an n x n mass matrix, $\underset{\sim}{x}$, $\dot{\underset{\sim}{x}}$ and $\ddot{\underset{\sim}{x}}$ are n-vectors of vehicle positions, velocities and accelerations, and $\underset{\sim}{g}$ is an n-vector of functional relationships.

In addition to solving Eq. (3.71), computer program DTAM performs several other functions, as suggested by its outputs which include:

- an echo of input data,
- time and speed of trains,
- brake line pressures and independent vehicle brake settings,

- mile post and distance reached by front locomotive,
- locomotive throttle setting and amperage reading,
- vehicle location with grade and curvature information,
- vehicle brake pipe and cylinder pressures,
- maximum and minimum, front and rear, vehicle-drawbar forces, and
- maximum drawbar force and its location in the train.

The user-supplied information required by the program includes:

- track data, e.g., station name, speed limit, curvature and profile data, rail condition,
- vehicle geometry and brake characteristics,
- number of vehicles in the train,
- brake pipe pressure,
- tractive effort and dynamic characteristics of non-standard power units,
- throttle and brake handling sequence, and
- integration time step.

3.8 SUMMARY

In this chapter, we described vehicle lateral stability, curve negotiation capability, and ride quality as the main dynamic performance criteria for rail vehicles. The deterministic and stochastic approaches commonly used to represent track irregularities were presented. Some insight into the mechanics of wheel/rail contact was gained by examining three creep-force models. The modeling of truck assemblies for locomotive, freight and passenger vehicles was examined. Three vehicle models and one train dynamics model were presented as specific examples for further use in the computer simulation case studies of Chapter 5: (a) a linear vertical dynamics model for determining the passenger vertical ride quality; (b) a freight car dynamic curving model for predicting the lateral dynamic performance of a single freight car negotiating spiral entries and exits, as well as tangent track and constant radius curves; (c) a nonlinear 29 degree of freedom model for determining the response to track irregularities of a six-axle locomotive on tangent track; (d) a detailed longitudinal train action model for predicting coupler forces, studying train make-up and handling, and designing brake systems and cushioning devices.

REFERENCES

1. AHLBECK D.R. (1974): Comparative Analysis of Dynamics of Freight and Passenger Rail Vehicles; *U.S. Dept. of Transportation*, Report FRA-ORD & D-74-39, Washington, D.C.
2. AHLBECK D.R. (1982): The Effects of Track Modulus on Vehicle-Track Dynamic Interaction; *ASME/IEEE Joint Railroad Conference*, Paper 82-RT-3, Erie, PA, U.S.A.
3. ANIS Z. and HEDRICK J.K. (1977): Characterization of Rail Track Irregularities; *U.S. Dept. of Transportation*, Final Report on Contract DOT-TSC-1206, Washington, D.C.
4. ANON. (1975): Track Safety Standards; *U.S. Dept. of Transportation*, Federal Railroad Administration, Washington, D.C.
5 ANON. (1978): Guide for the Evaluation of Human Exposure to Whole-Body Vibration; *International Standards Organization*, (2nd ed.), Ref. No. ISO 2631-1978(E), 15 January 1978.
6. ARSLAN A.V. (1980): The Application of Statistical Linearization to Nonlinear Rail Vehicle Dynamics; *Ph.D. Thesis*, Massachusetts Institute of Technology, Cambridge, U.S.A.
7. ARSLAN A.V. (1981): Nonlinear Mathematical Model of a Six-Axle Locomotive on Tangent Track; *Association of American Railroads*, Report R-498, Chicago, U.S.A.
8. BHATTI M.H. and GARG V.K. (1984): A Review of Railway Vehicle Performance and Design Criteria; *International Journal of Vehicle Design*, Vol. 5, (2), 232-254.
9. CARTER F.W. (1926): On the Action of Locomotive Driving Wheel; *Proceedings of the Royal Society*, Series A, Vol. 112, 151-157, London, U.K..
10. CLARK R.A., EIKHOFF B.M., and HUNT G.A. (1981): Prediction of the Dynamic Response of Vehicles to Lateral Track

Irregularities; *Proceedings of the 7th IAVSD-IUTAM Symposium on the Dynamics of Vehicles on Road and Tracks*, Cambridge, U.K., 7-11 Sept. 1981, 106-113.

11. COOPERRIDER N.K. and LAW E.H. (1976): Wheel/Rail Geometry for Five Wheel Profiles and Three Rail Profiles; *U.S. Dept. of Transportation*, Report ERC-R-76009, Washington, D.C.

12. COOPERRIDER N.K. and LAW E.H. (1983): Application of Mathematical Modeling Techniques to Rail Freight Car Dynamics; *Proceedings of the Symposium on the Economics and Performance of Freight Car Trucks*, Montreal, Canada, 29-44.

13. CORBIN J.C. and KAUFMAN W.M. (1975): Classifying Track by Power Spectral Density; in: *Mechanics of Transportation Suspension Systems*, PAUL B. (Ed.), ASME, Vol. 15, 1975, 1-20, New York.

14. DEAN F.E. and AHLBECK D.R. (1979): Criteria for High-Speed Curving of Rail Vehicles; Paper 79-WA/RT-12, *ASME Winter Annual Meeting*, New York, Dec. 1979.

15. GARG V.K. and DUKKIPATI R.V. (1984): Dynamics of Railway Vehicle Systems; *Academic Press*, Toronto.

16. HANES R.N. (1970): Human Sensitivity to Whole-Body Vibration in Urban Transportation Systems: A Literature Review; *The Johns Hopkins University*, Applied Physics Labortory Report APL/JK/U TPR 004, U.S.A.

17. HEDRICK J.K. (1977): Simulation and Analysis of Rail Vehicle Dynamics; *Proceedings of the Mini-Conference on Transportation*, University of Michigan, Ann Arbor, Vol. 7, (2), 127-135.

18. HEDRICK J.K., WORMLEY D.N., ARSLAN A.V., and CHIN R. (1980): Nonlinear Analysis and Design Tools for Rail Vehicles: Nonlinear Locomotive Dynamics; *Association of American Railroads*, Report R-463, Chicago, U.S.A.

19. HEDRICK J.K., WORMLEY D.N., KIM R.R., KAR A.K., and BAUM W. (1982): Performance Limits of Rail Passenger Vehicles: Conventional, Radial, and Innovative Trucks; *U.S. Dept. of*

Transportation, Federal Railroad Administration, Report DOT/RSPA/DPB-50/81/28, Washington, D.C.

20. HEDRICK J.K. and KROLEWSKI S.M. (1983): Freight Car Dynamic Curving Model: Volume 1 - Technical Documentation; *Association of American Railroads*, Report R-557, Chicago, U.S.A.

21. HELLER R. (1977): Analog and Digital Computer Simulation of Coulomb Friction; *U.S. Dept. of Transportation*, Federal Railroad Administration, Report FRA-ORD-78/07, Washington, D.C.

22. HELLER R. and COOPERRIDER N.K. (1977): Users' Manual for Asymmetric Wheel/Rail Contact Characterization Program; *U.S. Dept. of Transportation*, Federal Railroad Administration, Report FRA/ORD-78-05, Washington, D.C.

23. HOUBOLT J.C. (1961): Runway Roughness Studies in the Aeronautical Field; *Proceedings of the ASCE, Journal Air Transportation Division*, Vol. 87, (AT1), 11-31.

24. JOHNSON K.L. and VERMEULEN P.J. (1964): Contact of Non Spherical Bodies Transmitting Tangential Forces; *Journal of Applied Mechanics*, Vol. 31, 338-340.

25. KALKER J.J. (1967): On the Rolling Contact of Two Elastic Bodies in the Presence of Dry Friction; *Ph. D. Dissertation,* Delft University of Technology, Delft, The Netherlands.

26. KALKER J.J. (1980): Review of Wheel-Rail Rolling Contact Theories; in: *The General Problem of Rolling Contact*, BROWNE A.L. and TSAI N.T. (ed.), Vol. 40, 77-92, *ASME*, New York.

27. KLINGER D., COOPERRIDER N.K., HEDRICK J.K., and WHITE R.H. (1976): Guideway Vehicle Cost Reduction; *U.S. Dept. of Transportation*, Report DOT-TST-76-95, Washington, D.C.

28. KROLEWSKI S.M. (1982): Model Development for Freight Car Dynamic Curving Simulation; *Massachusetts Institute of Technology*, MS Thesis, Cambridge, U.S.A.

29. LAW E.H. and COOPERRIDER N.K. (1974): A Survey of Rail

Vehicle Dynamics Research; *Trans. of the ASME*, Journal of Dynamic Systems, Measurement and Control, Vol. 98, Series G, (2), 132-146.

30. LEATHERWOOD J.D. and DEMPSEY T.K. (1976): A Model for Prediction of Ride Quality in a Multifactor Environment; *NASA*, Report TMX-72842.
31. LYON D. and WEEKS R.J. (1983): The Design of Bogies for Low Wear of Wheel and Rail in Rapid Transit Applications; in *Contact Mechanics and Wear of Rail/Wheel Systems*, KALOUSEK J., DUKKIPATI R.V. and GLADWELL G.M.L. (Ed.), *University of Waterloo Press*, Waterloo, Canada.
32. MURRAY W., BAUM W., HEDRICK J.K., WORMLEY D.N., and KAR A.K. (1979): Performance Limits of Rail Passenger Vehicles: Evaluation and Optimization; *U.S. Dept. of Transportation*, Research and Special Programs Administration, Report DOT/RSPA/DPB-50/79/32, Washington, D.C.
33. RAIDT J.B., SHUM K.L., MARTIN G.C., and GARG V.K. (1977): Detailed Vertical Train Stability Model - Technical Documentation; *Association of American Railroads*, Report R-267, Chicago, U.S.A.
34. MARTIN G.C. and TIDEMAN H. (1977): Technical Documentation - Detailed Longitudinal Train Action Model; *Association of American Railroads*, Report R-221, Chicago, U.S.A.
35. SAYERS M.W. and HEDRICK J.K. (1977): Track Maintenance/Railcar Suspension Trade-offs to Obtain Acceptable Ride Quality; in *Passenger Vibration in Transportation Vehicles*, BERMAN A. and HANNIBAL A. (ed.), *AMD, ASME*, Vol. 24, 35-58, New York.
36. THOMAS R.L., MACMILLAN R.D., and MARTIN G.C. (1976): Quasi-Static Lateral Train Stability Model; *Association of American Railroads*, Report R-209, Chicago, U.S.A.
37. WICKENS A.H. (1965): The Dynamic Stability of Railway Vehicle Wheelsets and Bogies Having Profiled Wheels; *Int. J. Solids Structures*, Vol. 1, p. 319.
38. WHITE R.C., LIMBERT D.A., HEDRICK J.K., and COOPERRIDER N.K.

(1978): Guideway Suspension Trade-offs in Rail Vehicle Systems; *U.S. Department of Transportation*, Report DOT-OS-50107, Washington, D.C.

39. YOKOSE K. (1970): Experiment of Hunting Derailment with a One-Fifth Model Wheelset: *Railway Technical Research Institute*, Quarterly Reports, Vol. II, (4), 228-231.

4

PARAMETER IDENTIFICATION AND MODEL VALIDATION

4.1 INTRODUCTION

The 'validity' of a model is established by comparing simulation results with physical test results and accepting differences which are within specified error limits. A 'validated model' may be used to predict the behaviour of similarly-structured physical equipment under similar test conditions, provided the parameter values used in the new applications are correct within acceptable tolerances. The term 'parameter identification' is used here to signify the determination of correct parameter values for a given application. This chapter discusses parameter identification and model validation, two activities of fundamental importance to simulation in railway dynamics. The former is covered in Sections 4.2 to 4.7; the latter, in Section 4.8.

Some of the parameters in need of identification are: track data, component masses and moments of inertia, kinematic constraints, suspension stiffnesses and damping factors, geometric data, carbody bending and torsional characteristics, wheel/rail creep coefficients and train data. Masses and stiffnesses are the easiest parameters to determine accurately. Damping values are more difficult to obtain, especially when friction between pivots and slides or hysteresis of elastomeric bushings are involved. Moments of inertia and the positions of centers of mass are often estimated from design drawings. Some parameters may be determined measurements. However, such tests can be expensive and time consuming, or they may not exactly represent changing operating conditions. Therefore, to reduce the effort of characterization and represent the actual operating conditions more closely, parameter identification techniques which utilize vehicle response data taken during validation tests are becoming increasingly popular.

When simulation is invoked as a tool to predict the performance of new, or modified vehicles, it is usually sufficient to make use of typical track environments with known parameter values. Since track identification tests are expensive to perform, it is fortunate that the results of extensive tests have been published. Therefore, we will concentrate on the availability of track data, as opposed to the methods of track identification. Thereafter, since it is essential to know how to determine the parameter values of rolling stock, the emphasis will be shifted to vehicle and train identification methods.

4.2 TRACK PARAMETERS

The nominal geometry of a track and its irregularities in alignment, cross-level and vertical profile are the dominant forcing functions that cause a vehicle to respond dynamically. The availability of tangent and curved track data is discussed here.

4.2.1 Tangent Track

Various tolerances can be used to limit the irregularities during track construction and maintenance. The following three are employed extensively in the United States [31]: (1) maximum deviation from normal, (2) maximum allowable change in deviation in a given length, and (3) maximum allowable distance between the midpoint of a chord of given length touching the track at both chord-ends and the midpoint of the track between chord-ends.

The Federal Railroad Administration's (FRA) Track Class definitions for tangent track are summarized in Table 4.1. Class 6 track is the best; class 1, the worst. They limit the operating speed over a given section of track, based upon the above tolerances and a 18.9 m (62 ft) chord length [2]. Track irregularities are geometric parameters which can be categorized as periodic, aperiodic or random irregularities. Track compliance can also be a

Table 4.1. Condensed FRA safety standards for tangent track.

Safety Standard	Track Class 1	2	3	4	5	6
Speed Limits						
Freight, km/h	16	40	64	96	128	176
(mph)	(10)	(25)	(40)	(60)	(80)	(110)
Passenger, km/h	24	48	96	128	144	176
(mph)	(15)	(30)	(60)	(80)	(90)	(110)
Gauge, G						
$G \geq$ ... cm	142.24	142.24	142.24	142.24	142.24	142.24
(in)	(56.00)	(56.00)	(56.00)	(56.00)	(56.00)	(56.00)
$G \leq$ ... cm	146.68	146.05	146.05	146.05	144.78	144.15
(in)	(57.75)	(57.50)	(57.50)	(57.50)	(57.00)	(56.75)
Alignment*						
$\epsilon_a \leq$ ... cm	12.70	7.62	4.45	3.18	1.91	1.27
(in)	(5.00)	(3.00)	(1.75)	(1.25)	(0.75)	(0.50)
Rail Profile*						
$\epsilon_r \leq$ ... cm	7.62	6.98	5.72	5.08	3.18	1.27
(in)	(3.00)	(2.75)	(2.25)	(2.00)	(1.25)	(0.50)
Cross-level*						
$\epsilon_c \leq$ ... cm	7.62	5.08	4.45	3.18	2.54	1.27
(in)	(3.00)	(2.00)	(1.75)	(1.25)	(1.00)	(0.50)
$\delta_c \leq$ ... cm	7.62	5.08	4.45	3.18	2.54	1.59
(in)	(3.00)	(2.00)	(1.75)	(1.25)	(1.00)	(0.625)

*ϵ_a: Mid-offset of alignment from a 18.9 m (62 ft) chord touching the alignment at both chord-ends

ϵ_r: Mid-offset of a rail from a 18.9 m (62 ft) chord touching the rail at both chord-ends

ϵ_c: Deviation from zero cross-level at any point along the track

δ_c: Difference in cross-level between any two points less than 18.9 m (62 ft) apart along the track

significant parameter. These four types of track parameters are discussed separately.

(a) Periodic Irregularities: From Eqs. (3.11a) and (3.12), bolted-rail track with half-staggered joints can be seen to excite a vehicle with even harmonics in vertical profile (bounce) and the odd harmonics in crosslevel (roll). The coefficients of these equations are the harmonic amplitudes. Since the amplitudes decrease rapidly with harmonic number, only the first three Fourier coefficients in each case are typically used in practice.

To establish some resonable relationship between the rectified cosine wave used as a discrete input and the geometry deviations in the FRA Track Safety Standards (TSS), some random distribution of amplitudes must be assumed. If a Gaussian distribution of low joints is assumed, then the maximum allowed deviation in a 18.89 m (62 ft) chord length can perhaps be called the three-sigma value of the random distribution. This would imply that out of 10,000 rail joints, 27 would exceed the allowed maximum. The concept depends to a great extent on the method of track inspection, length of inspection sample, data reduction technique, and criterion for establishing the track class.

With the maximum allowed deviation equated to a statistical three-sigma value, the constant-amplitude rectified cosine wave representing the staggered-joint track can then be thought of as the one-sigma level, or the "most-expected" continuous track profile amplitude. The vehicle steady-state response can then be associated with this input amplitude, and the transient response to the occasional three-sigma low joint can be ignored. Based on the maximum allowed values for cross-level error, the input amplitudes were calculated by Ahlbeck et al. [8] and are given in Tables 4.2 and 4.3.

Since the response data generated by a linear model is scaled directly in proportion to the rectified cosine amplitude values, the vehicle response to any assumed track class can be easily determined once the response to any one track class input is computed.

An extensive database has been established for the statistical characterization of track irregularities in the United States. Approximately 800 km (500 miles) of typical track sections are represented. Two reports concerning the data acquisition efforts have been published (Corbin [12]; Hamid and Yang [26]).

Table 4.2. Equivalent RCW input amplitudes based on FRA track safety standards (adapted from Ahlbeck et al. [8]).

Track class	1	2	3	4	5	6
Cross-level						
deviation, cm	7.62	5.08	4.45	3.81	2.54	1.59
(in)	(3.00)	(2.00)	(1.75)	(1.50)	(1.00)	(0.625)
Equivalent RCW						
input, cm	2.54	1.70	1.47	1.07	0.84	0.53
(in)	(1.00)	(0.67)	(0.58)	(0.42)	(0.33)	(0.21)
FRA Speed Limit						
Freight, km/h	16	40	64	96	128	176
(mph)	(10)	(25)	(40)	(60)	(80)	(110)
Passegner, km/h	24	48	96	128	144	176
(mph)	(15)	(30)	(60)	(80)	(90)	(110)

(b) Aperiodic Irregularities: The analytical forms of the aperiodic irregularities listed in Table 3.1 (Chapter 3) are functions of two parameters: (i) an amplitude A, and (ii) a length-related parameter 1/k. A range of values of A and k for the various irregularities, as found in the track geometry data collected by FRA, are given in Tables 4.4 to 4.7. Typical occurrences of these isolated irregularities are summarized in Table 4.8.

(c) Random Irregularities: A survey and discussion of measured track PSDs is given by Corbin and Kaufman [13, Chapter 3]. Sayers and Hedrick [49] have developed a method of determining values of the PSD parameters A_a, A_v, Ω_s and Ω_c [Eqs. (3.10) to (3.12)] corresponding to the tolerance standards specified by the FRA for track classes 1 to 6 (Tables 4.1 and 4.2). They have established parameter values (Table 4.9) which yield a good agreement between the analytical expressions and measured data.

Table 4.3. Condensed track requirements from FRA track safety standards (adapted from Ahlbeck et al. [8]).

Requirement	Track Class 1	2	3	4	5	6
Maximum speed, km/h	16	40	64	96	128	176
(mph)	(10)	(25)	(40)	(60)	(80)	(110)
Maximum vertical profile*, cm	7.62	6.98	5.72	5.08	3.18	1.27
(in)	(3.00)	(2.75)	(2.25)	(2.00)	(1.25)	(0.50)
Maximum cross-level, cm	7.62	5.08	4.45	3.18	2.54	1.59
(in)	(3.00)	(2.00)	(1.75)	(1.25)	(1.00)	(0.625)
Maximum alignment*, cm	12.70	7.62	4.45	3.81	1.91	1.27
(in)	(5.00)	(3.00)	(1.75)	(1.50)	(0.75)	(0.50)

*maximum deviation in a 18.9 m (62 ft) chord length

Table 4.4. Parameters ranges for aperiodic gauge irregularities.

Irregularity	A cm	(in)	k m^{-1}	(ft^{-1})
Cusp	2.03-3.56	(0.8-1.4)	0.053-0.200	(0.016-0.061)
Bump	2.03-3.56	(0.8-1.4)	0.102-0.131	(0.031-0.040)
Jog	-	-	-	-
Plateau	2.03-3.30	(0.8-1.3)	0.095-0.262	(0.029-0.080)
Trough	-	-	-	-
Sine cycle	-	-	-	-
Damped cosine	1.27-2.54	(0.5-1.0)	-	-

Table 4.5. Parameter ranges for aperiodic alignment irregularities.

Irregularity	A cm (in)		k m^{-1} (ft^{-1})	
Cusp	1.27-7.62	(0.5-3.0)	0.036-0.338	(0.011-0.103)
Bump	1.27-7.11	(0.5-2.8)	0.029-0.272	(0.009-0.083)
Jog	1.27-8.38	(0.5-3.3)	0.019-0.082	(0.006-0.025)
Plateau	3.05-4.06	(1.2-1.6)	0.082-0.089	(0.025-0.027)
Trough	3.56-5.59	(1.4-2.2)	0.043-0.095	(0.013-0.029)
Sine cycle	2.03-3.05	(0.8-1.2)	0.043-0.066	(0.013-0.020)
Damped cosine	2.54-5.59	(1.0-2.2)	0.043-0.049	(0.013-0.015)

Table 4.6. Parameter ranges for aperiodic cross-level irregularities.

Irregularity	A cm (in)		k m^{-1} (ft^{-1})	
Cusp	2.29-7.62	(0.9-3.0)	0.102-0.312	(0.031-0.095)
Bump	2.25-7.62	(1.0-3.0)	0.056-0.102	(0.017-0.031)
Jog	4.06-7.11	(1.6-2.8)	0.066-0.164	(0.020-0.050)
Plateau	1.52-2.54	(0.6-1.0)	0.085-0.131	(0.026-0.040)
Trough	-	-	-	-
Sine cycle	-	-	-	-
Damped cosine	2.29-3.05	(0.9-1.2)	0.167-0.200	(0.051-0.061)

Table 4.7. Parameter ranges for aperiodic vertical-profile irregularities.

Irregularity	A cm	(in)	k m^{-1}	(ft^{-1})
Cusp	2.29-7.62	(0.9-3.0)	0.053-0.312	(0.016-0.095)
Bump	1.27-10.16	(0.5-4.0)	0.043-0.213	(0.013-0.065)
Jog	1.27-12.70	(0.5-5.0)	0.026-0.148	(0.008-0.045)
Plateau	2.29-7.62	(0.9-3.0)	0.029-0.108	(0.009-0.033)
Trough	1.78-5.08	(0.7-2.0)	0.066-0.082	(0.020-0.025)
Sine cycle	2.54-3.81	(1.0-1.5)	0.066-0.082	(0.020-0.025)
Damped cosine	-	-	-	-

Table 4.8. Typical Occurrences of Aperiodic Irregularities.

Irregularity	Occurrence
Cusp	Joints, turnouts, interlockings, sun kinks, buffer rails, insulated joints and splice bar joints in continously welded rail, piers at bridges
Bump	Soft spots, washouts, mud spots, fouled ballast, joints, spirals, grade crossings, bridges, overpasses, loose bolts, turnouts, interlockings
Jog	Spirals, bridges, crossings, interlockings, fill-cut transitions
Plateau	Bridges, grade crossings, areas of spot maintenance
Trough	Soft spots, soft and unstable subgrades, sprials
Sine cycle	Spirals, soft spots, bridges
Damped cosine	Spirals, turnouts, localized soft spots

Table 4.9. Track spectral-density parameter values (adapted from Sayers and Hedrick [49]).

Parameter	Track Class					
	1	2	3	4	5	6
A_v $cm^2 rad/m$	1.2107	1.0181	0.6816	0.5376	0.2095	0.0339
($in^2 rad/ft$)	0.0572	0.0481	0.0322	0.0254	0.0099	0.0016
A_a $cm^2 rad/m$	3.3634	1.2107	0.4128	0.3027	0.0762	0.0339
($in^2 rad/ft$)	0.1589	0.0572	0.0195	0.0143	0.0036	0.0016
Ω_s rad/m	0.6046	0.9308	0.8520	1.1312	0.8209	0.4380
(rad/ft)	0.1843	0.2837	0.2597	0.3448	0.2502	0.1335
Ω_c rad/m	0.8245	0.8245	0.8245	0.8245	0.8245	0.8245
(rad/ft)	0.2513	0.2513	0.2513	0.2513	0.2513	0.2513

For example, Fig. 4.1 is a graph of alignment and vertical profile spectral densities for class 6 track. Since there is negligible difference between the measured spectral densities, the following deductions apply to this case: (1) the identical form of Eqs. (3.10) and (3.11) is valid, and (2) the values of A_a and A_v are equal. Indeed, the track spectral-density parameter values of Table 4.9 give $A_a = A_v = 0.0339$ $cm^2 rad/m$ (0.0016 $in^2 rad/ft$) and $\Omega_c = 0.8245$ rad/m (0.2513 rad/ft) for class 6 track. The calculated curve of Fig. 4.1, obtained by inserting these parameter values into Eqs. (3.10) and (3.11), is seen to be a good approximation. In addition to the above parameter values, Table 4.9 gives

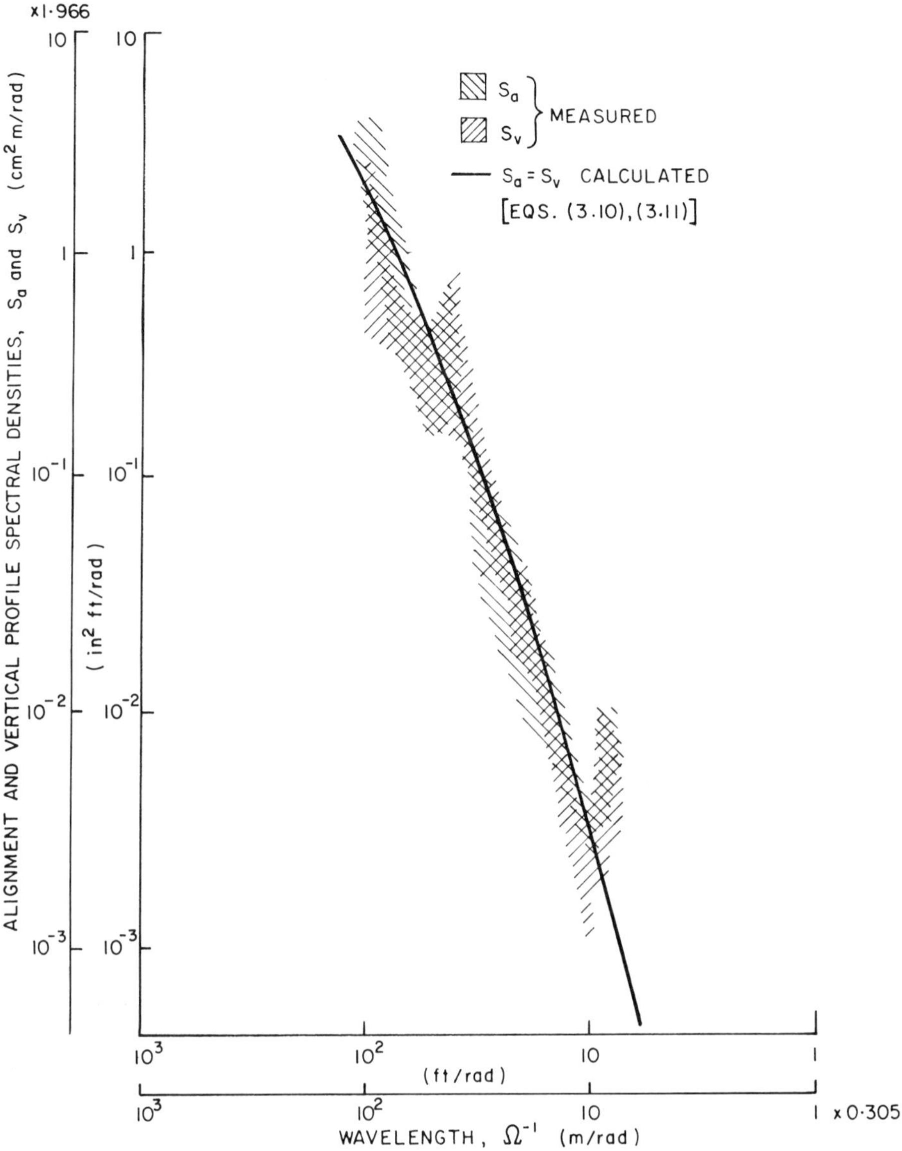

Fig. 4.1 Class 6 alignment and vertical profile spectral densities (adapted from Hedrick [31])

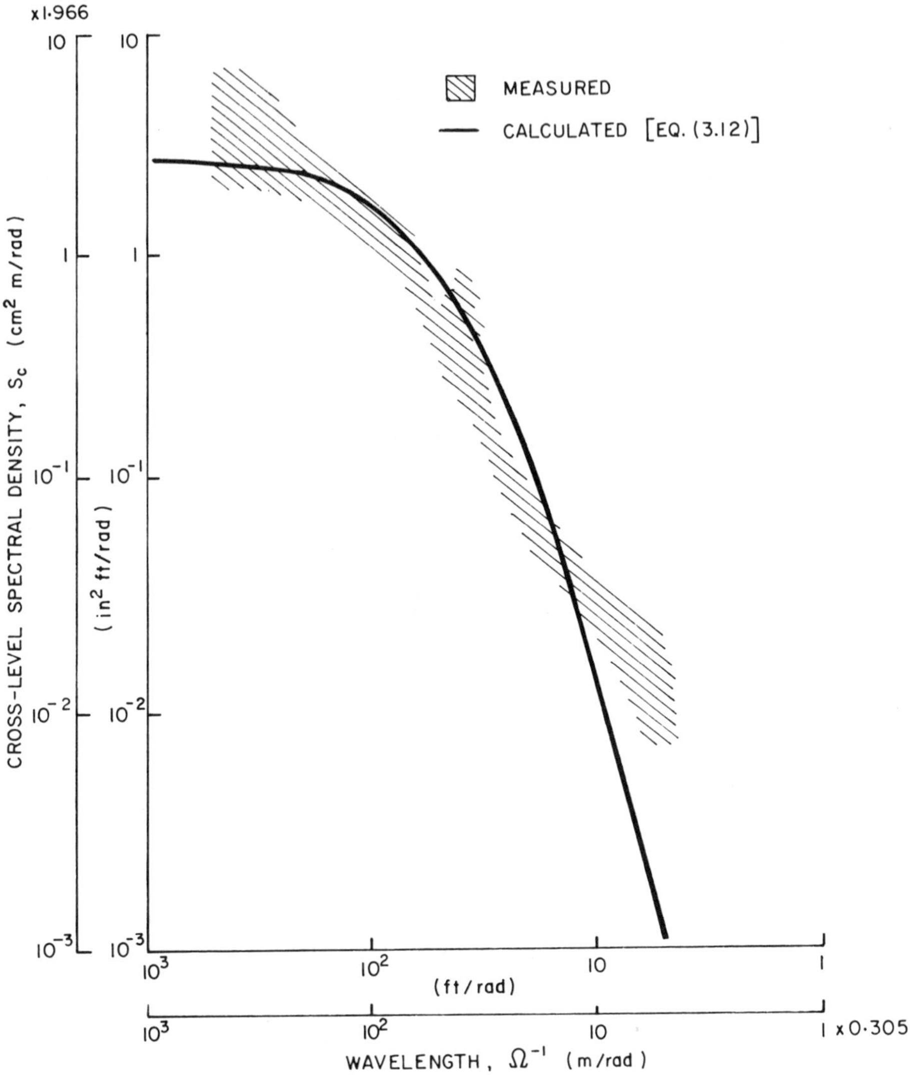

Fig. 4.2 Class 6 cross-level spectral density (adapted from Hedrick [31])

Ω_s = 0.4380 rad/m (0.1335 rad/ft) for the cross-level spectral density of class 6 track. Insertion of these four parameter values into Eq. (3.12) yields the calculated curve of Fig. 4.2, which is in fairly good agreement with the measured data. The parameter values applicable to track classes 1 to 5 are also listed in Table 4.9.

(d) Track Compliance: The lateral deflection of rails is a function of the lateral and vertical load components, as illustrated by the typical family of curves in Fig. 4.3 (Ahlbeck [7]). It is seen that the deflection depends nonlinearly on both compon-

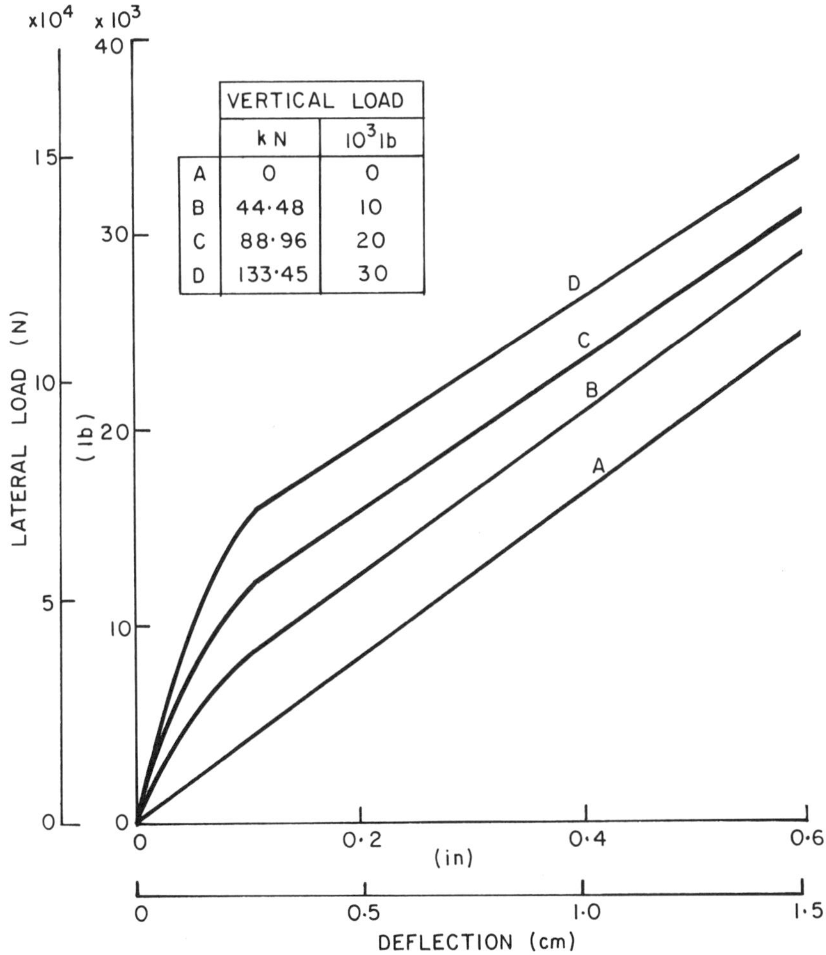

Fig. 4.3 Rail spring-rate curves (adapted from Hedrick [31])

ents. Furthermore, the characteristics vary with the type and condition of rails, ties, fasteners and roadbed.

4.2.2 Curved Track Geometry

The Federal Railroad Administration's (FRA) curved-track class definitions are summarized in Table 4.10. The geometry of track curvature is relatively the easiest of the track geometry measurements to obtain. The necessary data on track curvature and superelevation can generally be obtained from track charts. However, if accurate results are desired, a direct measurement of curvature and superelevation is recommended.

For the measurement of track curvature, for example, consider a vehicle traveling along Curve C in Figure 4.4. If the vehicle travels an infinitesimal distance, $d\ell$ (assuming R remains constant), a corresponding angle $d\theta$ is subtended by a chord drawn from point A to point B. The distance $d\ell$ is then given by

$$d\ell = \frac{R\pi d\theta}{180} = \frac{Rd\theta}{57.3} \tag{4.1}$$

where

R = radius of the curve, m (ft)

$d\theta$ = angle in degrees

If both sides of Eq. (4.1) are divided by dt, the infinitesimal time it takes the vehicle to travel from point A to point B, Eq. (4.2) results:

$$\frac{d\ell}{dt} = V = \frac{1}{K}\frac{d\theta}{dt} = \frac{1}{K}\dot{\theta} \tag{4.2}$$

where

V = velocity of the vehicle, m/s (ft/s)

$\dot{\theta}$ = angular rate, deg/s

K = 57.3/R

Therefore, curvature is given by:

$$K = \frac{\dot{\theta}}{V} \tag{4.3}$$

Multiplying K by 30.48 m (100 ft) gives the degrees subtended over 30.48 m (100 ft) of travel.

Table 4.10. Condensed FRA curved-track safety standards.

Safety Standard	Track Class 1	2	3	4	5	6
Speed Limits						
Freight, km/h	16	40	64	96	128	176
(mph)	(10)	(25)	(40)	(60)	(80)	(110)
Passenger, km/h	24	48	96	128	144	176
(mph)	(15)	(30)	(60)	(80)	(90)	(110)
Gauge, G						
$G \geq$ ... cm	142.24	142.24	142.24	142.24	142.24	142.24
(in)	(56.00)	(56.00)	(56.00)	(56.00)	(56.00)	(56.00)
$G \leq$ ... cm	145.68	146.68	146.68	146.05	146.05	144.78
(in)	(57.75)	(57.75)	(57.75)	(57.50)	(57.50)	(57.00)
Alignment*						
$\epsilon_a \leq$ ... cm	12.70	7.62	4.45	3.81	1.59	0.95
(in)	(5.00)	(3.00)	(1.75)	(1.50)	(0.625)	(0.375)
Rail Profile*						
$\epsilon_r \leq$ ... cm	8.89	7.62	5.08	3.81	2.54	1.27
(in)	(3.50)	(3.00)	(2.00)	(1.50)	(1.00)	(0.50)
$V_r \leq$ ... cm	4.45	3.81	3.18	2.54	1.90	1.27
(in)	(1.75)	(1.50)	(1.25)	(1.00)	(0.75)	(0.50)
Cross-level*						
$V_c \leq$ ... cm	5.08	4.45	3.18	2.54	1.91	1.27
(in)	(2.00)	(1.75)	(1.25)	(1.00)	(0.75)	(0.50)
$\epsilon_c \leq$ ... cm	7.62	5.08	4.45	3.18	2.54	1.27
(in)	(3.00)	(2.00)	(1.75)	(1.25)	(1.00)	(0.50)
$\delta_c \leq$ ... cm	7.62	5.08	4.45	3.18	2.54	1.59
(in)	(3.00)	(2.00)	(1.75)	(1.25)	(1.00)	(0.625)

*ϵ_a: Deviation of midordinate of a 18.9 m (62 ft) chord

ϵ_r: Runoff in any 9.45 m (31 ft) of rail at end of a raise

V_r: Deviation from designated elevation on spirals

V_c: Variation in cross-level on spirals in any 9.45 m (31 ft)

ϵ_c: Deviation from zero cross-level at any point from designated elevation on curves between spirals

δ_c: Difference in cross-level between any two points less than 18.9 m (62 ft) apart on curves between spirals.

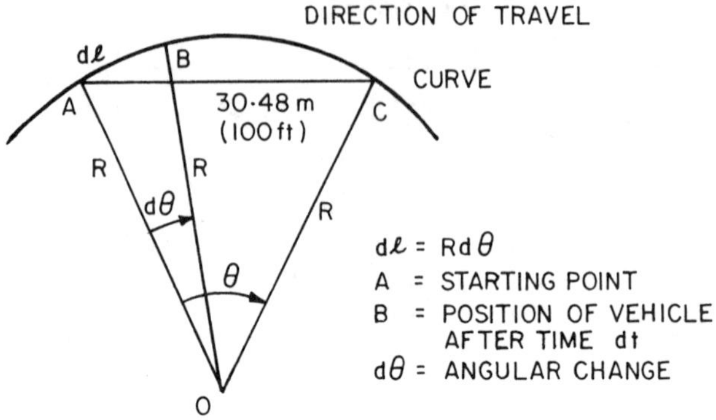

Fig. 4.4 Method of measuring track curvature

Although the definition of track curvature considers the arc subtended by a 30.48 m (100 ft) chord, the use of the 30.48 m (100 ft) of actual travel introduces a negligible error. Referring to Fig. 4.4, we have:

Chord AC = $2R\sin(\theta/2)$

Arc AC = $R\theta$

where θ is in radians.

Therefore, the percentage error introduced is:

$$\% \text{ error } \left(\frac{\text{Arc AC} - \text{chord AC}}{\text{Arc AC}}\right) 100 = \left[1 - \frac{2\sin(\theta/2)}{\theta}\right] 100 \tag{4.4}$$

For a track curvature of 12 degrees (considered large), we have:

$$\% \text{ error } = \left[1 - \frac{2(0.10453)}{0.20944}\right] 100 = 0.1814\%$$

A survey of the measurement of track geometry and track irregularities by specialized geometry cars using various techniques are well described by Cooperrider and Law [11], Shladover [55], and Gostling and Cooperrider [24]. These techniques are grouped as follows: (a) manual measurements including the use of land surveying techniques; (b) automated measurements using track geometry measurement vehicles; (c) highly specialized equipment such as the FRA Track Geometry Survey Device (TGSD); and (d) data acqusition vehicles. Each of these methods has certain advantages and disadvantages [14].

4.3 WHEEL-RAIL PARAMETERS

There are two kinds of wheel-rail contact parameters to identify: geometry and creepage. They are discussed in the following sections.

4.3.1 Contact Geometry

The Measurement of wheel and rail-head profiles, using various devices which can accurately record left and right profiles, simultaneously has been reported by Cooperrider and Law [11], Shladover [55], and Cooperrider [24].

A technique was also developed for fast and efficient digitization of graphical wheel and rail profile data [10]. These digitized data are required as inputs to the wheel-rail data-reduction programs described below.

A data-reduction program, WHRAILA, developed by Cooperider et al. [10,32] accepts digitized wheel-rail data to produce geometric variables of interest, defined in Section 3.4.4, as functions of lateral wheelset position y_w. The geometric variables are: contact positions z_L and z_R, rolling radii r_L and r_R, contact angles δ_L and δ_R, and wheelset roll angle ϕ; also, the differences $(r_L - r_R)$ and $(\delta_L - \delta_R)$. The negligible dependence of these variables on the wheelset yaw angle ψ was omitted. Case studies exemplifying the use and typical results of this program are presented in Section 5.2.

4.3.2 Creep Parameters

Kalker's nonlinear theory of creep [34,35] is regarded as the most complete and accurate theory available. The nonlinear creep force characteristics given by this theory compare well with laboratory experiments [20]. Furthermore, in addition to predicting the longitudinal and lateral forces due to relative longitudinal and lateral velocities between wheel and rail, Kalker's linear theory includes the dependence of the lateral creep force on the relative angular velocity, or spin, between wheel and rail. This force component, termed in the lateral-spin creep force, is usually less than the lateral creep force component due to lateral creepage when the wheel/rail contact remains on the wheel tread. However, as the contact point moves towards the flange and the wheel/rail contact angle increases, the lateral-spin component becomes dominant.

It has been observed that the creepages corresponding to a given creep force may be considerably larger in the field than those given by the theory. Furthermore, these relationships are functions of wheel and rail surface conditions.

A computer program which calculates the linear creep coefficients of Kalker's linear theory (Section 3.4.1) was prepared and made available by Haque et al. [27]. Since the coefficients (Eq. 3.36) are functions of the wheel-rail geometry, normal force and material properties, these parameters are inputs to the program.

Another program was translated to FORTRAN, by Goree and Law [22] from the original ALGOL written by Kalker. The program gives an approximate solution for the resultant tangential creep forces and spin moment acting between two bodies of equal linearly elastic material properties. Hertz' contact theory was used and two additional simplifying assumptions were made, resulting in a significant reduction in computation time.

For application in nonlinear models, such as the dynamic curving program of a rail vehicle either the nonlinear creep force model of Johnson and Vermeulen (Section 3.4.2) or the heuristic nonlinear model of White et al. (Section 3.4.3) are used as built-in subprograms, to compute the creep forces and moments.

4.4 TRUCK PARAMETERS

In this section, we subdivide the parameters of trucks into geometry, inertia, stiffness and damping groups. It will be seen, however, that some identification methods and devices are applicable to more than one of these groups.

4.4.1 Truck Geometry

The geometry of a truck can be readily determined from its design drawings. A procedure has been proposed to determine the relative motions (clearances) between truck components. It consists of a series of drawing board layouts depicting the possible clearances and resulting motions. A high degree of accuracy in measurements is possible by using full-scale layouts. The procedure applied to each layout consists of the following steps: (a) Determine the type of relative motion possible (translation or rotation); (b) Determine motion limits; (c) Assemble the layout with nominal maximum and minimum dimensions (tolerances); (d) Superimpose the desired motion on the layout (e.g., bolster rotation relative to sideframe); and (e) Compute, or measure, extremes of motions for nominal, maximum and minimum clearances.

Since a wheelset angular-misalignment of only a few miliradians has been found to be significant to the dynamic performances of trucks, it follows that the measurement accuracy of this parameter is important. Elkins and Allen [14] have reported an optical method which is accurate within less than one milliradian.

Truck Wheel Base

Truck wheel base, the distance between the centers of the outside axles of a truck, can have a substantial effect on curving performance, especially on tight curves. Long wheel bases increase the lateral forces required to move through a curve or spiral. However, in operation on tangent track, a long wheel base can reduce the probability of truck hunting, as shown by parameter b in Eq. (4.9). The length classification and typical truck wheel base values are given in Table 4.11.

Number of Axles per Truck

The addition of a third or fourth axle to a truck greatly increases the geometric constraints imposed during a curve or spiral negotiation. However, such trucks tend to be less prone to hunting.

Table 4.11. Sample truck wheel base values (adapted from Ehrenbeck and Polcari [13]).

Classification m(ft)	Example m (ft)
Short ≤ 2.134 (7)	70-ton freight truck, 1.728 (5.67); 125-ton freight truck, 1.829 (6).
Long > 2.134 (7)	GSI swing hanger passenger truck, 2.667 (8.75); GM-EMD six-axle HTC locomotive truck, 4.176 (13.7).

4.4.2 Truck Inertias

The mass of a truck component can be determined by its weight; the cg position, by its geometry. Given two masses, m_1 and m_2 and corresponding cg positions, p_1 and p_2, the resultant mass and cg position are:

$$m = m_1 + m_2 \tag{4.5}$$

and

$$p = p_1 + \left(\frac{m_2}{m_1+m_2}\right)\left(p_2-p_1\right) \tag{4.6}$$

respectively. Thus, the masses and cg positions of individual components can be combined two at a time, in this way, to obtain the net mass and cg position of the whole truck.

The moment of inertia of a truck component can often be determined by calculations based on its geometry and specific mass. For a bolster, the roll and yaw moments of inertia are usually significant, whereas the yaw and pitch moments of inertia of a sideframe are dominant.

An experimental method of determining moments of inertia is also available. The method consists of suspending the castings individually on a free pivot and treating the resulting system as a compound pendulum. The arrangement should maintain the castings in a single plane of oscillation perpendicular to the desired inertial axis. The period of oscillation T is obtained by angularly displacing the suspended castings and stopwatch timing the motion through a specific number of free cycles; the total time divided by the number of cycles yields T. The timing procedure can be repeated 3 to 5 times to obtain an average value of T. The test specimen weight is determined through simple scale weighing. Corrections can be made to account for the small weights of the pivot pins. The pendulum arm length, L, is tape measured.

For small angles, the moment of inertia about the pivotis calculated using the following formula:

$$I_p = \frac{T^2 WL}{4\pi^2} \tag{4.7}$$

where

T = period of oscillation

W = weight of test specimen

L = distance from pivot to specimen cg

From the parallel axis theorem, the moment of inertia about the specimen cg is given by:

$$I_{cg} = I_o - \frac{WL^2}{g} \tag{4.8}$$

where g is the acceleration due to gravity.

The yaw moment of inertia of a truck is measured about a vertical axis through its center plate. This parameter is determined by the weight of the truck and its distribution about the center of yaw rotation. A high yaw moment of inertia will increase the dynamic wheel/rail forces during dynamic curving or spiral negotiation and slightly reduce the tendency toward truck hunting. A truck yaw moment of inertia may be classified as low, medium or high, as shown in Table 4.12.

Table 4.12. Sample truck yaw moment of inertias (adapted from Ehrenbeck and Polcari [13]).

Classification $N\ s^2\ m$ ($lb\ s^2\ in$)	Examples $N\ s^2 m$ ($lb\ s^2\ in$)
Low < 3390 (30000)	Pioneer III passenger truck, 2712 (24000)
Medium 3390-5085 (30000-45000)	50-ton freight truck, 3435 (30400) 70-ton freight truck, 4062 (35950) 100-ton freight truck, 4952 (43820)
High > 5085 (45000)	Four-axle WH GE locomotive, 9040 (80000)

Various inertia data for conventional three-piece freight trucks can be found in the literature, e.g., Hasselman and Johnson [28], the American Steel Foundaries [36,37,46] and the Budd Company [53].

Wheelset Weight (Effective Lateral Wheelset Mass)

The inertial mass or weight of an individual wheelset is relevant to the lateral wheel/rail forces generated during the negotiation of relatively short alignment perturbations. In many truck designs, there is sufficient lateral clearance between the axle and other components, such as sideframes and journal boxes, to allow rapid accelerations of the wheelset without immediate coupling to the larger truck or carbody masses. Traction motors may often be excluded from the effective lateral wheelset mass. In other cases, tight lateral clearances, such as those in tapered roller bearings, make the effective lateral dynamic wheelset mass substantially greater than that of the wheelset alone. Most freight and passenger trucks fall into the "light" category. A low wheelset effective mass is generally beneficial for all performance issues. Each design in question must be reviewed and some portion of the weight of all attached components may be added to the wheelset weight, based on geometry and clearances. Typical values of wheelset weights are given in Table 4.13.

4.4.3 Truck Stiffnesses

Generally, the parameter values of suspension elements, such as springs and friction devices, are relatively easy to determine individually; however, they may perform differently in their installed environment. The British Rail Technical Centre [41] and the Transportation Test Center of the Association of American Railroads [13] use air tables, hydraulic jacks and other equipment to measure displacement; special fixtures are used to measure primary and secondary suspension stiffnesses and breakaway friction levels. The French National Railways at their test centre at Vitry sur Seine [3] have several test devices to measure suspension

Table 4.13. Sample wheelset weight values (adapted from Ehrenbeck and Polcari [13]).

Classification N(lb)	Example N (lb)
Low ≤ 22249 (5000)	50 ton freight truck, 8597 (1932)
	70 ton freight truck, 9630 (2164)
	100 ton freight truck, 11516 (2588)
	125 ton freight truck, 13928 (3130)
	Pionner III passenger truck, 15295 (3437)
Heavy > 22249 (5000)	GM-EMD four-axle locomotive truck, 22855 (5136)

parameters and test fully-assembled vehicles. There is a machine that rotates an entire truck, relative to the carbody of a fully-assembled vehicle, to measure the yaw stiffness and friction of the interconnection. Lyon and Remedios [42] have described methods of identifying suspension parameters directly from track tests; thus, realistic motions are automatically achieved without having to simulate them in a laboratory. Cooperrider and Fallon [15] used parameter identification techniques to determine freight-car coupler characteristics and snubber friction levels.

Truck shear stiffness is a measure of a truck's resistance to the parallel displacement of its wheelsets. The shear stiffness (Fig. 4.5) of one wheelset of a truck with respect to another can be determined by dividing the lateral force exerted experimentally on a given axle by the resulting lateral displacement relative to the other. Similarly, the bending stiffness can be obtained by dividing the yaw torque applied to a given axle by the relative yaw angular displacement between the axles.

In general, increasing the shear stiffness of a truck will improve its hunting performance. This is especially significant for conventional (non-radial) truck designs. Radial truck designs are likely to be relatively rigid in shear due to structural elements such as cross-bracing. Curving performance may be reduced slightly by increases in shear stiffness.

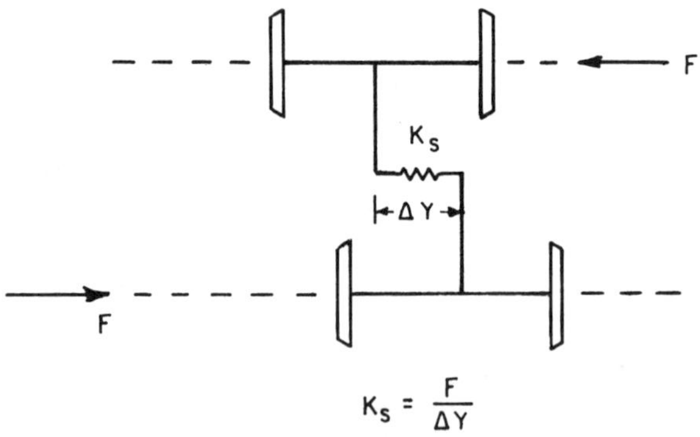

Fig. 4.5 Truck shear stiffness measurement

Truck dynamic curving and hunting behaviour vary inversely in many truck designs and are highly dependent on shear stiffness and axle yaw bending stiffness. An approximate formula, developed by Wickens [58], is qualitatively useful to show the influence of these parameters on the critical hunting speed, V_c:

$$V_c = \frac{C}{m\lambda a}\left[K_B + (a^2+b^2)K_S - \left((K_B + (a^2+b^2)K_S)^2 - 4b^2K_SK_B\right)^{\frac{1}{2}}\right]^{\frac{1}{2}} \qquad (4.9)$$

where

C is a constant

m is the effective hunting mass, which is, in general, the mass of the wheelsets

a is the track semi-gauge

b is one-half of the wheel base

λ is an effective wheel/rail contact conicity

K_B is the truck axle yaw bending stiffness

K_S is the truck shear stiffness

Since dynamic truck performance is not solely a function of shear stiffness, but involves other truck design parameters, major groups of designs, namely, steerable and nonsteerable freight and passenger trucks, are defined and ranges of shear stiffness established for each. Typical low, medium and high shear stiffness ranges have been established for these truck types (Table 4.14).

Truck axle yaw bending stiffness is a measure of a truck's resistance to relative yaw between two wheelsets, as shown schematically in Fig. 4.6. This stiffness is particularly relevant to curving performance. Low bending stiffness allows the wheelsets of a truck to take positions radial to a curve, thereby reducing the wheelset-to-rail angle of attack. Steerable, or radial, truck designs allow low bending stiffness through the structural interconnections of the wheelsets, such as cross-bracing or special shear pads. Lower bending stiffness allows sharper curving before lateral forces, due to the angle of attack, become a problem. Conversely, low bending stiffness can increase the tendency for wheelset hunting.

Table 4.14. Sample truck shear stiffness values (adapted from Ehrenbeck and Polcari [13]).

Classification N/m (lb/ft)	Example N/m (lb/ft)
Nonsteerable freight trucks	
Low ≤ 29.19×10^7 (2×10^7)	Conventional three-piece 70-ton freight truck, 27.73×10^7 (1.9×10^7) cannot
Medium > 29.19×10^7 (2×10^7)	become much softer due to pinned joints
High: not applicable	and cannot become much stiffer without becoming a radial design.
Steerable freight trucks	
Low < 87.56×10^5 (6×10^5)	70-ton radial design freight truck,
Medium, 87.56×10^5–143.02×10^5 (6×10^5–9.8×10^5)	109.46 (7.5×10^5) Significant changes to shear stiffness are unlikely.
High > 143.02×10^5 (9.8×10^5)	
Nonsteerable passenger trucks	
Low < 116.75×10^4 (8×10^4)	
Medium, 116.75×10^4–18.97×10^5 (8×10^4–1.3×10^5)	Budd-Pioneer III truck, 14.59×10^5 (10^5) some design variations are possible.
High > 18.97×10^5 (1.3×10^5)	
Steerable passenger trucks	
Low: not applicable	
Medium < 43.78×10^6 (3×10^6)	GSI-Scheffel, 54×10^6 (3.7×10^6)
High ≥ 43.78×10^6 (3×10^6)	Significant design variations are unlikely.

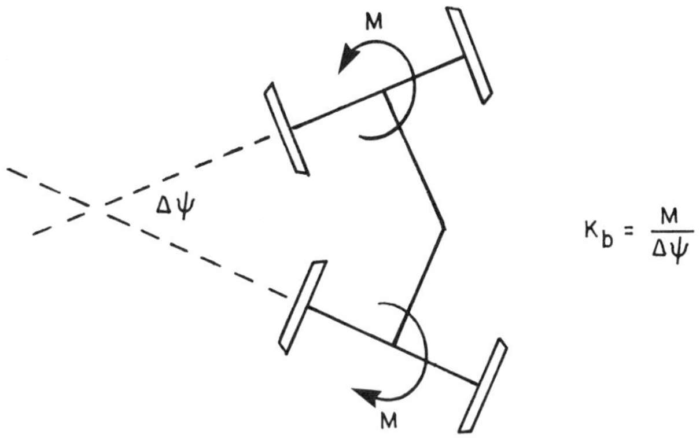

Fig. 4.6 Truck bending stiffness measurement

As with shear stiffness, the absolute value of bending stiffness is meaningful only within the prescribed design groups of steerable and non-steerable passenger and freight trucks, as shown in Table 4.15.

Suspension stiffnesses in the various translational (vertical, lateral, longitudinal) and rotational (roll, pitch, yaw) directions are usually determined by applying a range of loads in a prescribed direction and measuring the resulting deflections. For example, the test set-up for lateral stiffness is shown in Fig. 4.7; in this case, for a 70-ton ride-control spring group. Figure 4.8 shows lateral spring-rate curves for various coil heights, obtained by vertical load variations. Although minimized by a greased teflon sheet, the non-zero frictional resistance had the effect of shifting the spring-rate curves vertically without changing their slopes. The lateral stiffness, given by the slope, is seen to increase with vertical load.

A method of measuring the primary longitudinal stiffness was reported by Elkins and Allen [14]. The axles of a truck were supported by an air-bearing table to allow the axles to float freely and a longitudinal force was applied at each wheel by means

Table 4.15. Sample truck yaw bending stiffness values (adapted from Ehrenbeck and Polcari [13]).

Classification Nm/rad (lb ft/rad)	Example Nm/rad (lb ft/rad)
Nonsteerable freight trucks	
Low - not applicable	
Medium ≤ 1.36×10^{8} (10^{8})	
High > 1.36×10^{8} (10^{8})	Conventional 70-ton three-piece truck, 5.44×10^{8} (4×10^{8}) stiffness is fixed by design.
Steerable freight trucks	
Low < 1.36×10^{6} (10^{6})	
Medium, 1.36×10^{6}-2.72×10^{6} (10^{6}-2×10^{6})	
High > 2.72×10^{6} (2×10^{6})	Radial design 70-ton truck, 1.77×10^{8} (1.3×10^{8}). Some design modifications are feasible.
Nonsteerable passenger trucks	
Low < 2.72×10^{6} (2×10^{6})	Budd-Pioneer III truck, 3.26×10^{6} (2.4×10^{6})
Medium, 2.72×10^{6}-5.44×10^{6} (2×10^{6}-4×10^{6})	
High > 5.44×10^{6} (4×10^{6})	
Steerable passenger trucks	
Low < 11.56×10^{5} (8.5×10^{5})	GSI-Scheffel radial truck, 8.16×10^{5} (6×10^{5}). Some bending stiffness changes may be significant.
Medium > 11.56×10^{5} (8.5×10^{5})	
High: not applicable	

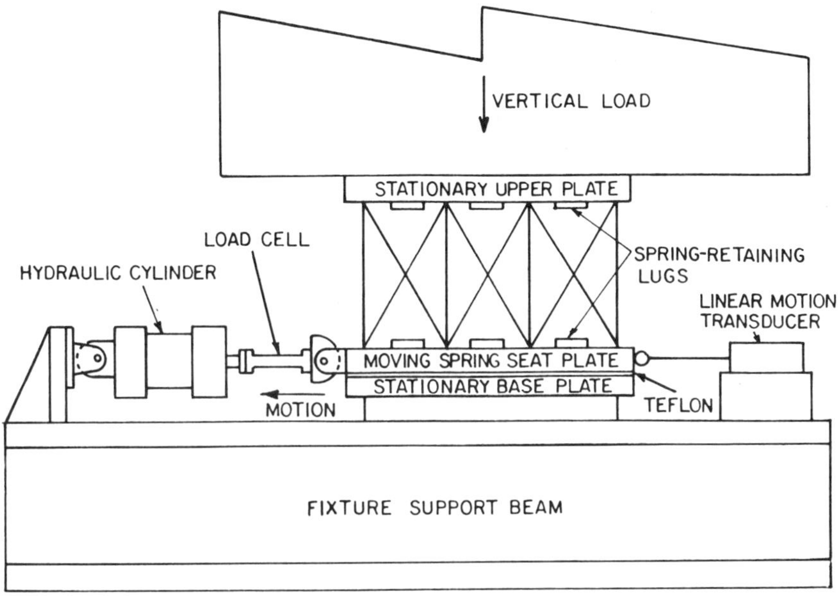

Fig. 4.7 Lateral spring-rate test set-up for a 70 ton Ride-Control Spring Group (adapted from ORR [46])

of a friction brake system. The normal force of the brake shoe was measured by a load cell between the brack actuator and the shoe. Linear variable-displacement transformers were used to measure the longitudinal displacements between the axles and sideframes.

The primary lateral stiffness of a truck whose wheelsets are anchored to ground can be obtained by applying a lateral force to a sideframe. The lateral force can be indicated by a load cell while displacement transducers measure the lateral displacement between the axles and sideframes.

4.4.4 Truck Damping

Damping requires energy dissipation, which can involve solid- and fluid-friction processes. The former will be referred to as friction damping, Coulomb friction, solid friction or, simply, friction; the latter will be called viscous damping.

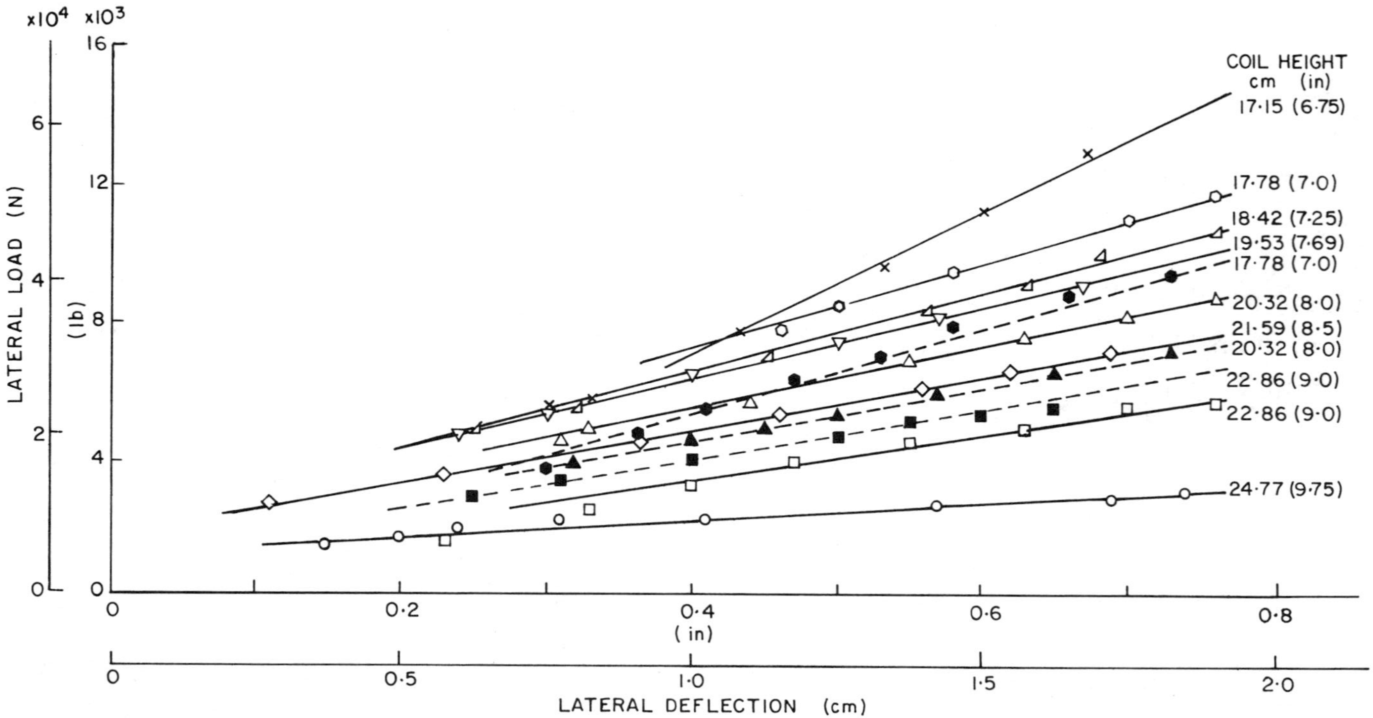

Fig. 4.8 Lateral spring-rate curves for a 70 ton Ride-Control Spr.ng Group (adapted from ORR [46])

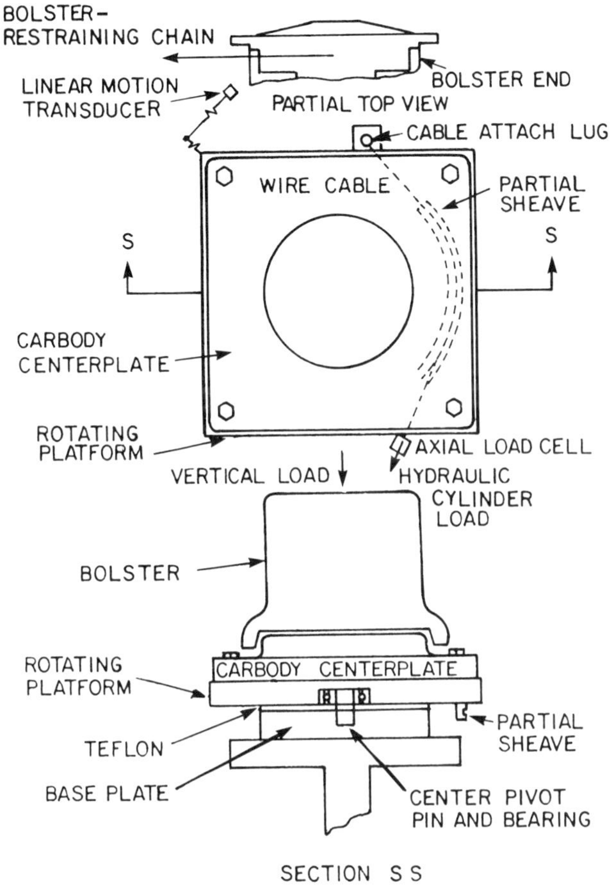

Fig. 4.9 Centerplate torsional friction test set-up (adapted from ORR [46])

Consider the torsional friction between a bolster and carbody. Figure 4.9 shows a test set up used to turn a carbody centerplate relative to a restrained bolster, in the presence of an adjustable normal load [30]. A greased teflon sheet was used to minimize friction between the turning platform and base plate. A twisting force was applied to the carbody centerplate and platform assembly by a hydraulic cylinder and cable system acting through a partial sheave. The cable load was measured by a load cell between the cylinder and cable clevis. An oscillograph recorded the load and displacement signals.

The test method consisted of zeroing the position of the carbody centerplate, applying a vertical load and, then, a cable tension controlled in such a way as to produce a constant velocity until cylinder run-out. This gave the break-away (static friction) load, followed by the dynamic friction load.

The slope of a plot of torque as a function of normal force P gave the coefficient of friction. Thus, the static and dynamic coefficients of friction, μ_s and μ_d, could be obtained from two such plots, yielding:

$$T_s = \mu_s P \tag{4.10}$$

and

$$T_d = \mu_d P \tag{4.11}$$

for the static (break-away) torque T_s and the dynamic torque T_d, respectively. Furthermore, it was found that, for a given case, the dynamic coefficient of friction μ_d was related to the inner and outer carbody-centerplate radii, R_1 and R_2, by:

$$\mu_d = \frac{2}{3}\left(R_2^3-R_1^3\right)/\left(R_2^2-R_1^2\right) \tag{4.12}$$

A secondary suspension model, including three springs and a friction element, was presented in Fig. 3.12. A simplified model

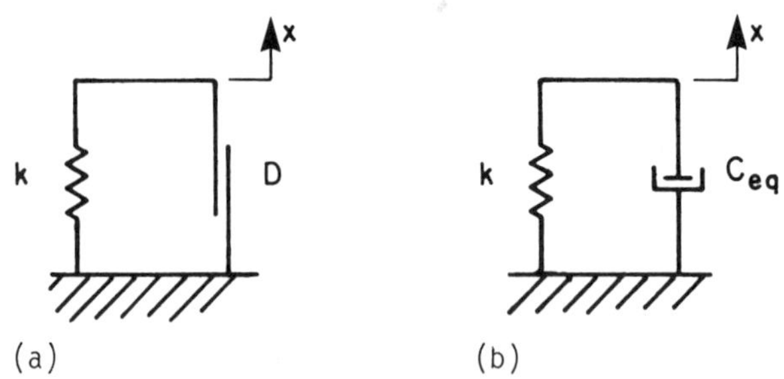

Fig. 4.10 (a) Nonlinear truck suspension model, (b) Linear truck suspension model

is given in Fig. 4.10(a), where there is one spring in parallel with a friction element. The friction element is characterized by a constant force D. Under certain assumptions made by Fallon et al. [16], the friction element can be replaced by an equivalent viscous damping element [Fig. 4.10(b)]. Then, the damping coefficient C_{eq}, as determined by assuming a sinusoidal motion of amplitude x_o and frequency f_o and matching the energy dissipated per cycle in both models, is given by:

$$c_{eq} = 2D/\left(\pi^2 f_o x_o\right) \tag{4.13}$$

In the following, the procedure adopted in Ref. [46] for determining the vertical frictional damping forces developed by a 6"×6" Ride Control truck bolster/friction shoe/sideframe system for a column load of 13,434 N (3,020 lb), as a function of relative bolster-to-sideframe motion and velocity is described. Various displacements and frequencies were used within definition to satisfy the object of determining the friction characteristics as a function of displacement and velocity. The test consisted of a reasonable number of motion cycles with a specific peak-to-peak displacement and cyclic frequency. Within limitations of the test apparatus, the amplitude and frequency variations were directed toward encompassing typical service phenomena.

Figures 4.11(a) and 4.11(b) illustrate the general damping characteristics associated with this test. A comparison of the two force/displacement oscilloscope photographs leads to the conclusion that the load magnitudes are roughly equal. Since the two cases represent different amplitude/frequency combinations, it follows that vertical damping force is independent of both amplitude and velocity. Although in light of known friction laws the stated independence is not surprising, the establishment at this point simplifies the energy modeling of purely vertical damping since it may now be based on a constant friction force acting through some given distance. Indeed, an expended energy comparison of Figs. 4.11(a) and 4.11(b), using a planimeter, relates them almost exactly through the ratio of their amplitudes of motion. That is,

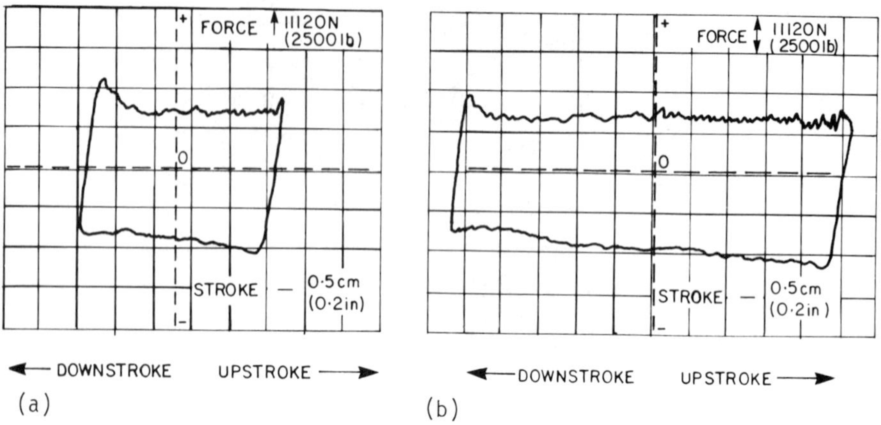

Fig. 4.11 (a) Vertical damping cycle at 0.25 Hz and 836.1 Nm (7400 in lb) energy dissipation per cycle (adapted from ORR [46])
(b) Vertical damping cycle at 0.1 Hz and 1728.6 Nm (15,300 in lb) energy dissipation per cycle (adapted from ORR [46])

Fig. 4.11(b) with a two-inch peak-peak stroke gives roughly twice the energy absorption as Fig. 4.11(a) with a one-inch stroke.

Figure 4.12 is the result of many tests with various combinations of motion amplitude and frequency. Loads are plotted on the abscissa as functions of the calculated maximum velocities. It is seen that, whereas the loads were not exactly constant for a given displacement amplitude, variations were not too extensive in most cases. Since the departures from the average lines show no obvious trend, the averages, 3100 lbs (μ_s = .49) and 3140 lbs (μ_s = .5) for the down- and up-strokes, respectively, could be used to characterize the system. Furthermore, a single value, $\mu_s = 0.5$, would probably be adequate for both directions of motion.

Truck Yaw Friction

Truck yaw friction, or break-away torque, is defined as the peak torque required to initiate rotation of a truck relative to the carbody. A high level of yaw friction tends to compound the problems associated with curving. On the other hand, it tends to increase the critical truck-hunting speed. Most trucks fall into

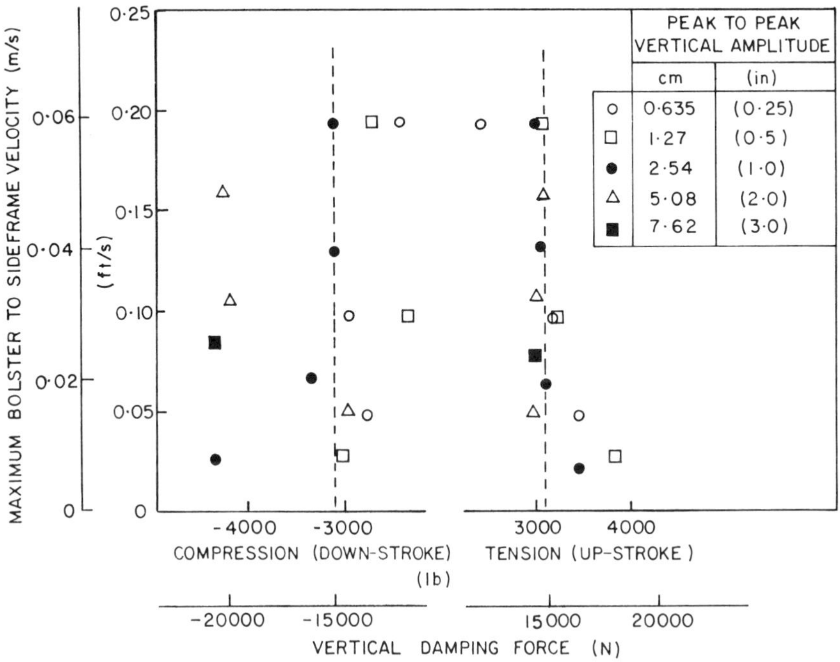

Fig. 4.12 Vertical damping force vs. maximum boster to sideframe velocity (adapted from ORR [46])

the medium or low yaw-friction categories, as defined in Table 4.16.

4.5 VEHICLE PARAMETERS

The lumped parameter rail vehicle dynamic models used for most analyses require characterization of the component mass and inertia properties, the geometry of the interconnections between components, and the force-motion laws governing the suspension interconnections. When body bending and torsion are included in the model, the frequencies and shapes of the bending and torsional modes must also be determined. Thus, in addition to the wheel-rail and track parameters, the carbody parameters discussed in this section are required for a vehicle model.

Table 4.16. Sample truck yaw friction or break-away torque values (adapted from Ehrenbeck and Polcari [13]).

Classification N m/N (lb in/lb)	Example N m/N (lb in/lb)
Low < 0.0381 (1.5)	Teflon surface, 0.0104 (0.41)
Medium, 0.0381-0.0762 (1.5-3.0)	50-ton freight truck, dry:0.0533 (2.1) 70-ton freight truck, dry:0.0609 (2.4) 125-ton freight truck, dry:0.0711 (2.8)
High > 0.0762 (3.0)	-

4.5.1 Vehicle Geometry

The spacing between truck centers, the overall distance between coupler pulling faces and the height of a vehicle's center of gravity are important.

Truck-Center Spacing

Track geometry inputs to the carbody are filtered, spatially, by the truck-center spacing. Specifically, track geometry alignment and profile perturbations which have wavelengths close to the truck-center spacing (or one-half multiples of it) have particularly strong effects on carbody dynamics compared to other-wavelength inputs. Typical revenue-track geometry contains a spectrum of wavelengths and amplitudes. However, the most severe inputs for carbody response are those with regular repeated cycles. On American railroads, the common use of 11.89 m (39 ft) staggered jointed-rail segments make vehicles with truck-center spacings of about 11.89 m (39 ft) and 17.68 m (58 ft) ($1\frac{1}{2}$ rail lengths) particularly sensitive to track geometry inputs (Table 4.17).

Table 4.17. Sample truck center spacing values (adapted from Ehrenbeck and Polcari [13])

Classification m (ft)	Example m (ft)
11.89 (39) range	70 ton box car, 12.19 (40)
10.97-12.80 (36-42)	Six-axle locomotive, 12.19 (40)
	100 ton coal Gondola car, 12.344 (40.5)
17.68 (58) range	70 ton flat car, 17.68 (58)
16.76-18.59 (55-61)	Passenger cars, 18.29 (60)
other < 10.97 (36)	70 ton tank car, 8.839 (29)
	F 7/9 locomotive, 9.144 (30)
	GP 7/9 locomotive, 9.144 (30)
> 12.80 (42)	100 ton covered Hopper car, 14.02 (46)
< 16.76 (55)	
> 18.59 (61)	70 ton vehicular flat car, 19.81 (65)

Distance Between Coupler Pulling Faces

The overall length of a vehicle affects spiral negotiation due to: (1) the continuously increasing warp over increasing lengths of spiral, and (2) the lateral force components of axial buff and draft forces (Table 4.18).

Center of gravity

The height of a vehicle's center of gravity has a pronounced effect on its response to roll inputs. Dynamic roll inputs are produced by cross-level variations on tangent or curved track. Static roll inputs are due to operation at other than balance speeds on curved track. The center of gravity height is classified

Table 4.18. Sample length between coupler pulling faces (adapted from Ehrenbeck and Polcari [13]).

Classification m (ft)	Example m (ft)
Short ≤ 22.86 (75)	100-ton box car, 20.73 (68) 70-ton refrigerated box car, 17.68 (58) 100-ton covered hopper car, 16.46 (54) Six-axle locomotive, 19.81 (65)
Long > 22.86 (75)	High cube box cars, 27.43 (90) TOFC/COFC flat cars, 27.43 (90) 70 ton auto rack cars, 28.65 (94)

Table 4.19. Sample center of gravity height values (adapted from Ehrenbeck and Polcari [13]).

Classification m (in)	Example m (in)
Low < 2.286 (90)	70 ton covered hopper car, 2.032 (80) 70 ton box car, 1.778 (70) Six-axle locomotive, 2.032 (80)
High ≥ 2.286 (90)	100 ton tank car, 2.337 (92) 100 ton coal hopper car, 2.489 (98)

as either low (≤ 2.286 m (90 in)) or high (≥ 2.286 m (90 in)) as shown in Table 4.19.

One method of determining the height of a carbody's cg is described in [46]. It involves placing four jacks at each jacking pad of the carbody, with load cells in between. The carbody is then jacked up until free of trucks and in a level position. The jack loads are adjusted to be equal on both sides of each car end. The jacks are then adjusted to tip the car with height increments of 5.334 cm (2.1 in), until a maximum height of 26.67 cm (10.5 in) is reached. The load cell readings are recorded at each tipped position. The following formula is used to calculated the cg height of the carbody relative to its jacking pad (Fig. 4.13):

$$h = \frac{\Delta W}{W} \cdot \frac{\ell^2}{S} \tag{4.14}$$

where

h = cg height relative to jacking pad, m (in)

ℓ = distance between two jacking pads, m (in)

S = tip height, m (in)

W = total weight of empty car, N (lb)

ΔW = weight change in tip position, N (lb)

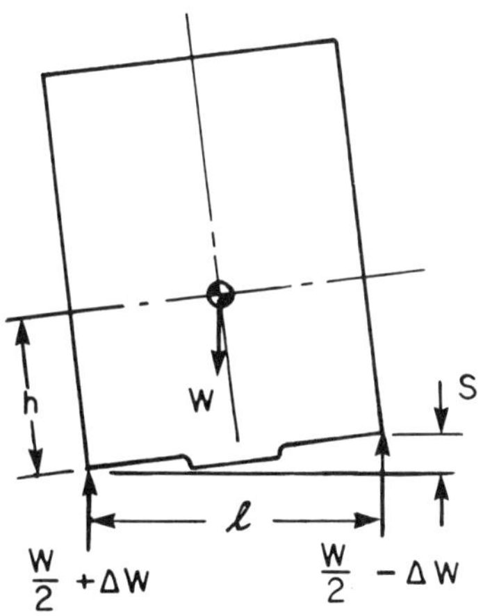

Fig. 4.13 Carbody center of gravity height determination

The cg height of the carbody relative to the rail is obtained by adding the height of the jacking pad. The truck centers of gravity are then taken into account to determine the entire vehicle's cg height.

4.5.2 Carbody Inertias

All three carbody moments of inertia are important. However, only the pitch and roll moments of inertia are considered here, leaving it to the reader to extend the methods to the yaw degree of freedom.

Pitch Moment of Inertia

A carbody's pitch moment of inertia may be obtained by a dynamic test. In this test, the car is hinge-supported at the side sills directly beneath its cg, as a pivot point for pitch motion. Truck snubbers are made inactive. The car is then excited in pitch at one end and the pitch response motions are recorded on an oscillograph. The resonant frequency is used in the following equation to obtain the pitch inertia of the car about its cg (Fig. 4.14):

$$I_p = \frac{K \frac{C^2}{2} - Wh}{\omega_p^2} - \frac{W}{g} h \qquad (4.15)$$

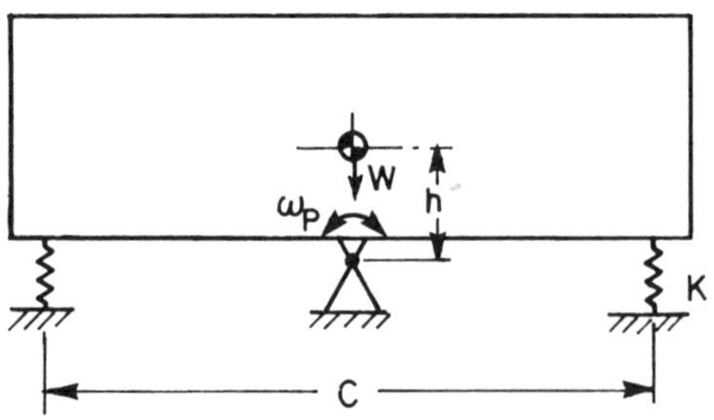

Fig. 4.14 Determination of carbody inertia in pitch motion

where

g = acceleration due to gravity, 9.81 m/s^2 (32.2 ft/s^2)

h = pivot point to cg, m (ft)

I_p = pitch inertia, m N s^2 (ft lb s^2)

K = spring constant per truck, N/m (lb/ft)

C = truck center distance, m (ft)

W = empty car weight, N (lb)

ω_p = measured resonant pitch frequency, rad/s

Roll Moment of Inertia

In a roll moment of inertia test, a carbody car is hinged at its center sill ends to produce roll pivot points. Shims are provided at the side bearings to prevent relative movement between carbody and truck bolster; therefore, when the car rolls, the carbody and truck bolster roll together. The car is excited at the middle of the side sill to determine resonance. The motion is recorded on an oscillograph. The resonant frequency is used in the following equation to obtain the carbody's roll inertia (Fig. 4.15):

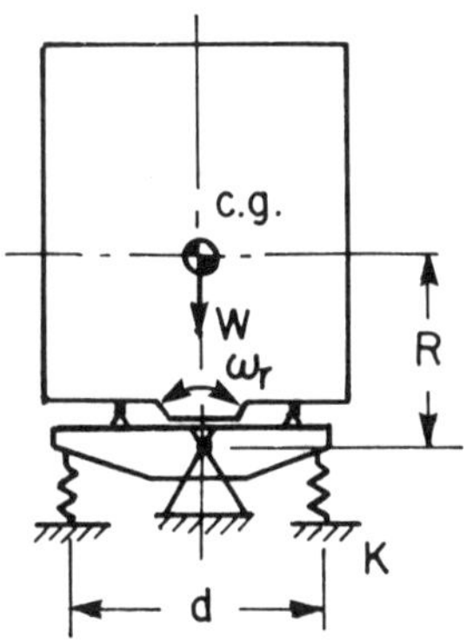

Fig. 4.15 Determination of carbody inertia in roll motion

$$I_r = \frac{\frac{K}{2} d^2 - WR}{\omega_r^2} - \frac{W}{g} R^2 \qquad (4.16)$$

where

d = distance between truck spring group centers, m (ft)

g = acceleration due to gravity, 9.81 m/s^2 (32.2 ft/s^2)

I_r = roll inertia, m N s^2 (ft lb s^2)

K = spring constant, N/m (lb/ft)

R = distance of cg to the pivot point, m (ft)

W = empty car weight, N (lb)

ω_r = measured resonant roll frequency, rad/s

4.5.3 Carbody Stiffnesses

The stiffness or rigidity of a carbody in torsion or flexure are considered in this section.

Torsional Stiffness

A torsionally stiff carbody can exhibit wheel lift when negotiating a sharp spiral or any other track geometry with relatively large cross-level differences. Torsional stiffness can be defined as the amount of torque on the carbody necessary to produce a certain amount of twist. It can be measured by the following procedure.

Support the carbody in a level position on load cells at each of its four corners as shown in Fig. 4.16. Lower the support at one of the corners by one inch and measure the resulting forces (F). Repeat until the force at the low corner is less than 10% of the force at the high corner at the same end. Then

Torsional stiffness = $[(f_{high} - F_{low})(\text{width})]/(\text{angular deflection})$

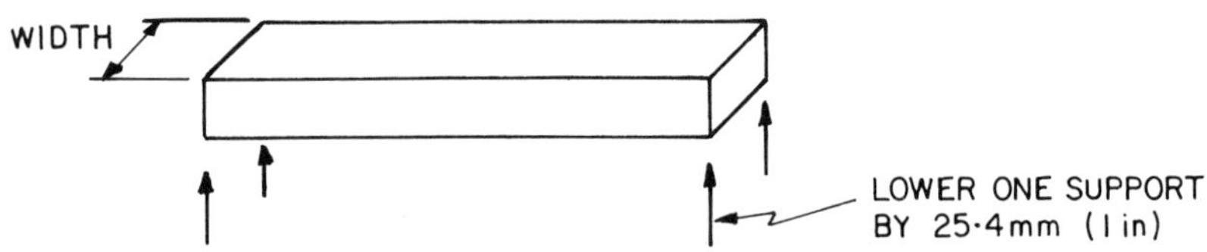

Fig. 4.16 Carbody torsional rigidity determination

Table 4.20. Sample carbody torsional stiffness values (adapted from Ehrenbeck and Polcari [13])

Classification Nm/rad (lb in/rad)	Example Nm/rad (lb in/rad)
Low stiffness ≤ 0.113×10^9 (10^9)	15.24m (50 ft) 70-ton box car, 4.633×10^7 (41×10^7) 12.19 m (40 ft) 50-ton box car, 19.323×10^7 (171×10^7)
High stiffness > 0.113×10^9 (10^9)	10,000 gallon 50-ton tank car, 237.865×10^7 (2105×10^7); Locomotives

where

$$\text{Angular deflection} = \tan^{-1}\left(\frac{\text{displacement of support}}{\text{width}}\right) \tag{4.17}$$

Typical values for carbody torsional stiffnesses are presented in Table 4.20.

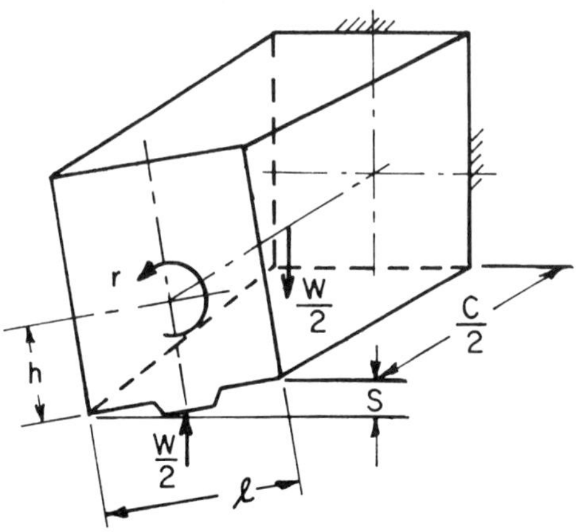

Fig. 4.17 Carbody torsional rigidity determination

In another procedure, (Fig. 4.17), two diagonal corners are jacked up equal amounts with 0.635 cm (0.25 in) increments until the other two corners are lifted off the jacks. The height of all four corners and the load cell readings are recorded. The procedure is repeated by jacking up the opposite diagonal corners.

The following formula can then be used to obtain the torsional rigidity:

$$G_T = \frac{T + \frac{Wh}{2}\frac{S}{2\ell}}{\frac{S}{\ell}} \frac{C}{2} \tag{4.18}$$

where

G_T = Torsional rigidity, N m^2/rad (lb ft^2/rad)

h = height of cg relative to jacking pad, (m) (ft)

ℓ = distance between jacking pads, m (in)

S = average elevation of the jacked-up corners relative to the corner at the other side, m (in)

T = average of applied torque at two bolsters, N m (ft lb)

C = truck-center distance, m (ft)

W = empty car weight, N (lb)

Flexural Rigidity

The flexural rigidity of hopper cars can be obtained analytically by the following formula since the moment of inertia of the side girders is easy to calculate:

$$\text{Flexural Rigidity} = \frac{EI}{L}, \text{ N m rad (ft lb rad)} \tag{4.19}$$

where E is the modulus of elasticity, Pa (lb/in^2), I is the moment of Inertia, m^4 (in^4), and L is the length of the beam, m (in).

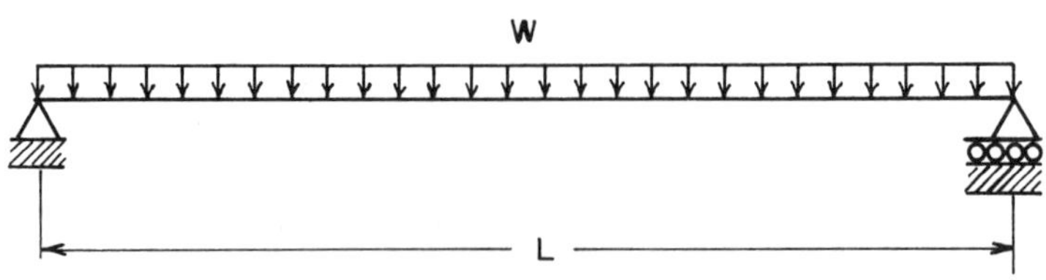

Fig. 4.18 Carbody flexural rigidity determination

For box cars, due to the doorway, an equivalent flexural inertia can be calculated by assuming the carbody to be a simple supported beam in bending (Fig. 4.18):

$$I = \frac{5WL^4}{384\ E\Delta_{max}}\ ,\ m^4\ (in^4) \tag{4.20}$$

where Δ_{max} is the maximum centerline deflection, m (in), and W is the uniformly distributed load N/m (lb/in).

4.5.4 Carbody Damping

The damping of a carbody reduces its displacement amplitude. Low damping can allow large carbody displacements to occur, especially in the case of repeated inputs. The damping ratio may be classified as low (≤ 0.20) or high (> 0.20). The 70-ton box car, for instance, is classified as low for all modes. The following test procedure may be used to calculate the damping ratio: Displace the carbody and release; record carbody displacements during free response; measure oscillation amplitudes; for each consecutive pair of oscillations, calculate the logarithmic decrement by:

$$\delta = \ln (x_1/x_2) \tag{4.21}$$

where x_1 and x_2 are consecutive amplitudes (Fig. 4.19); then, calculate the damping ratio by

$$\zeta = \left(\frac{\delta^2}{\delta^2 + 4\pi^2}\right)^{\frac{1}{2}} . \tag{4.22}$$

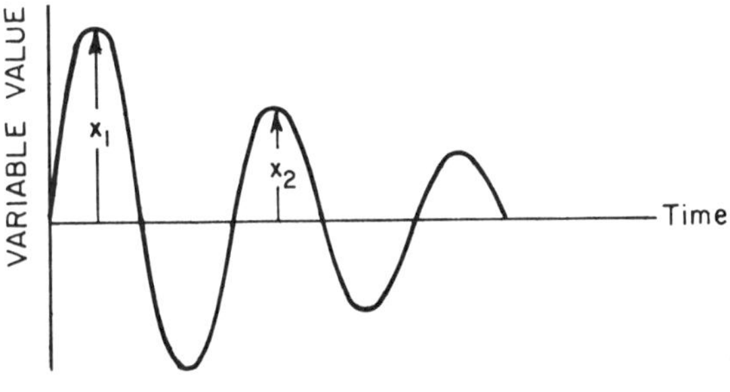

Fig. 4.19 Logarithmic decrement method

4.5.5 Carbody Resonant Frequencies

Railroad vehicles are often modeled as simple "spring-mass" systems, with the mass being the carbody mass and the spring being the lateral or vertical suspension system between the carbody and the rail. Such a spring-mass system will have displacement-amplitude gains dependent on the frequency of the input. The greatest gain comes at the resonant frequency. The frequency of the input is determined by the wavelength of a repeated track geometry perturbation and the vehicle speed. Although there is a continuum of perturbation wavelengths in typical tracks, the most common are due to the use of 11.89 m (39 ft) rail. Vehicles with resonant frequencies in the range encountered during normal operating speeds over typical tracks will be especially susceptible to twist and roll, pitch and bounce, and yaw and sway. The carbody resonant frequencies may be determined using the modern methods of modal analysis [41,42] and are classified as low (≤ 6 Hz) and high (> 6 Hz). Under conditions of "free response" to an initial displacement, a carbody will oscillate at its natural (resonant) frequency.

Knowing the suspension stiffness K, N/m (lb/ft), and mass M, kg (lb) of a carbody, its natural frequency ω_n, rad/s, is given by

$$\omega_n = (K/M)^{\frac{1}{2}} \tag{4.23}$$

The following test procedure may also be used. Apply a static displacement; release and record the free response; then, count the oscillations per second.

4.6 TRAIN PARAMETERS

Many computer programs have been developed in the track/train dynamics area [52-57]. The inputs to these models may be broadly divided into track and train data sets. Most of the parameters are obtained from track layout charts and vehicle drawings, coupler parameters, tractive effort curves and brake data are usually obtained from the manufacturers.

Track parameters were discussed in Section 4.2. Curvature data includes degree of curve, length of spiral and curve, and superelevation. Track profile data includes mileage and elevation. Most track layouts are sequenced tangent-spiral-curve-spiral-tangent. The standard inputs usually assume this but various combinations, including reverse and compound curves, are possible.

4.6.1 Train Makeup and Operation

Some of the train makeup and operation data typically required by train dynamics computer programs are listed:

(1) number of vehicles;
(2) vehicle identification numbers;
(3) vehicle types;
(4) vehicle weights;
(5) distance between pulling faces;
(6) weight of sliding sills;
(7) initial lateral bolster displacements at front and rear of each vehicle;
(8) centre of gravity height for each vehicle;
(9) drawbar force acting at the rear of each vehicle;
(10) net lateral load at leading outer wheel of each vehicle (if L/V ratios are to be computed);
(11) speed at which locomotive experiences transition;
(12) original throttle position;
(13) initial speed of train;
(14) number of automatic brake operations;
(15) number of independent brake operations;
(16) sand on the rail (yes/no);
(17) wet rail (yes/no).

4.6.2 Coupler and Draft Gear Parameters

A train can be regarded as a spring-mass system composed of cars (masses) and couplers with their draft gears (springs) can undergo longitudinal oscillations under certain conditions of braking and accelerating. Longitudinal train action programs are used to investigate this situation. The primary deterrent to oscillations is damping present in a train, including damping of the draft gears. Some of the energy absorbed during a cycle of compression and extension is converted into heat and not returned to the oscillating system. This energy loss can be measured by laboratory or field tests in which the forces generated during one cycle of the draft gear are measured. The energy returned during extension is subtracted from the energy required for compression. This differnce is the energy dissipated due to damping. It can be measured graphically as the area between the load and release curves of a force versus displacement plot such as Fig. 4.20.

The amount of energy which can be stored by the draft gear during compression is dependent on its longitudinal stiffness and the amount of deflection allowed. If the coupler/draft gear, even when deflected to its limit, cannot absorb the kinetic energy of the impacting car, a high longitudinal force can result, which may cause the car to pitch significantly and be susceptible to damage. For the purpose of evaluating the performance of a coupler and draft gear during an impact, one method is to compare the total compressive energy which can be stored by the draft gear with the energy required to decelerate a 100-ton vehicle from 12.87 km/h (8

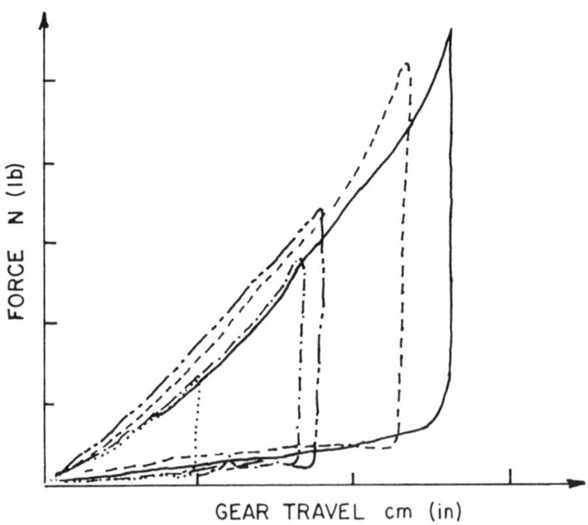

Fig. 4.20 Coupler/draft gear longitudinal energy absorption

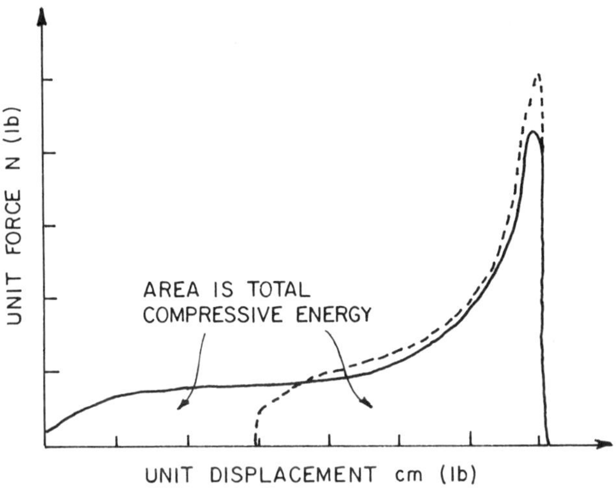

Fig. 4.21 Coupler/draft gear compressive energy

mph) to standstill. During laboratory or field testing, the loads may be applied slowly, if necessary, to establish conservative estimates (Fig. 4.21).

4.6.3 Traction and Brake Data

Tractive effort is the force exerted by a locomotive on the track for the movement of a train. This tractive effort is usually not constant. For example, a basic characteristic of a diesel-electric locomotive is its high starting tractive effort, which decreases with speed. The relationship of throttle setting to tractive effort for a typical high-horsepower locomotive is shown in Fig. 2.23.

Train braking is handled in a manner to provide flexibility in the control of train speed, slack action, and also to hold wheel damage and brake shoe wear to a minimum as well as to reduce the dynamic forces between track and train.

Figure 4.22 shows typical dynamic braking effort curves achieved as functions of speed. Investigations are being made into the possible redesign of dynamic brakes to give a smoother function, as in curve B of the same figure. Figure 4.22 also shows the

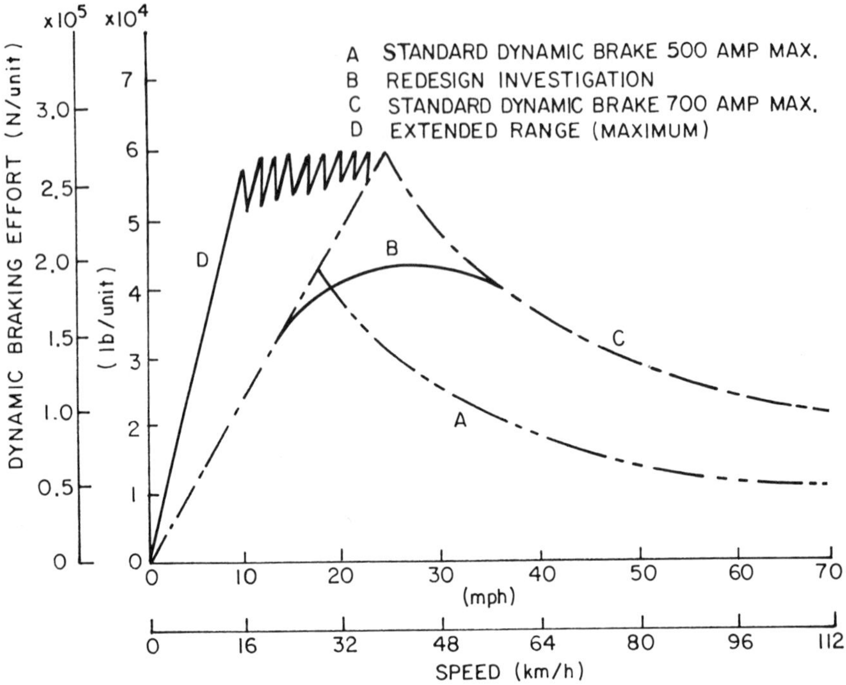

Fig. 4.22 Typical dynamic braking effort curves (adapted from [1])

relationship for the standard dynamic brake, curve C, and the extended range dynamic brake, curve D. Other brake data often required by programs include: (a) pressure at which brake retainers are set, N/m^2 (psi); (b) maximum pressure in the independent brake pipe, N/m^2 (psi); (c) automatic brake pipe pressure, N/m^2 (psi); (d) leakage, Pa/min (psi/min); and (e) initial wheel tread temperature, °C (°F)

4.7 SYSTEM IDENTIFICATION METHODS

System identification is defined here to mean the determination of the parameter values of a system by measurements of its input and output variables. The determination of wheel-rail, truck and vehicle parameters directly from dynamic test results offers great savings in time, effort and costs over laboratory testing of individual components. Moreover, the efforts involved in the application of system identification methods to an entire vehicle model constitute a model validation technique in themselves.

4.7.1 System Identification Background

The system identification concept is illustrated in the information flow diagram of Fig. 4.23. Given postulated set of model equations and a measurement of input and output variables, the parameter values may be determined by various system identification techniques. An advantage of these techniques compared with typical laboratory methods is that they use measurements under dynamic loading conditions. The general procedure may be summarized as follows: (a) Hypothesize a model representing the relationship between input and output variables; (b) Measure and record the input and output variables; (c) Condition the recorded data for analysis, including filtering, integration, differentiation, etc., as required; (d) Apply parameter estimation methods; (e) Analyse, correlate and interpret the results; and (f) If results are unsatisfactory, consider changing the model.

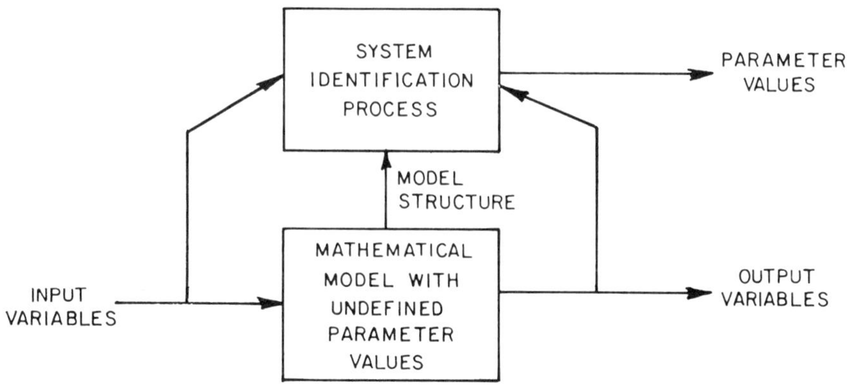

Fig. 4.23 (a) Parameter estimation technique - relationship between input and output
(b) Parameter estimation technique - determination of parameters c and x of equation $m\ddot{y} + c\dot{x} + kx = 0$

System identification methods have been applied to process control, communications, and space craft dynamics for some time but have only recently found application in rail vehicle dynamics. A survey of the application of identification methods to rail vehicle dynamics was given by Hull et al. [33] and Shladover [54]. An update was contributed by Fries [17]. Sage and Melsa [48] presented the identification theory and Gustavsson [25] compiled a survey of applications to other engineering areas. Hasselman and Johnson [29] discussed the problems and pitfalls in the identification process.

These identification methods are not always successful, in that they sometimes fail to converge to any meaningful set of parameter values. These failures can usually be attributed to measurement and model structure inaccuracies.

Fries [17], Scheffel [50] and Broerson [9] have identified methods which minimize output errors (i.e. the difference between measured and predicted outputs) in rail vehicle identification processes. Gilan et al. [19] and Scheffel [51] have indicated that the best success in convergence has been obtained with the Bayesian formulation of the output error problem. In this approach, the method involves minimization of a performance index given by

$$J = (\underset{\sim}{Y} - \underset{\sim}{H})^T U (\underset{\sim}{Y} - \underset{\sim}{H}) + (\underset{\sim}{r}_o - \underset{\sim}{r})^T W (\underset{\sim}{r}_o - \underset{\sim}{r}) \qquad (4.24)$$

where

$\underset{\sim}{Y}$ = vector of dynamic system measurements

$\underset{\sim}{H}$ = vector of corresponding model outputs

$\underset{\sim}{r}$ = vector of parameters (to be identified)

$\underset{\sim}{r}_o$ = a priori estimates of the parameters

U = observation of weighting matrix

W = parameter weighting matrix

The dynamic system measurements may be time series data, or processed data in the form of power spectral densities (PSD's), transfer functions, autocorrelation functions, etc. The parameter weighting matrix is often taken as the observation error covariance matrix. Gilan and Hedrick [19] have applied this method to identify creep coefficients and the primary suspension parameters of a cross-braced truck from field test data for a rail vehicle. They have minimized the difference between computed and measured PSD's. Convergence was obtained using a single truck dynamic model and a good fit was obtained between model and experimental PSD's. Fries [17] used the same Bayesian output error approach for identification of creep coefficients and suspension parameters of a transit vehicle on the full-scale roller rig at the Pueblo Test Center. In this case, excitation was provided by hydraulic actuators between the ground and the trucks, permitting experimental determination of response transfer functions. Successful identification of creep coefficients and suspension parameters was obtained and a good fit with additional test results was also found.

4.7.2 System Identification Example

The application of a system identification method to the secondary lateral suspension of a high-traction-coefficient (HTC) truck of an SDP-4DF locomotive is presented here as an example [13]. Figure 4.24 shows schematically the instrumentation arrays: two accelerometers to measure the lateral acceleration directly above the center plates of each truck and two displacement transducers to measure the relative lateral displacement between the truck bolsters and frames.

The pure yaw response mode can be isolated from the sway mode using an input-output relationship of the form:

$$M\ddot{z} + c(\dot{x}) + k(x) = 0 \tag{4.25}$$

where

$M = I/2\ell^2$

I = moment of inertia

2ℓ = distance between accelerometers placed symmetrically with respect to the carbody center of gravity

$z = \ddot{y}_1 - \ddot{y}_2$

$x = \delta_1 - \delta_2$

The terms $c(\dot{x})$ and $k(x)$ may represent either linear or nonlinear viscous damping and spring forces, respectively. Thus, assuming a

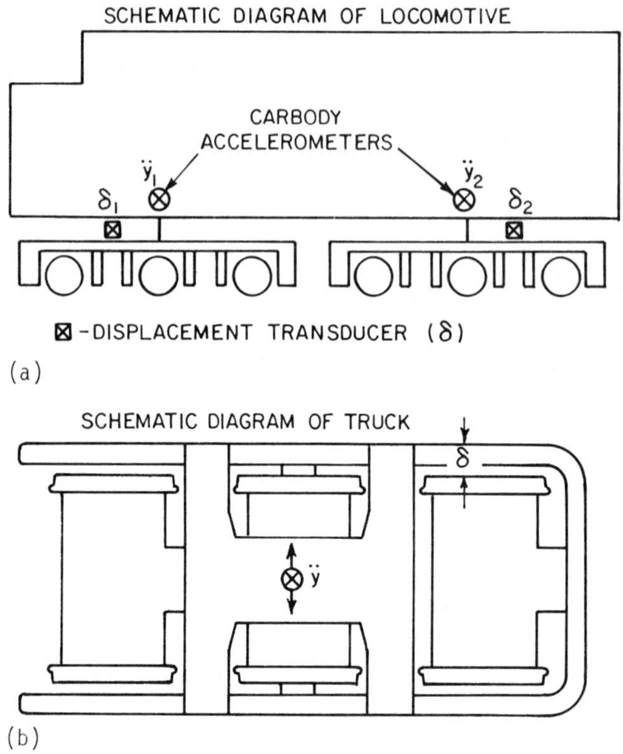

Fig. 4.24 (a) Schematic diagram of locomotive and instrumentation location
(b) Schematic diagram of locomotive truck and instrumentation location

nonlinear function for both elements, the moment balance equation can be written as:

$$M\ddot{z} + c_1\dot{x} + c_3\dot{x}^3 + k_1x + k_3x^3 = 0 \qquad (4.26)$$

Assuming M can be calculated, the parameter estimation technique involves the estimation of coefficients c_1, c_3, k_1 and k_3. The model structure is the equation, the input variables are δ_1 and δ_2, and the output variables are $\ddot{y}_1$ and $\ddot{y}_2$.

Before proceeding with the estimation of the parameters, c_1, c_3, k_1 and k_3, it was necessary to derive the relative velocity $\dot{x}$, from the relative displacement measurements using numerical differentiation. Since the rigid body yaw response mode of a typical railway vehicle is below 5 Hz, data were filtered using a Fast Fourier Transform (FFT) filter with a cut-off frequency of 5 Hz. The velocity was computed by taking the first finite difference divided by the sample interval. Estimation was then sought of the parameters.

The tests were conducted on a tangent zone designed with piecewise linear alignment perturbations, as shown in Fig. 4.25. These perturbations were specifically intended to excite the yaw mode of the locomotive. The instrumented vehicles were run over the test zone at eleven speeds between 64 and 113 Km/h (40-70 mph), as shown in Table 4.21. The sample rate used to digitize the data was 256 Hz per channel. Thus, a 64 Km/h (40 mph) run was 6.6 seconds long and the related data stream consisted of 1,700 samples per channel. A 96 Km/h (60 mph) run produced almost 1,000 samples per channel.

Table 4.22 gives the representative results for a 101.4 Km/h (63 mph) test. In addition to the values of the suspension characteristics are given an F value, a coefficient of determination R^2 and a standard error s^2 for the estimate. Without going into details an F value is the statistic which measures the strength of the relationship between two quantities. A large F value (>3) indicates that the regression model explains a significant amount of variations. The coefficient of determination R^2 provides an

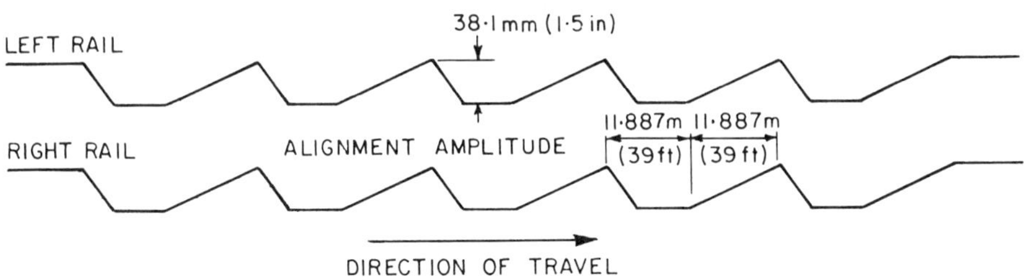

Fig. 4.25 Plan view of Perturbated Test Track (PTT) Test Section 8 (adapted from Ehrenbeck and Polcari [13])

Table 4.21. Test matrix (adapted from Ehrenbeck and Polcari [13]).

Run number	Test speed km/h (mph)
1	64.4 (40)
2	88.5 (55)
3	112.7 (70)
4	64.4 (40)
5	88.5 (55)
6	69.2 (43)
7	101.4 (63)
8	75.6 (47)
9	96.6 (60)
10	80.5 (50)
11	112.7 (70)

indication of what portion of output variations are explained by the model. The F value in Table 4.22 confirms that the indicated results are not just by chance, but that the hypothesized model does explain the observations. The R^2 value of 0.99 indicates that the model accounts for a significant portion of variations.

The results indicate that the suspension has a "softening" spring since the k_3 coefficient is negative. This means that the natural frequency of the system would be slightly lower for large displacements than for small displacements. This result is typical of the type of used elastomeric shear springs in this locomotive. The damping coefficient is similarly nonlinear. As the suspension velocity increases, the effective damping decreases slightly. This is typical of the force-limited type of dampers installed on the SDP-40F. The dampers have a pressure relief which limits the maximum damping force to 8,009 N (1,800 lb).

Table 4.22. Suspension characteristics (adapted from Ehrenbeck and Polcari [13]).

k_1 N/m (lb/in)	k_3 N/m^3 (lb/in^3)	c_1 N s/m (lb s/in)	c_3 Ns^3/m^3 (lb s^3/in^3)	F	R^2	s^2
613×10^4 (3.5×10^4)	-17.1×10^7 (-630)	175×10^3 (1000)	-2.98×10^5 (-1.1)	2.9×10^5	0.9994	0.0001

4.8 MODEL VALIDATION

The process of validating a model consists of verifying that the model yields a valid representation of the real system. It would be impossible to set absolute standards for model validation that would be applicable to all models under all circumstances. Whether a particular model is valid or not depends on its intended use. The process of validation attempts to correlate test and model results. The model validation process is summarized in the flowchart of Fig. 4.26. The main steps are:

(1) defining the validation criteria;

(2) comparing the test and model results; and

(3) making adjustments to the model.

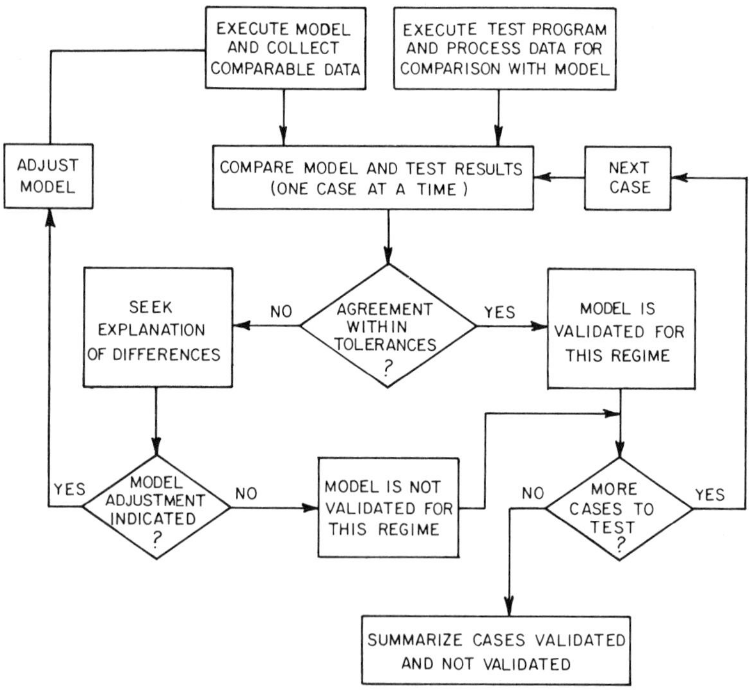

Fig. 4.26 Model validation process (adapted from Ehrenbeck and Polcari [13]).

Upon completion of the validation process, the results are summarized by stating that the particular model was or was not validated and defining the vehicle or class of vehicle(s), the terrain type and the operating conditions for which it is intended. Then, subject to these conditions, the prediction of vehicle performance can be achieved with the model in place of physical testing. The model cannot be regarded as having been validated for any other conditions.

The process of planning, testing, data reduction and analysis needed for a model's validation are tedious and require substantial time, effort, and expense. The trade-off between the effort needed to validate a model and the additional tests that would otherwise be required determines the cost-effectiveness of model validation. The cost depends on: (1) the nature of the model, (2) its intended purpose, (3) the type of vehicle or train required for tests, (4) the operating conditions and terrain to be considered, (5) the required instrumentation, and (6) difficulties involved in data acquisition and reduction. Most of the cost elements that are involved in a model validation project also exist in a well-designed performance testing program that is meant to evaluate the performance of a particular vehicle under specific operating conditions without any consideration for applications of the results to other vehicles or conditions.

However, more careful measurements of track geometry and a more thorough analysis of data are often required for model validation tests compared with performance tests.

4.8.1 Performance Criteria

The many diverse rail vehicle dynamic models can be categorized into three types: quasi-static, frequency domain and time domain. Certain performance criteria used for validation tend to be associated with each type. Let us consider them briefly.

Quasi-Static Models

The steady-state models designed to predict the dynamic equilibrium performance of vehicles often use quasi-static techniques. Some steady-state curving models for freight cars fall into this category. Because each output density is a single number which can be compared easily with a similar single number obtained during a test, these types of models are the simplest to validate.

Frequency-Domain Models

Although frequency-domain models are usually based on linear assumptions, they can be applied to certain nonlinear systems by the use of quasi-linearization techniques such as describing functions. The natural frequencies, damping ratios, and mode shapes of a vehicle's response are determined by the eigenvalues and eigenvectors. The critical speed and damping of each response mode for the onset of hunting can be predicted. The hunting speed can be used as a performance criterion for validation purpose. Transfer functions are often used to calculate vehicle response spectra as performance criteria.

Time-Domain Models

Time-domain models are applicable to linear and nonlinear dynamic systems. A time history of each response variable and various auxiliary variables are typical outputs of such models. However, validation processes which use these time responses directly as performance criteria tend to be the most difficult.

Regardless of the specific performance criteria used, the ultimate objective of a vehicle model is to predict the vehicle's performance pertaining to one or more of the following: hunting or lateral stability, lateral forced response (yaw and sway), vertical forced response (pitch and bounce), rock and roll, steady-state curve negotiation capability, and dynamic curving behaviour, including spiral negotiation. In the case of trains, there is longitudinal train action and impact response, as well as steady-state buff and draft behaviour.

4.8.2 Validation Criteria

The validation criteria are a set of relations that are uniquely defined and used to decide whether a model's predictions are close enough to test results for the model to be regarded as valid. Before one proceeds with the process of validation, the validation criteria must be selected based on the purpose of the model. The selection of validation criteria involves the following considerations: (1) Which variables should be compared? (2) Which measurements, or statistics, of each variable should be used? (3) What level of tolerance should be applied to each measurement? (4) What mathematical form should be used to combine the individual measurements?

The chosen variables should be the primary variables that are most important for a model's ultimate use and the intermediate variables that can reveal its deficiencies. The choice of intermediate variables is more involved and depends on an understanding of the model and the physical behaviour of the target vehicle. If the model does not represent the important intermediate variables well, then there is an inherent structural defect in the model and erroneous predictions of the primary variables under different conditions can be expected. Some examples of primary and intermediate response variables that could be used as validation criteria for models of different dynamic processes are shown in Table 4.23.

There are many different ways, or statistics, in which a variable can be characterized. The chosen statistics should be in line with the analytical methods used in the model. Some examples of the statistics which can be used for frequency- and time-domain models are summarized by Garg and Dukkipati [18].

The tolerances for the statistics of a single number, such as the mean, rms, and maximum, are easy to define, but it can be much more complicated for the statistics of a series of numbers, such as spectral densities, correlations, and frequency response functions. The tolerances may be different for different variables. The measured values may be compared with the predicted values of these statistics by one or more of the following mathematical relations: (a) maximum amplitude difference, (b) fit error, (c) phase, or time, or frequency difference, and (d) weighted sums of the preceding comparisons.

The validation criterion consists of combining all separate measures of the validity of a model into a unique mathematical expression. In regard to the relative importance of the separate

Table 4.23. Response Variables for Model Validation Criteria (adapted from Garg and Dukkipati [18]).

Parameters	Primary response variables	Intermediate response variables
Hunting (lateral stability)	Critical speed and critical frequencies, truck lateral and yaw displacements, and damping ratios	Wheel-rail contact angles and forces, body accelerations, and suspension deflections
Rock and roll	Body roll angle, wheel-lift duration, suspension deflection, and L/V ratios	Body accelerations, wheel/rail forces, centerplate and side bearing loads, natural frequencies, and damping ratios
Lateral forced response (yaw and sway)	Wheel-rail displacements, wheel-rail forces, body lateral yaw, and accelerations	Suspension deflections, truck accelerations, L/V ratios, natural frequencies, and damping ratios

Vertical forced response (pitch and bounce)	Body vertical and pitch accelerations, L/V ratios, and wheel unloading	Suspension deflections, truck accelerations, centerplace and side bearing loads, natural frequencies, and damping ratios
Steady-state curve negotiation	Wheel/rail displacements and forces, angle of attack, suspension deflections and wear indices	Body roll angle and truck
Dynamic cuving behaviour	Body accelerations, truck lateral and yaw displacements, wheel/rail forces, L/V ratios, and time durations	Suspension deflections, and angle of attack
Longitudinal train action	Buff and draft forces, L/V ratios, times of buff and draft transitions	Body accelerations and coupler angles
Longitudinal impact	Buff and draft forces and draft gear displacements	Body accelerations and suspension deflections
Steady-state buff and draft	Coupler lateral forces, coupler angles and L/V ratios	Draft gear deflections

validity measures, each different mathematical expression is based on different assumptions. A few of the mathematical forms for model validation are: threshold on weighted sum of errors, threshold on weighted sum of tolerance exceedances, threshold on weighted product of absolute values of erros, and no tolerance exceedances. An example of the validation of a six-axle locomotive nonlinear model in response to tangent track irregularities is given in Chapter 5 (Section 5.5).

4.9 SUMMARY

In this chapter, we discussed methods of determining the values of parameters associated with systems of railway vehicles, or trains, traveling on tracks. Track, wheel-rail, truck, vehicle and train parameters were involved. Specifically, there were tangent and curved track parameters, wheel-rail contact geometry and creep coefficients, truck and carbody geometry, inertias, stiffnesses and damping, train make-up data, coupler and draft gear parameters, and traction and brake data. We introduced the concept of system identification, which is based on input-output measurements. Finally, we described the use of performance and validation criteria in the model validation process.

REFERENCES

1. ANON. (1973): Track-Train Dynamics to Improve Freight Train Performances; *Association of American Railroads*, Report R-185, Chicago, U.S.A.
2. ANON. (1975): Track Safety Standards; *U.S. Department of Transportation*, Federal Railroad Administration, Report FRA/ORD-77-04, Washington, D.C.
3. ANON. (1978): Vitry-sur-Seine Test Centre; *SNCF*, Rolling Stock Department, Test Division.
4. ANON. (1980): Car and Locomotive Encyclopedia; *Simmons-Boardman*, Centennial Edition, New York, U.S.A.
5. ABBOT P.W. (1976): Comparison of the Nonlinear Dynamic

Characteristics of Barber S-2 and ASF Ride Control Freight Car Trucks; *Martin Marietta Corporation*, Bethesda, Maryland, U.S.A.

6. ABBOT P.W. (1977): Track Train Dynamics, Test Results, HT-C Truck Static Test; *Martin Marietta Corporation*, Report MRC-77-100, Bethesda, Maryland, U.S.A.
7. AHLBECK D.R. (1982): Effects of track modulus on vehicle track dynamic interaction; *ASME/IEEE Joint Railroad Conference*, Paper 82-RT-3, Erie, Pennsylvania, U.S.A.
8. AHLBECK D.R., PRAUSE R.H., DAY J.B. and MEACHAM H.C. (1974): Comparative Analysis of Dynamics of Freight and Passenger Rail Vehicles; *U.S. Dept. of Transportation*, Federal Railroad Administration, Report FRA-ORD&D-74-39, Washington, D.C.
9. BROERSON P.M.T. (1977): Railway Contact Forces: Coherence Functions and Model Verification; *Proceedings of 5th IAVSD* and *2nd IUTAM Symposium on the Dynamics of Vehicles on Roads and Tracks*, Swets and Zeitlinger B.V., Amsterdam, The Netherlands, 500-508.
10. COOPERRIDER N.K., LAW, E.H., HULL, R., KADALA, P.S. and TUTEN, J.M. (1975): Analytical and Experimental Determination of Nonlinear Wheel/Rail Geometric Constraints; *U.S. Dept. of Transportation*, Federal Railroad Administration, Report FRA-OR&D-76-244, Washington, D.C.
11. COOPERRIDER N.K. and LAW, E.H. (1978): A Survey of Rail Vehicle Testing for Validation of Theoretical Dynamic Analyses; *Transactions of the ASME, Journal of Dynamic Systems, Measurement, and Control*, Vol. 100, 238-251.
12. CORBIN J.C. (1980): Statistical Representation of Track Geometry; *U.S. Dept. of Transportation*, Federal Railroad Administration, Report FRA/ORD-80-22, Washington, D.C.
13. EHRENBECK R. and POLCARI S. (ed.) (1984): Vehicle/Track Interaction Assessment Technqiues; *U.S. Department of Transportation*, Federal Railroad Administration, Report DOT/FRA/ORD-84/01.3, Washington, D.C.
14. ELKINS J.A. and ALLEN, R.A. (1982): Verification of a Transit

Vehicle's Curving Behaviour and Projected Wheel/Rail Wear Performance; *Transactions of the ASME, Journal of Dynamic Systems, Measurement, and Control*, Vol. 104, (3), 247-255.

15. FALLON W.J. (1977): An Investigation of Railcar Model Validation; *M.S. Thesis*, Arizona State University, Tempe, Arizona, U.S.A.
16. FALLON W.J., COOPERRIDER, N.K. and LAW, E.H. (1978): An Investigation of Techniques for Validation of Railcar Dynamic Analyses; *U.S. Dept. of Transportation*, Federal Railroad Admnistration, Report FRA/ORD-78/19, Washington, D.C.
17. FRIES R.H. (1983): Estimation of Transit Rail Vehicle Parameters From Roll Rig Tests; *Ph.D. Thesis*, Arizona State University, Tempe, Arizona, U.S.A.
18. GARG V.K. and DUKKIPATI, R.V. (1984): Dynamics of Railway Vehicle Systems; *Academic Press*, Toronto, Canada.
19. GILAN A. and HEDRICK, J.K. (1982): Validating Rail Vehicle Dynamic Models: A Case Study; *Joint ASME/IEEE Railroad Conference*, Erie, Pennsylvania, Paper 82-RT-6, U.S.A.
20. GILCHRIST A.O. and BRICKLE, B.V. (1976): A Re-examination of the Proneness to Derailment of a Railway Wheelset; *Journal of Mechanical Engineering Science*, Vol. 18, (3), 131-141.
21. GOGINENI B.R. and GARG, V.K. (1978): A Dynamic Model for Longitudinal Train Action; *Association of American Railroads*, Report R-297, Chicago, U.S.A.
22. GOREE J.G. and LAW, E.H. (1977): Users' Manual for Kalker's Simplified Nonlinear Creep Theory; *U.S. Department of Transportation*, Report FRA/ORD-78/60, Washington, D.C.
23. GOREE J.G. (1978): Users' Manual for Kalker's Exact Nonlinear Creep Theory; *U.S. Department of Transportation*, Federal Railroad Administration, Report FRA/ORD-78/50, Washington, D.C.
24. GOSTLING R.J. and COOPERRIDER, N.K. (1983): Validation of Railway Vehicle Lateral Dynamic Models; *Vehicle System Dynamics*, Vol. 12, 179-202.
25. GUSTAVSSON I. (1973): Survey of Applications of Identifica-

tion in Chemical and Physical Processes; *Proceedings of the 3rd IFAC Symposium*, Paper S-3, 67-85, The Hague, The Netherlands.

26. HAMID A., RASMUSSAN K., BALUJA, M., and YANG, T.L. (1983): Analytical Descriptions of Track Geometry Variations; *U.S. Dept. of Transportation*, Federal Railroad Administration, Office of Research and Development, Report DOT/FRA/ORD-83/03.1, Washington, D.C.
27. HAQUE I., LAW, E.H. and COOPERRIDER, N.K. (1979): Users' Manual for Program for Calculation of Kalker's Linear Coefficients; *U.S. Dept. of Transportation*, Federal Railroad Administration, Report FRA/ORD-78/71, Washington, D.C.
28. HASSELMAN T.K. and JOHNSON, L.K. (1979): Validation and Verification of Rail Vehicle Models; *ASME Winter Annual Meeting*, Paper 79-WA/DSC-8, New York, U.S.A.
29. HASSELMAN T.K. and JOHNSON, L. (1979): Validation and Verification of Rail Vehicle Models; *ASME Winter Annual Meeting*, Paper 79-WA/DSC-8, U.S.A.
30. HAWTHORNE V.T. (1979): Truck Hunting in the Three-Piece Freight Car Truck; *ASME Winter Annual Meeting*, Paper 79-WA/RT-14, U.S.A.
31. HEDRICK J.K. (1977): Simulation and Analysis of Rail Vehicle Dynamics in: *Overview of the Simulation in Highway Transportation*, Vol. 7, (2), Ch. 11, 127-135, Society for Computer Simulation, LaJolla, California, U.S.A.
32. HELLER R. and COOPERRIDER, N.K. (1977): Users' Manual for Asymmetric Wheel/Rail Contact Characterization Program; *U.S. Dept. of Transportation*, Federal Railroad Administration, Report FRA/ORD-78/05, Washington, D.C.
33. HULL R., TRANKLE, T.K. and KLINGER, D.L. (1979): Application and Evaluation of System Identification Techniques to Rail, Vehicle Dynamics; *Systems Control Incorporated*, Report TR5307-100, U.S.A.
34. KALKER J.J. (1967): On the Rolling Contact of Two Elastic

Bodies in the Presence of Dry Friction; *Ph.D. Dissertation*, Delft University of Technology, Delft, The Netherlands.

35. KALKER J.J. (1980): Review of Wheel-Rail Rolling Contact Theories; in *The General Problem of Rolling Contact*, Browne, A.L. and Tsai, N.T. (ed.), ASME-AMD Vol. 40, 77-92, ASME, New York.
36. LIND E. (ed.) (1974): Harmonic Roll Series - Volume 4; *Association of American Railroads*, Track-Train Dynamics, Report R-184, Chicago, U.S.A.
37. LIND E.F. (1976): Truck and Carbody Characterization; *Association of American Railroads*, Track-Train Dynamics Phase I, Report R-186, Chicago, U.S.A.
38. LOW E.M. and GARG, V.K. (1977): Track Train Dynamics Report-Technical Documentation-Train Operation Simulator; *Association of American Railroads*, Report R-269, Chicago, U.S.A.
39. LOW E.M. and GARG, V.K. (1978): Validation of the Train Operation Simulator; *Association of American Railroads*, Report R-335, Chicago, U.S.A.
40. LUTTRELL N.W., GUPTA, R.K., LOW, E.M. and MARTIN G.C. (1976): Train Operation Simulator-Users' Manual; *Association of American Railroads*, Report R-198, Chicago, U.S.A.
41. LYON D. and REMEDIOS, N.C. (1979): Resonance Testing of Railway Vehicles; *Proceedings of the 6th IAVSD-IUTAM Symposium on the Dynamics of Vehicles on Roads and Tracks*, Berlin, 233-247.
42. LYON D. and REMEDIOS, N.C. (1979): Determination of Vehicle Parameters from Track Testing Data; *Proceedings of the 6th IAVSD-IUTAM Symposium on the Dynamics of Vehicles on Roads and Tracks*, Berlin, 249-258.
43. MARTIN G.C. and HAY, W.W. (1967): Method of Analysis for Determining the Coupler Forces and Longitudinal Motion of a Long Freight Train in Over-the-Road Operation; *University of Illinois*, Department of Civil Engineering, Urbana, Illinois, U.S.A.

44. MARTIN G.C. and TIDEMAN, H. (1977): Detailed Longitudinal Train Action Model - Technical Documentation; *Association of American Railroads*, Report R-221, Chicago, U.S.A.

45. MARTIN G.C., PLOUFFE, W.E., AHMED, S., ANTEZAK, H. and TIDEMAN, H. (1977): User's Manual - Detailed Longitudinal Train Action Model; *Association of American Railroads*, Report R-220, Chicago, U.S.A.

46. ORR D.G. (ed.) (1974): Track Train Dynamics Report - Harmonic Roll Series - Volume 1; *Association of American Railroads*, Report R-172, Chicago, U.S.A.

47. RAIDT J.B., SHUM, K.L., MARTIN, G.C. and GARG, V.K. (1979): Detailed Vertical Train Stability Model - Technical Documentation; *Association of American Railroads*, Report R-267, Chicago, U.S.A.

48. SAGE A. and MELSA, J.L. (1971): Estimation Theory with Applications to Communication Control; *McGraw-Hill Company,* New York.

49. SAYERS M.S. and HEDRICK, J.K. (1977): Track Maintenance/ Railcar Suspension Trade-Offs to Obtain Acceptable Ride Quality; *ASME Design Engineering Technical Conference*, Chicago, U.S.A. 35-58.

50. SCHEFFEL H. (1974): New Design Approach for Railway Vehicle Suspension; *Rail International*, Vol. 10, 638-651.

51. SCHEFFEL H. (1974): The Hunting Stability and Curving Ability of Railway Vehicles; *Rail International*, Vol. 5, No. 2, 154-176.

52. SCHWAM S.A. (1983): The Hunting in the Three-Piece Freight Car Truck; *Proceedings of the Conference on the Economics and Performance of Freight Car Trucks*; 77-91, Montreal, Canada.

53. SHAPIRO S.M. (1978): Engineering Data on Selected High Speed Passenger Trucks; *U.S. Department of Transportation, Federal Railroad Administration*, Report FRA/ORD-78/29, Washington, D.C., Railroad Administration, Report FRA/ORD-78/71, Washington, D.C.

54. SHLADOVER S.E. and HULL, R. (1981): Rail Vehicle Dynamics Model Validation; *Systems Control Incorporated*, Internal Report.

55. SHLADOVER S.E. (1982): Anthology of Rail Dynamics Research; *U.S. Department of Transportation*, Federal Railroad Administration, Report FRA/ORD-8332/37, Washington, D.C.

56. SHUM K.L., SINGH, S.P. and CHANG, E.H. (1979): Track Train Dynamics Report - Track Structures - Vol. III. Study of Effects of Spiral Length on Lateral Stability on Simplecurve Negotiation; *Association of American Railroads*, Report R-355, Chicago, U.S.A.

57. THOMAS R.L., MacMILLAN, R.D. and MARTIN, G.C. (1976): Quasi-Static Lateral Train Stability model - Technical Manual; *Association of American Railroads*, Report R-209, Chicago, U.S.A.

58. WICKENS A.H. (1965): The Dynamic Stability of Railway Vehicle Wheelsets and Bogies Havinng Profiled Wheels; *Int. J. Solids and Structures*, Vol. 1, p. 319.

5

CASE STUDIES

5.1 INTRODUCTION

Wheel/rail geometric constraints, passenger vehicle vertical dynamics, freight car curving dynamics, locomotive response to track irregularities, and longitudinal train action were discussed in Chapter 3. Several of the analytical models presented there were chosen because of their availability as computer programs for use in the case studies of this chapter. Therefore, the contents of Chapter 3, including nomenclatures of the various models, are prerequisites for these case studies.

The wheel/rail geometric constraint variables, which have a strong influence on the lateral dynamics of railway vehicles, are nonlinear functions of the lateral displacement of wheelsets, as was described in Section 3.4.4. Results of a parametric study with various typical wheel and rail profiles are presented in Section 5.2 to determine the effects of wheel and rail wear on the geometric constraint variables.

A vertical dynamics study of a passenger vehicle equipped with a conventional secondary suspension is conducted in Section 5.3. The study utilizes the model of Section 3.6.1 and a complete set of baseline parameter values based on published Amcoach carbody data and typical passenger truck data. Results are obtained for the baseline vehicle and parametric variations about it.

In Section 5.4, examples of the effects of wheel/rail geometry, curve entry geometry, rail stiffness, Coulomb friction and suspension design parameters on the dynamic response of a freight vehicle during curve entry and exit are studied, based on the curving model of Section 3.6.2.

The nonlinear model of a six-axle locomotive's response to tangent track irregularities, described in Section 3.6.3, is evaluated in Section 5.5 by comparing its predictions of wheel/rail loads and carbody accelerations with field test data.

The Detailed Longitudinal Train Action Model (DTAM), presented in Section 3.7, analyzes the longitudinal behaviour of a whole train, or the vehicles within it. Validation results on emergency braking operations with a freight train carrying steel coils are given in Section 5.6.

5.2 WHEEL-RAIL GEOMETRIC CONSTRAINTS

The results of the influence of variations in wheel and rail head profiles on the wheel/rail geometric constraint variables are presented in this section. An analytical procedure based on the describing function technique was developed by Cooperrider et al. [5,12,14] to determine the contact positions, contact angles, difference in contact angles, rolling radii, difference in rolling radii and wheelset roll angle as functions of the wheelset lateral position.

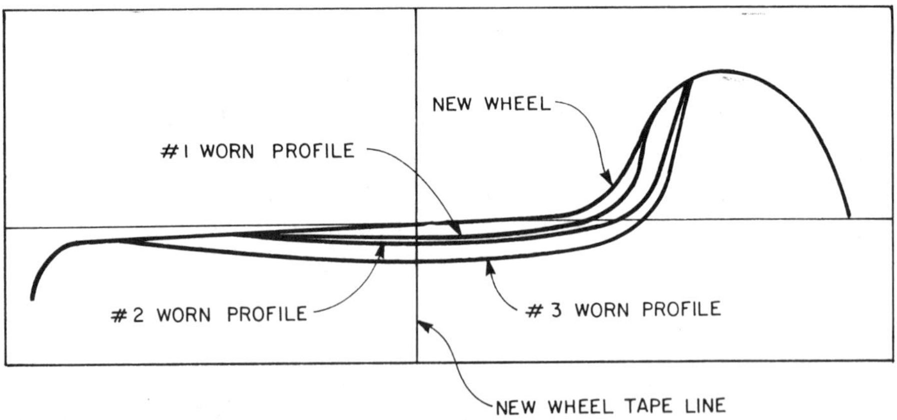

Fig. 5.1 Wheel profiles from a 63.50-tonne (70-ton) hopper car (adapted from Cooperrider et al. [5])

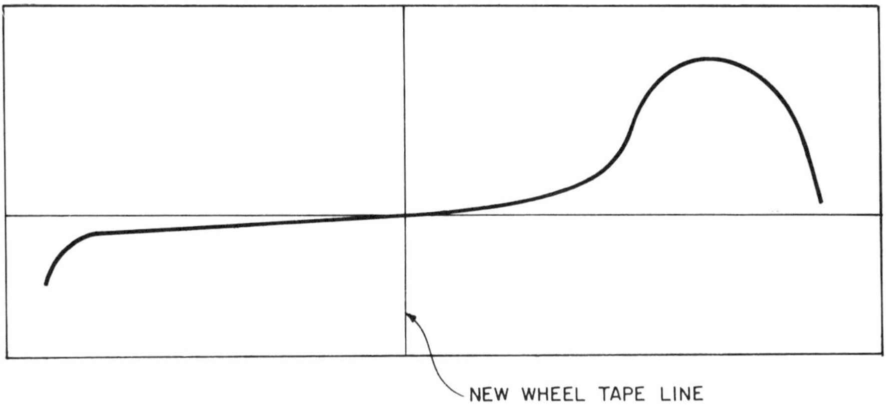

Fig. 5.2 Modified Heumann wheel profile (adapted from Cooperrider et al. [5])

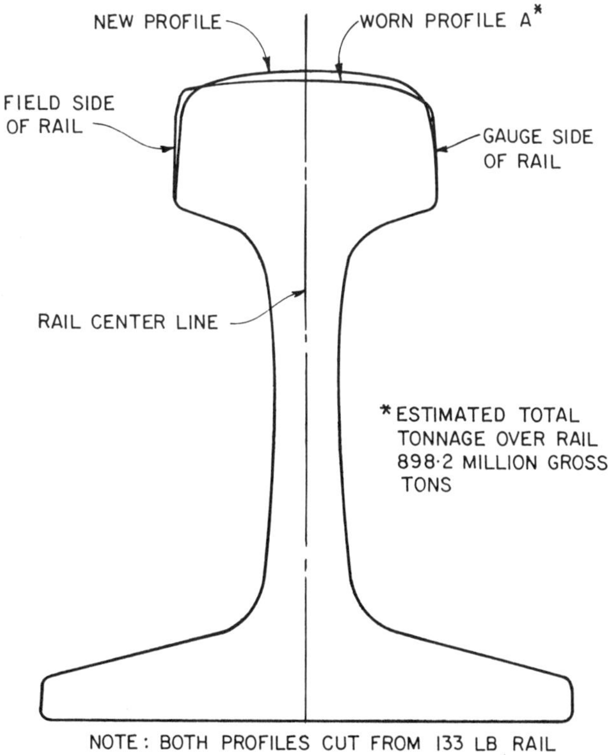

Fig. 5.3 Rail head profiles (from Ref. [5])

The effect of various combinations of the wheel and rail profiles shown in Figs. 5.1 to 5.3 on the wheel/rail constraint variables are presented in the following sections, based on Cooperrider et al. [5].

5.2.1 Wheel Wear Effects

Wheel/rail analysis results for five different wheel profiles with a specific worn rail profile are presented and discussed below. The wheel profiles dealt with are those of a new wheel, a wheel in three stages of wear and a wheel with a modified Heumann profile.

(a) New Wheels: The wheel/rail contact positions and constraint variables for new wheels on worn rails at nominal track gauge are shown in Fig. 5.4. The contact position on the rail remains nearly stationary at a position just to the inside of the rail center until flange contact occurs. The slight jump in contact point at a lateral displacement of about 0.6858 cm (0.27 in), attributed to computational errors, of the profiles in the data generation process or during the curve fit has negligible effect on the constraint variables shown in Fig. 5.4 c, d, e, f, and g.

The constraint variables are all nearly linear over the range of wheelset lateral positions excluding flange contact. Accordingly, in this range, the normalized rolling radii difference may be approximated by

$$\frac{r_L - r_R}{2a_r} = \lambda_e\left(\frac{y}{a_w}\right) \tag{5.1}$$

Likewise,

$$\frac{\delta_L - \delta_R}{2} = \Delta\left(\frac{y}{a_w}\right) \tag{5.2}$$

Fig. 5.4 New wheels on worn rails at nominal gauge: (a) wheelset contact position; (b) rail contact position; (c) wheelset roll; (d) normalized rolling radii difference; (e) rolling radii; (f) one half contact angle difference; (g) contact angles; (adapted from Cooperrider et al. [5])

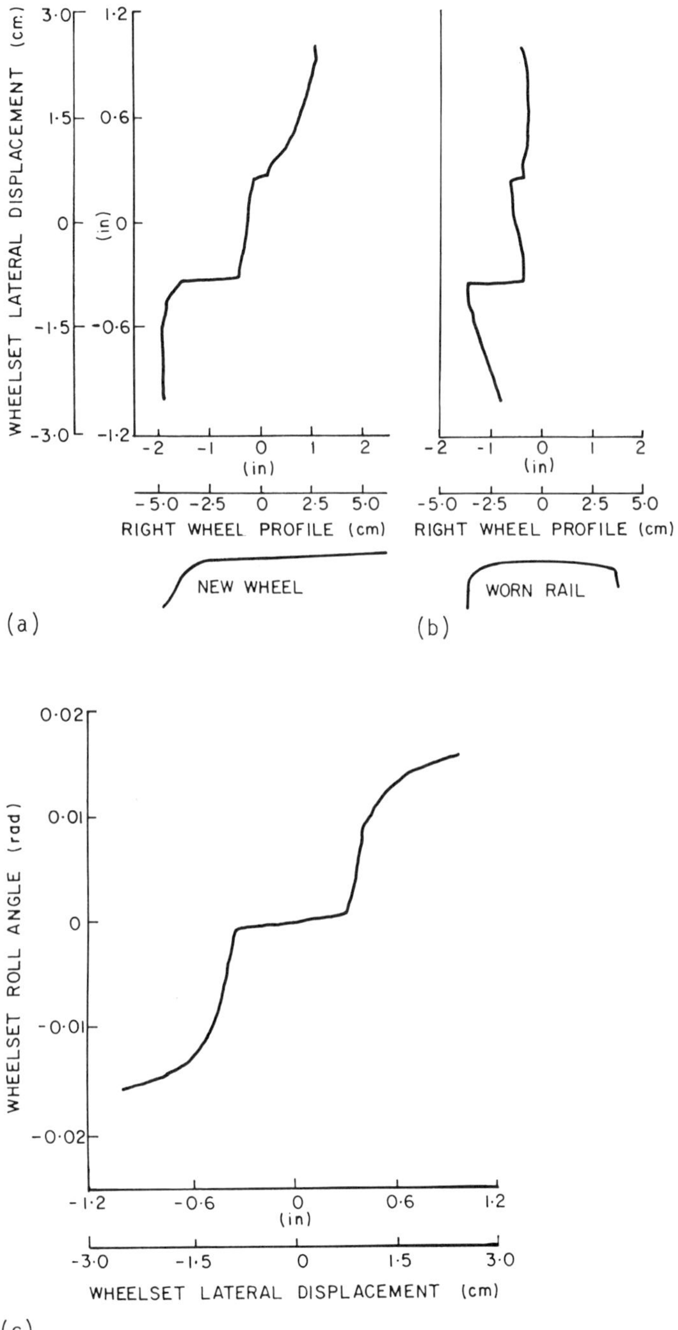
WHEELSET LATERAL DISPLACEMENT (cm)
3·0
1·5
0
-1·5
-3·0
1·2
0·6
(in) 0
-0·6
-1·2
-2 -1 0 1 2
(in)
-5·0 -2·5 0 2·5 5·0
RIGHT WHEEL PROFILE (cm)
NEW WHEEL
(a)
-2 -1 0 1 2
(in)
-5·0 -2·5 0 2·5 5·0
RIGHT WHEEL PROFILE (cm)
WORN RAIL
(b)
WHEELSET ROLL ANGLE (rad)
0·02
0·01
0
-0·01
-0·02
-1·2 -0·6 0 0·6 1·2
(in)
-3·0 -1·5 0 1·5 3·0
WHEELSET LATERAL DISPLACEMENT (cm)
(c)

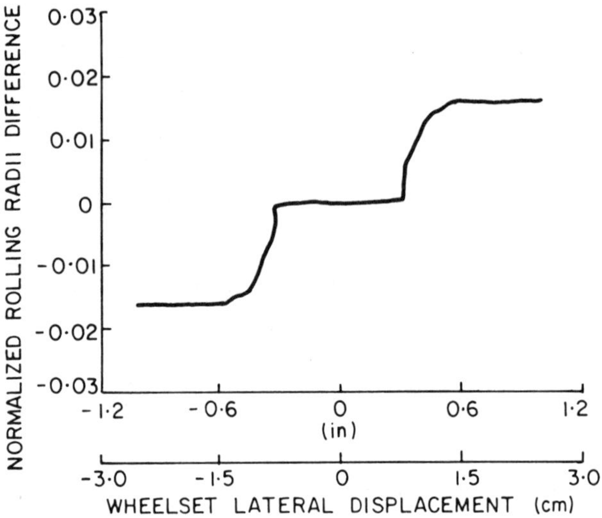

(d)

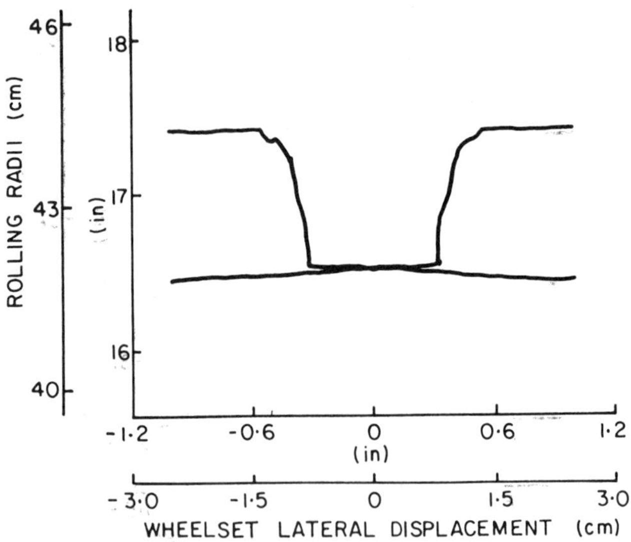

(e)

Fig. 5.4 (continued)

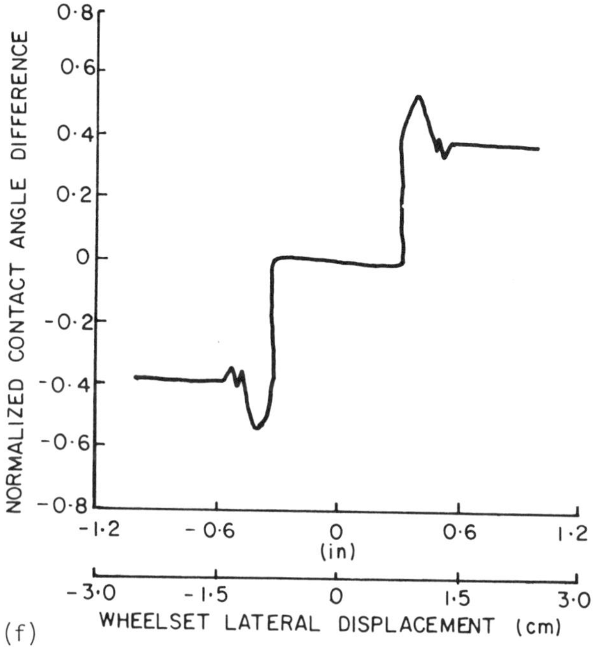
0·8
0·6
0·4
0·2
0
-0·2
-0·4
-0·6
-0·8
NORMALIZED CONTACT ANGLE DIFFERENCE
-1·2
-0·6
0
0·6
1·2
(in)
-3·0
-1·5
0
1·5
3·0
WHEELSET LATERAL DISPLACEMENT (cm)

(f)

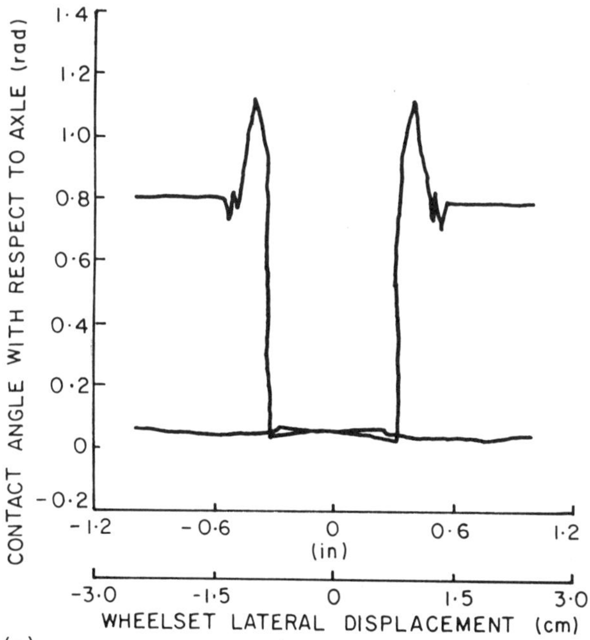
1·4
1·2
1·0
0·8
0·6
0·4
0·2
0
-0·2
CONTACT ANGLE WITH RESPECT TO AXLE (rad)
-1·2
-0·6
0
0·6
1·2
(in)
-3·0
-1·5
0
1·5
3·0
WHEELSET LATERAL DISPLACEMENT (cm)

(g)

is the one-half contact angle difference, and

$$\phi = a_1\left(\frac{y}{a_w}\right) \tag{5.3}$$

is the wheelset roll angle. The slopes λ_e, Δ and a_1 are given by the describing function values within this range. Thus at $\left(\frac{y}{a_w}\right) = 0.30$ approximate values for the linear coefficients in Eqs. (5.1), (5.2) and (5.3) obtained from the describing functions yield:

$\lambda_e = 0.0446$

$\Delta = -0.921$

$a_1 = 0.0643$

The corresponding values for an ideal new wheel on a new rail are $\lambda_e = a_1 = 0.05$ and $\Delta = 0$. The above deviations from these "ideal" values is due to the effect of the worn rails.

The contact point moves to the inside edge of the rail, and remains there whenever the flange contacts the rail. The contact point on the wheel also jumps from the tread region to the flange. This contact jump is reflected in jump discontinuities in the rolling radii difference and contact angle difference. Further increase in lateral displacement of the wheelset moves the contact point on up the flange and eventually down the other side when the displacement exceeds 1.778 cm (0.70 in).

(b) Slightly Worn Wheels: Similarly, it can be shown that a slightly worn wheel (profile #1 in Fig. 5.1) on a worn rail (profile A in Fig. 5.3), at nominal rail gauge, causes the contact points on the wheel and rail to move in a series of small jumps across a wider band of the wheel and rail profiles than with a new wheel. Furthermore, there exists a range of y in which contact is made on the outside half of the rail.

Both the rolling radii and contact angle differences are found to be nearly zero in a narrow ± 0.127 cm (± 0.05 in) band about the wheelset's centered position. The rolling radii difference and wheelset roll angle appear to be nearly linear over the full

±0.9398 cm (± 0.37 in) range before flange contact occurs. The linear coefficients in Eqs. (5.1), (5.2), and (5.3) obtained from the describing functions at $\left(\frac{y}{a_w}\right) = 0.30$ are:

$\lambda_e = 0.0885$

$\Delta = 6.9530$

$a_1 = 0.7870$

The roll coefficient value is larger than for a new wheel. Also, the larger value of conicity λ_e indicates that a worn wheel tends to be more unstable than a new one. The contact angle difference, which contributes to the gravitational stiffness is also larger for this wheel. Flanging occurs at about the same wheelset lateral position as for a new wheel. However, since the flange is steeper on a worn wheel, both the contact angle and rolling radii differences increase faster, during flanging than with a new wheel.

(c) Moderately Worn Wheels: For a moderately worn wheel (profile #2 of Fig. 5.1), the contact position shifts to the outside (field side) over much of the wheelset's lateral range. This moves the contact point to a portion of the wheel profile with negative slope when the wheel moves away from the rail. The contact angle is then negative and both the rolling radii and contact angle differences have negative slopes in this region.

In a small region about the wheelset's centered position, this moderately worn wheel yields negative values of coefficients λ_e, Δ, and a_1. On the other hand, during flange contact, i.e., $y \geq 0.9398$ cm (0.37 in), the values of wheel/rail geometric constraint variables are very similar to the preceding case of slightly worn wheels on worn rails.

(d) Severely Worn Wheels: The wheel/rail behaviour of the most severely worn wheel (profile #3 of Fig. 5.1) is quite similar to that of the moderately worn wheel. The rolling radii difference, contact angle difference and wheelset roll angle have negative slopes of larger magnitude than for the moderately worn wheel in the region preceding flange contact. The lateral wheelset travel to the flange is quite large with this severely worn profile. The clearance is 0.3048 cm (0.12 in) more than for the moderately worn wheel, and nearly 0.5080 cm (0.20 in) more than the new wheel value, for a total of almost 1.524 cm (0.60 in) of flange clearance. Once flange contact is made, the geometric constraint variables assume positive values.

(e) Heumann Profile: The original Heumann profile was designed to obtain single point contact between the wheel and rail at all wheelset positions. The modified Heumann profile of Fig. 5.2 on worn rail with nominal gauge closely approximates single point contact over the full range, including a smooth transition onto the flange. This wide range of continuous single-point contact produces a more uniform wheel wear than with a standard wheel profile.

The wheel/rail constraint variables for the modified Heumann profile are also well behaved. The difference in rolling radii is approximately linear over a wide range. Even at small wheelset lateral displacements, the value of λ_e lies in the range (0.20, 0.40). Conicities of this magnitude are strongly destabilizing. However, this is counter-balanced to some extent by the steep slopes of the difference in contact angles and wheelset roll angle. At large wheelset lateral displacements, the wheel/rail constraint variables reach levels comparable to those of new, slightly worn and moderately worn wheels.

The relative influence of the various wheel profiles on rail vehicle dynamics can be determined by comparing their describing functions [5]. Figure 5.5 shows the describing function of the effective conicity λ_e pertaining to the normalized difference in rolling radii for the five wheels discussed above. It can be seen that, in the range excluding flange contact, it is largest for the modified Heumann profile, somewhat smaller for the slightly worn wheel, small but positive for the new wheel, and negative for the two most worn wheels. The large conicity of the Heumann profile may cause hunting at relatively low speeds unless the vehicle is

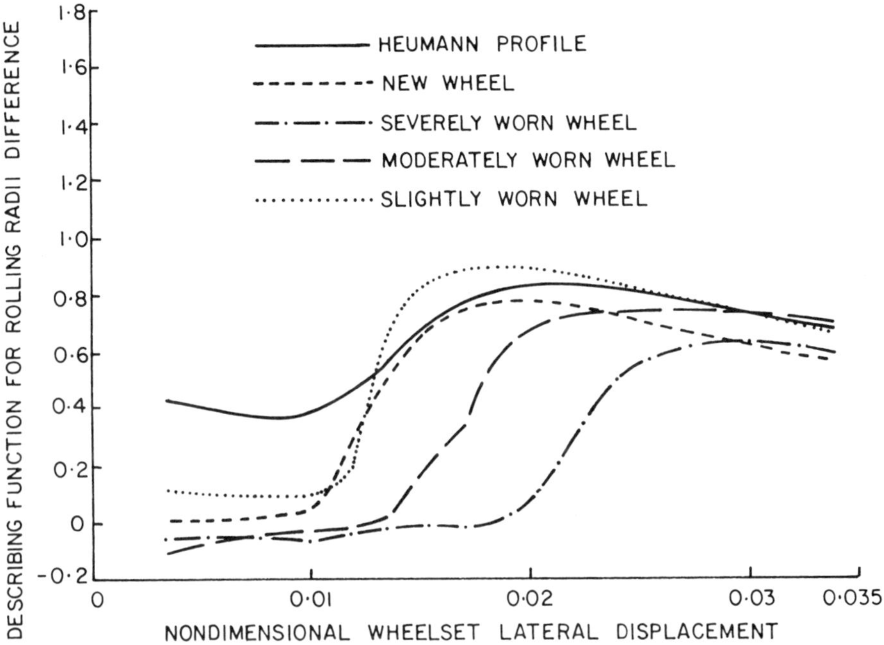

Fig. 5.5 Describing function for rolling radii difference vs. nondimensional wheelset lateral displacement on worn rails at nominal gauge (adapted from Cooperrider et al. [5])

specifically designed to accommodate these characteristics. The slightly worn or new wheels tend to promote more stable behaviour. However, the negative conicity exhibited by the two most worn wheels would probably cause vehicles to diverge from their centered position on the rails at all speeds and settle on either a stable position or hunting oscillation with flange contact.

The describing function pertaining to one-half the difference in contact angles is shown in Fig. 5.6 for the five wheel profiles. At small lateral wheelset displacements, it is greatest for the modified Heumann and slightly worn wheel profiles. The other wheel profiles have relatively small contact angle differences until flange contact occurs. Referring to the approximate gravitational stiffness force equation (Eq. (3.26)), it can be expected that the

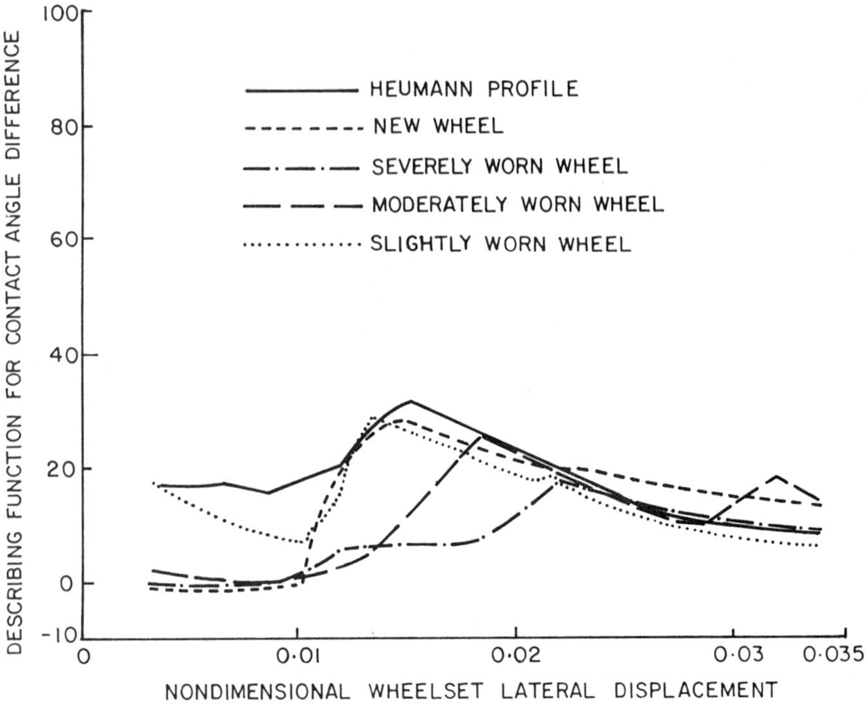

Fig. 5.6 Contact angle difference describing function vs. nondimensional wheelset lateral displacement on worn rails at nominal gauge (adapted from Cooperrider et al. [5])

gravitational stiffness effect on a heavily loaded Heumann or slightly worn wheel may be large enough to counter the destabilizing effect of the relatively large conicity of these same wheels and allow stable running over a reasonable speed range. For the new wheel and the two worn wheels the gravitational stiffness is not large enough to affect the dynamic behaviour during small lateral wheelset excursions. When the wheel/rail contact moves up the flange, the gravitational stiffness is, of course, the primary stabilizing influence.

The describing function pertaining to the wheelset roll angle is shown in Fig. 5.7 for the same wheel profiles. Only the Heumann

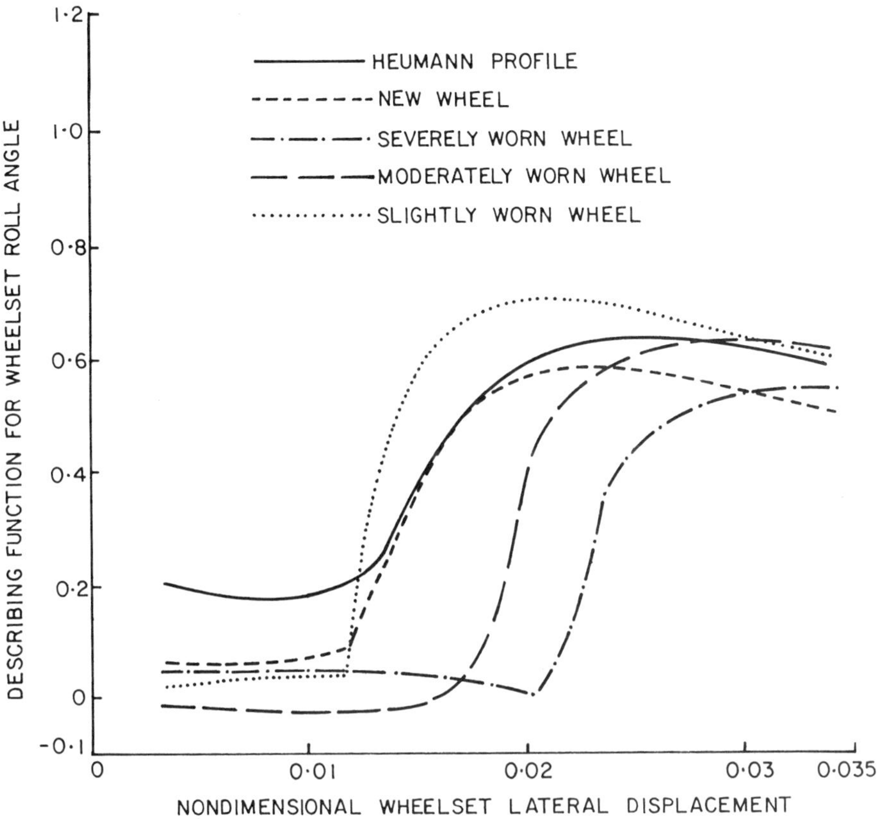

Fig. 5.7 Describing function for wheelset roll vs. nondimensional wheelset lateral displacement on worn rails at nominal gauge (adapted from Cooperrider et al. [5])

profile has a large describing function value for small wheelset lateral displacements. At larger displacements, when the contact position moves onto the wheel flange, the describing function is about the same for all profiles, although the wheelset lateral position for flange contact differs.

Summarizing, the wheel profiles discussed above can be grouped into three categories. Heumann and slightly worn wheels are typified by large conicities and gravitational stiffnesses. New wheels, by comparison, have small conicities and gravitational stiffnesses. Highly worn wheels are characterized by negative conicities at small wheelset lateral displacements. Each of these three groups has a distinct influence on vehicle behaviour.

5.2.2 Rail Wear Effects

The influence of rail wear on the wheel/rail geometric constraint variables for new, severely worn, and modified Heumann wheel profiles is presented here [5].

(a) New Wheels: A comparison of the contact position graphs for a new wheel on new rail at nominal gauge with those of Fig. 5.4 for a new wheel on worn rail shows that the contact point on the new rail remains essentially fixed until the flanging occurs, whereas it shifts slightly on the worn rail as the wheelset moves laterally. Thus, the small jumps that appear in the geometric constraint variables for worn rails do not occur in the case of new rails. Furthermore, the slope of the rolling radii difference and roll angle functions in the range before flange contact occurs is somewhat larger in the new rail case.

(b) Severely Worn Wheels: Similarly, for severely worn wheels, it can be shown that the contact point on a new rail remains in a fixed position over a wider range of wheelset lateral motion than on a worn rail, and that this position is closer to the center of the rail. Although the final jump in contact from the wheel tread to the wheel flange occurs at about the same lateral wheelset position in both cases, an intermediate jump occurs at a larger wheelset displacement on the new rail.

The influence of the difference in rail head profile on the wheel/rail constraints is quite minimal. The rolling radii difference, contact angle difference and roll angle functions are found to remain flatter over a wider range on the new rail than on the worn rail.

(c) Modified Heumann Profile: When the modified Heumann profile on new and worn rails is examined, it is found that the contact position on the wheel and rail have jumps of about the same magnitude in both cases, but they occur at different wheelset lateral positions. The rail contact position covers a wider range on the new rail than on the worn one, but the wheel contact position covers the same range in the two situations.

These small contact position differences have a large impact on the geometric constraint variables near the centered wheelset position. For instance, at a lateral position of 0.7620 cm (0.30 in) the rolling radii difference, contact angle difference and roll angle are all about 50% less on the new rail. At larger lateral displacements, this difference between the new and worn rail cases becomes small until, at 1.524 cm (0.60 in), it is approximately zero.

The following conclusions may be drawn with respect to new and worn rails:

(1) New wheel profiles have identical behaviour at large lateral wheelset displacements, but their conicity on new rails is larger and nearly constant. Thus, the stability of vehicles with new wheels may be slightly better on worn rails.

(2) Conicity and roll coefficient values for the modified Heumann profile are greater by 50% on worn rails.

(3) The describing functions for the severely worn wheel were substantially unaffected, even at large displacements.

(4) Rail wear does not affect the geometric constraint variables as strongly as the wheel wear. Indeed, some wheel profiles, such as severely worn wheel profile #3 of Fig. 5.1, are unaffected by rail wear. Rail wear can be either stabilizing or destabilizing, depending on the applicable wheel profile.

5.3 PASSENGER VEHICLE VERTICAL DYNAMICS

The results of studies conducted by Hedrick et al. [10,11] to evaluate the influence of parametric variations on vehicle performance are discussed in this section. The studies were based on the analytical model described in Section 3.6.1 and the baseline parameter values summarized in Table 5.1. These values correspond to a standard Amcoach vehicle, with estimates of the carbody's first bending mode and first two torsional modes.

Table 5.1. Baseline Amcoach Vehicle Parameters.

Vehicle Geometry: m (ft)			
Truck	b	= 1.295	(4.25)
	d_p	= 0.585	(1.92)
Carbody	d_s	= 1.143	(3.75)
	ℓ_c	= 25.908	(85)
	ℓ_s	= 9.068	(29.75)
Masses: kg (slug)			
Truck	M_t	= 3094	(212)
Carbody	M_c	= 41739	(2860)
Moments of Inertia: kg m^2 (slug ft^2)			
Truck	I_{tx}	= 2366	(1745)
	I_{ty}	= 498.9	(368)
Carbody	I_{cx}	= 103500	(76342)
	I_{cy}	= 2.391×10^6	(1.764×10^6)
Suspension Stiffness: N/m (lb/ft)			
Primary	k_{pz}	= 2.824×10^7	(1.935×10^6)
Secondary	k_{sz}	= 324000	(22200)
Suspension Damping: N s/m (lb s/ft)			
Primary	c_{pz}	= 21890	(1500)
Secondary	c_{sz}	= 29190	(2000)
Miscellaneous Parameters:			
	V	= 177 km/h	(110 mph)
	x_d	= 9.068 m	(29.75 ft)
	z_d	= 1.143 m	(3.75 ft)
Carbody torsion	ω_{1t}	= 12 Hz	
Carbody bending	ω_{1b}	= 7.1 Hz	

The response of the vertical railcar model is investigated by evaluating transfer functions between track irregularities and carbody displacements at representative passenger locations, and calculating accelerations in response to track profile and cross-level spectral densities. The acceleration data for Class 6 track traversed at 177 km/h (110 mph), described in Chapter 4, are used to determine the effect of variations in primary and secondary suspension parameters on passenger ride comfort.

5.3.1 Baseline Vehicle

The baseline vehicle is investigated in three parts: (a) natural modes of vibration, dealing with carbody displacement transfer functions, (b) track input spectra, describing vertical profile and cross-level track irregularities, and (c) forced response, where carbody vertical acceleration spectra are computed from the transfer functions and track input spectra.

(a) Natural Modes of Vibration: Insight into the baseline vehicle's dynamic characteristics was obtained by investigating a histogram of its natural response, where damping ratio was plotted against natural frequency for various modes of the baseline vertical model. It was found that, with a conventional secondary suspension, the rigid carbody modes are moderately damped and have natural frequencies in the neighbourhood of 1.0 Hz, whereas the flexible carbody modes are lightly damped with natural frequencies in the intermediate frequency range (5-11 Hz), where vehicle accelerations have a relatively large effect on passenger ride comfort. The rigid truck modes are slightly less damped than the rigid carbody modes and their natural frequencies are above the sensitive range for ride comfort. The truck pitch modes have no effect on carbody accelerations.

Transfer functions yielding carbody vertical displacement as a function of track profile and cross-level irregularities respectively, were obtained for two representative passenger locations in the vehicle: 1) the carbody center of gravity, and 2) a point located at 9.068 m (29.75 ft) behind and 1.143 m (3.75 ft) to the

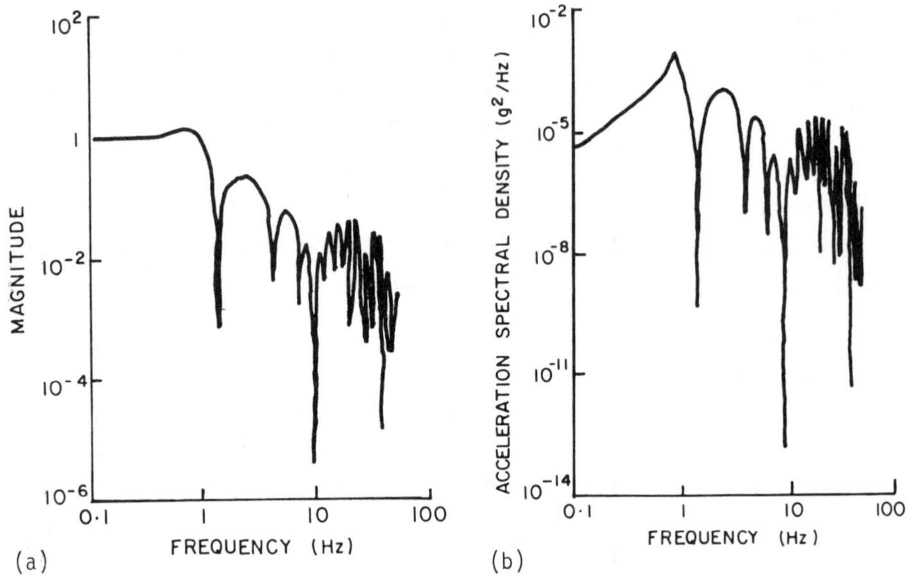

Fig. 5.8 Displacement transfer function for track profile input: (a) at cg; (b) at rear passenger point (adapted from Hedrick et al. [10])

right of the vehicle centre of gravity, over which the righthand rear secondary suspension elements are attached to the carbody.

The resonance peaks seen in Fig. 5.8(a), for the track profile irregularities to carbody center of gravity displacement transfer function shown are due to the carbody and truck heave natural frequencies (0.89 Hz and 30.5 Hz, respectively). Dips or dropouts in this frequency response occur at 1.4 Hz, 4.1 Hz, 6.8 Hz, 9.5 Hz, 12.3 Hz, etc. These are approximately the frequencies for which the truck wheelbase and/or the distance between truck centers is an odd multiple of half-wave-lengths of the profile irregularity input, which results in zero heave for a symmetrical carbody. Similarly, the pitch is zero for frequencies where the distance between truck centers is an integral number of wave-lengths). At certain speeds these dropout frequencies coincide with vehicle resonant frequencies, causing marked attenuations of the transfer functions and corresponding accelerations. For example, the carbody bending resonance peak is significantly attenuated by the dropout at 6.8 Hz in Fig. 5.8(a). For the aft passenger location displacement transfer function shown in Fig. 5.8(b), a resonance peak is present due to the carbody pitch natural frequency (1.06 Hz) in addition to those peaks present in the center of gravity displacement transfer function. Again, the carbody bending resonance peak is very nearly cancelled by the dropout at 6.8 Hz.

The track cross-level to carbody vertical displacement frequency response (not shown) displays two prominent resonance peaks: a low-frequency peak resulting from the carbody roll natural frequency (0.64 Hz) and an intermediate-to-high-frequency group of peaks due to the carbody fundamental frequency in torsion (12.1 Hz), the truck roll natural frequency (20.5 Hz), and the carbody second natural frequency in torsion (24.0 Hz). The same dropout frequencies are associated with the carbody roll due to track cross-level as with carbody heave due to vertical profile; and, in the profile irregularity to carbody vertical displacement transfer function, some resonance peaks are attenuated due to these dropouts (e.g., the carbody roll response peak decreases faster than it should because of the dropout at 1.36 Hz and the higher frequency group of resonance peaks are somewhat attenuated due to the dropouts at 12.1 Hz, 20.5 Hz, and 23.2 Hz).

(b) Track Input Spectra: The track inputs to the dynamic model are the vertical profile, cross-level and lateral alignment, but only the first two affect the vertical response. Figure 5.9(a) shows these spectra as a function of frequency ($\omega = V\Omega$) evaluated at V = 177 km/h (110 mph) over the range of Ω (cycles/m). It is important to note that, at this speed, the most important frequency range is between 1 and 2 Hz.

(c) Forced Response: Carbody vertical accelerations are computed from the transfer functions and track profile and/or cross-level spectra as described in Section 3.6. The vertical acceleration data for the two representative passenger locations in the baseline vehicle are presented in three ways: (1) the combined

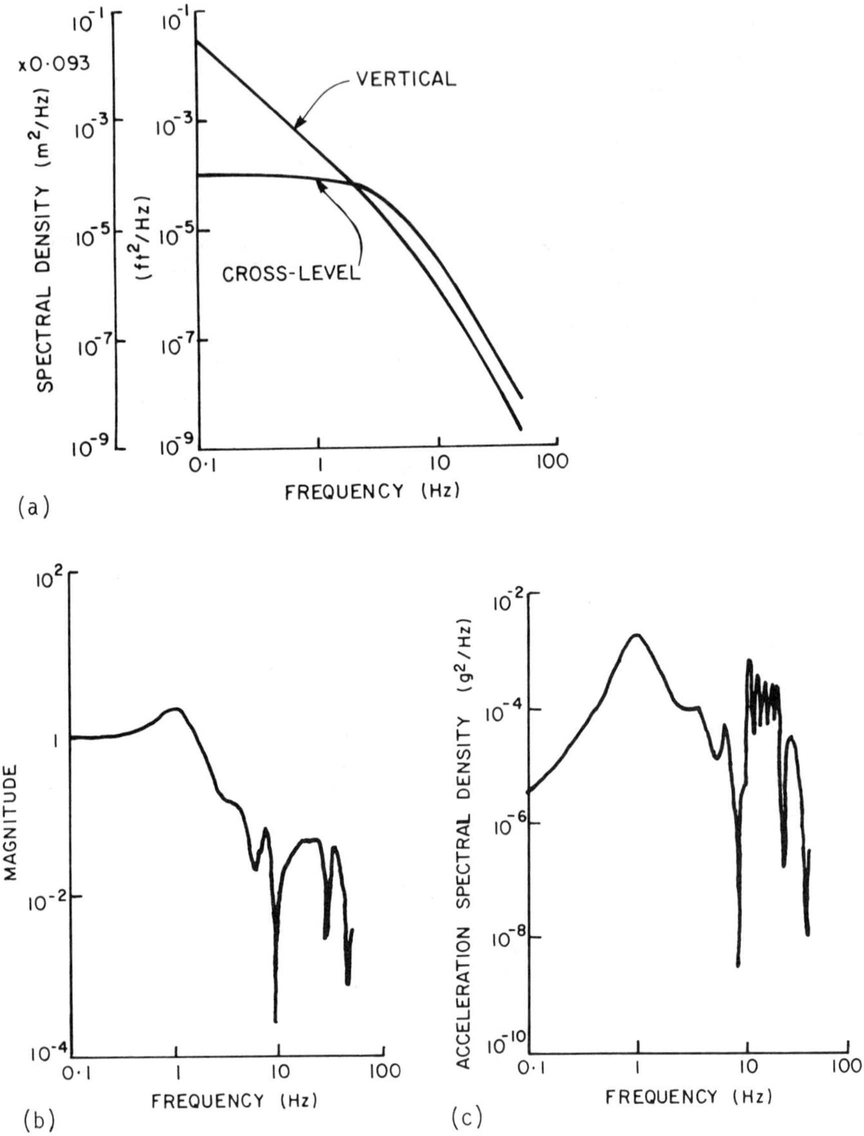

Fig. 5.9 (a) Track profile and cross-level inputs at 176 km/h (110 mph), (b) Acceleration spectral densities at cg for the baseline vehicle with conventional secondary suspension, (c) Acceleration spectral densities at rear passenger point for the baseline vehicle with conventional secondary suspension; (redrawn from Hedrick et al. [10])

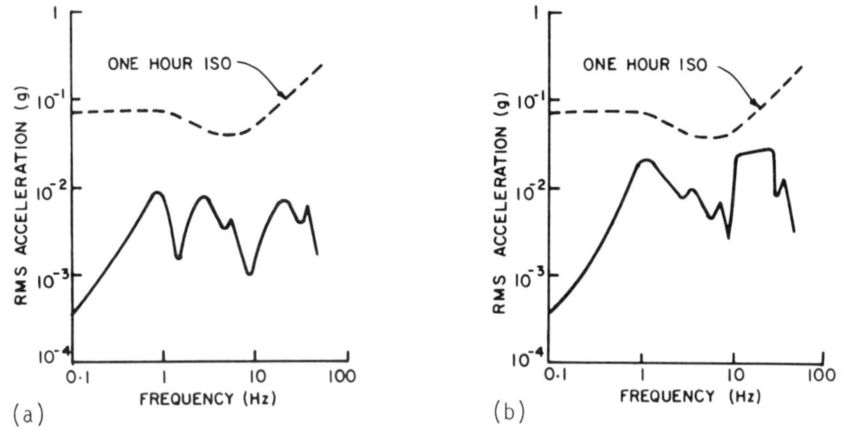

Fig. 5.10 One-third octave band rms acceleration plots for the baseline vehicle with conventional secondary suspension: (a) at cg; (b) at rear passenger point; (redrawn from Hedrick et al. [10])

total mean square acceleration spectra (Figs. 5.9(a) and (c)), (2) the combined total rms acceleration per one-third octave band compared with the one-hour ISO reduced-comfort boundary (Fig. 5.10), and (3) the rms acceleration and secondary stroke in the 0.1-50 Hz frequency range (Table 5.2).

A comparison of the general shape of the acceleration spectrum of Fig. 5.9(c) and the transfer function of Fig. 5.8(b) shows that the response due to vertical profile irregularities is the dominant component in the total acceleration response at the rear passenger location for all frequencies except the 10-25 Hz range where the response due to cross-level is relatively significant. Since the acceleration response below 10 Hz is dominated by the component due to vertical profile track irregularities, it follows that the ride quality is chiefly determined by that component; however, the contribution due to cross-level irregularities is significant. For example, the baseline vehicle's poorest ride quality in any single one-third octave band (see Fig. 5.10(b)) occurs in the band with a center frequency of 11.4 Hz, which is in that portion of the

Table 5.2. RMS Acceleration and RMS Secondary Stroke Data for Primary Suspension Stiffness and Damping Variations (from Hedrick et al. [10]).

PRIMARY STIFFNESS N/m (lb/ft)	PRIMARY DAMPING N s/m (lb s/ft)	DESCRIPTION	RMS ACCELERATION AT REAR POINT g	RMS SECONDARY STROKE (mm) (in)
2.8239×10^{7} (1.935×10^{6})	21891 (1500)	baseline	0.0736	9.5555 (0.3762)
1.1295×10^{8} (7.74×10^{6})	21891 (1500)	increased stiffness	0.0684*	9.4869 (0.3735)
7.0598×10^{6} (4.8375×10^{5})	21891 (1500)	decreased stiffness	0.0575	9.6317 (0.3792)
2.8239×10^{7} (1.935×10^{6})	58376 (4000)	increased damping	0.0674	9.5326 (0.3753)
2.8239×10^{7} (1.935×10^{6})	131345 (9000)	increased damping	0.0616	9.5123 (0.3745)

*does not include contribution due to truck heave resonant peak (60.8 Hz)

response affected by cross-level irregularities. Another measure of ride comfort is the rms roll acceleration at a passenger location. At the aft passenger location of the baseline vehicle, the rms roll acceleration (the quotient of the rms vertical acceleration due to cross-level inputs and y_d) is 0.446 rad/s², which is within the range of the first discomfort curve (DISC = 1) of Fig. 3.4.

In summary, the acceleration component due to track profile dominates the carbody acceleration response over most of the frequency range from 0.1 Hz to 50 Hz. Although the frequency range where passenger ride comfort is most sensitive to vehicle accelerations is dominated by the acceleration response component due to track profile, cross-level does influence ride comfort.

5.3.2 Parametric Study

(a) Primary Stiffness Variations: Two variations in the primary suspension stiffness k_{pz} were made. First, it was increased by a factor of four, then decreased by the same factor relative to the baseline value. In each case, the responses to track irregularities were compared with the baseline responses (Fig. 5.11). For the vehicle with stiffened primary suspensions, the combined total mean square acceleration spectral density at both the center of gravity and aft passenger locations and by comparison with the baseline response at these locations is somewhat diminished in the 10 - 25 Hz range while the response above 30 Hz is significantly increased. The rms accelerations for the centre of gravity and aft passenger point are 0.0221g and 0.0684g as compared to 0.0239g and 0.0736g for the baseline response (Tables 5.2 and 5.3). These rms accelerations decreased because the resonance peak due to the truck heave natural frequency, shifted to 60.8 Hz from the baseline 40.8 Hz, is not in the frequency range for which the above rms accelerations are calculated. The acceleration spectra are integrated from 0.1 Hz to 50 Hz to obtain the corresponding rms acceleration. Thus, due to the increased distance from the one-hour ISO curve in the 10-25 Hz range, the ride comfort is improved by the stiffened primary suspensions.

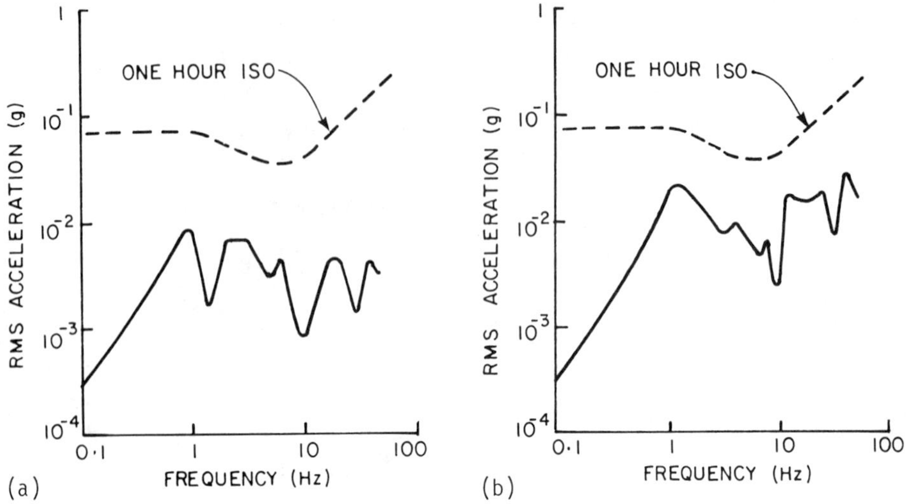

Fig. 5.11 One-third octave band rms acceleration plots for the baseline vehicle with stiffened primary suspension: (a) at cg; (b) at rear passenger point; (redrawn from Hedrick et al. [10])

On the other hand, the softened primary suspensions lead to greater filtering of accelerations transmitted to the truck and, thereby, the carbody. The total rms accelerations for the center of gravity and aft passenger point are 0.0230g and 0.0575g, compared to 0.0239g and 0.0736g for the baseline response. However, the ride comfort is only slightly improved because the reduction in rms acceleration occurs in the high-frequency region where vehicle accelerations have a relatively small effect on passenger ride comfort.

(b) Primary Damping Variations: The primary damping coefficient, c_{pz}, was increased successively by factors of 2.67 and 6 times the baseline value. The total rms accelerations for the first and second increases in primary damping are 0.0674g and 0.0616g , respectively, as compared to 0.0736g for the baseline vehicle; however, as shown by the one-third octave band rms acceleration plots in Fig. 5.12, the decreases in total rms acceleration

Table 5.3. RMS Acceleration and Secondary Stroke Data for Secondary Suspension Damping Ratio Variations (from Hedrick et al. [10]).

DAMPING RATIO FOR CARBODY HEAVE	PASSENGER LOCATION	COMBINED TOTAL rms ACCELERATION g	COMBINED TOTAL rms SECONDARY STROKE mm (in)
0.25	cg	0.0239	———
(baseline value)	right rear	0.0736	9.5555 (0.3762)
0.375	cg	0.0289	———
	right rear	0.0886	7.5158 (0.2959)
0.50	cg	0.0350	———
	right rear	0.1038	6.4262 (0.2530)
0.707	cg	0.0453	———
	right rear	0.1266	5.4559 (0.2148)

due to the increases in primary damping occur in the high-frequency portion of the response where passenger ride comfort is not as sensitive to vehicle accelerations.

A summary of the primary suspension parameter variations and the resulting rms acceleration and rms secondary stroke is presented in Table 5.2. It can be concluded from this parametric study that the secondary stroke was essentially unaffected by primary suspension variations in the parameter ranges considered. Furthermore, although primary suspension parameter variations caused reductions of as much as 25% in the rms acceleration of the carbody, ride quality was only slightly improved.

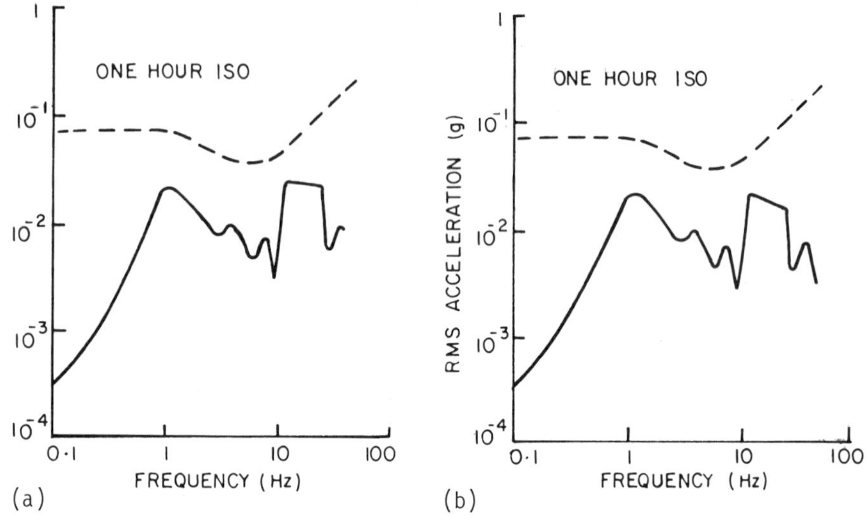

Fig. 5.12 One-third octave band rms acceleration at rear passenger point for baseline vehicle with increased primary damping
(a) c_{pz} = 58,376 N s/m (4,000 lb s/ft)
(b) c_{pz} = 131,346 N s/m (9,000 lb s/ft)

(c) Natural Frequency Variations: The natural frequencies of the carbody rigid-body modes (heave, pitch and roll) were varied without changing their damping ratios. This was achieved by suitable variations in the secondary suspension stiffness, k_{sz}, and damping, c_{sz}. The effect of halving and doubling the baseline natural frequencies was investigated.

For the reduced carbody natural frequencies, it was found that, although the shape of the combined total acceleration responses at the centre of gravity and aft passenger location was similar, their level level was significantly reduced for frequencies greater than approximately 1 Hz because the softer secondary suspension transmits less high-frequency force to the carbody. Thus, the ride comfort was substantially improved. The total rms accelerations for the center of gravity and aft passenger location were 0.0129g and 0.0395g compared with the baseline values of 0.0241g and

0.0747g, respectively, while the rms secondary stroke length was 1.50876 cm (0.594 in) compared with 0.95504 cm (0.376 in) for the baseline case.

When the carbody natural frequencies were doubled, it was found that a resonance peak near 2 Hz, due to the carbody rigid-body modes, was a large factor in the acceleration response at both locations. Because the stiffened secondary suspension transmitted more high-frequency force to the carbody, the acceleration spectra were higher than the baseline spectra for all frequencies above approximately 2 Hz. The total rms accelerations at the center of gravity and rear passenger location were 0.0586g and 0.1190g, respectively, compared with 0.0241g and 0.0747g for the baseline vehicle, and the rms secondary stroke was 0.5334 cm (0.210 in) compared with the baseline value of 0.9550 cm (0.376 in). The higher carbody natural frequencies reduced the ride comfort substantially as seen in the one-third octave band rms acceleration plot (Fig. 5.13).

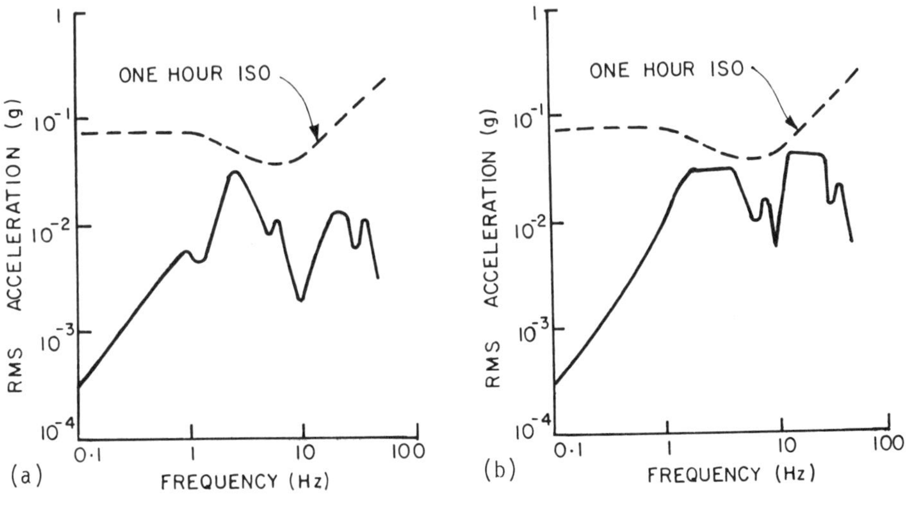

Fig. 5.13 One-third octave band rms acceleration plots for the baseline vehicle with increased carbody rigid-body mode natural frequencies: (a) at cg; (b) at rear passenger point; (redrawn from Hedrick et al. [10])

(d) Damping Ratio Variations: The effect of damping on carbody acceleration and passenger ride comfort was studied through three variations in the secondary suspension damping coefficient, c_{sz}, such that the carbody's heave damping ratio, ξ_h was increased from the baseline value of 0.25 to 0.375, 0.50 and 0.707, in turn. Corresponding proportional increases in the carbody's pitch and roll damping ratios occured in conjunction with the changes in ξ_h because the damping ratio for each of the carbody rigid-body modes is linearly dependent upon c_{sz}.

Comparing with the acceleration response of the baseline vehicle (Fig. 5.9), it was found that, as secondary damping was increased, the acceleration response at both passenger locations decreased in the frequency range below 2 Hz because of the reduced resonance peaks associated with the more heavily damped carbody rigid-body modes, and increased above 2 Hz because the more highly-damped suspension transmitted more high-frequency force to the carbody. In Table 5.3, it is seen that increases in the amount of damping caused significant increases in total rms acceleration and substantial reductions in rms secondary stroke.

Due to the increased acceleration levels in the sensitive frequency range for passenger ride comfort, damping ratio increases were associated with ride comfort deteriorations.

5.4 FREIGHT CAR CURVING DYNAMICS

The application of the freight car dynamic curving model (Section 3.6.2) to simulate the dynamic response of a vehicle during curve entry and exit is presented in this section. Following an evaluation of the model, simulation results are used to determine the effects of parameter variations in wheel/rail geometry, geometry, curve entry geometry, Coulomb friction and suspension design.

5.4.1 Model Evaluation

Preliminary evaluation studies were conducted by Krolewski [13] to compare the present model with a nonlinear steady-state curving model for passenger vehicles, developed by Bell [4], and a linear model of a freight car on tangent track, developed by Hadden [9].

For comparison with the steady-state curving results of the Bell's model, the present model was used, with consistent initial conditions, to simulate operation along a constant radius track. Equilibrium forces and other results were in close agreement.

For comparison with the Hadden model, nonlinearities were removed from the present model and simulation results for tangent track operation were obtained. Creep force saturation, Coulomb friction, and nonlinear wheel/rail profiles were some of the nonlinearities involved. System parameters were chosen to correspond to a given linear induction motor rail vehicle (LIMRV) [9]. The LIMRV vehicle had a soft primary suspension, resulting in a critical speed of 42.37 m/s (139 ft/s). The modified model was run at incremental vehicle speeds to determine the critical velocity. The results obtained were consistent with the eigenvalue analysis of the Hadden model [9].

5.4.2 Parametric Study

Computer simulation studies by Krolewski [13] on dynamic curving performance are presented here. The influence of vehicle parameters, track layout, wheel/rail profiles and rail flexibility is examined. The curving performance, as determined by the lateral forces, L/V ratios, angles of attack and work in the contact regions, is an indirect measure of the tendency for wheel-climb, rail rollover, lateral track shift and wheel/rail wear to occur. The graphical results are presented as time histories of state variables, wheel/rail forces and work.

(a) Baseline Vehicle Performance: A set of baseline vehicle parameters for an unloaded hopper freight car are listed in Table 5.4 [9]. The baseline vehicle was made to travel, by simulation, along a 45.72 m (150 ft) transition spiral into a 2.5 degree curve at a balanced running speed of 15.24 m/s (50 ft/s). The results were used to evaluate the effects of changing operating conditions. Tabulated wheel/rail profiles developed by Cooperrider and Law [6] and the geometric relationships for new wheels on worn

Table 5.4. Baseline Freight Car Data (from Krolewski [13]).

Vehicle Geometry: m (ft)

a = 0.7498 (2.46)

b = 0.8534 (2.8)

d = 0.9906 (3.25)

h_C = 0.9126 (2.994)

h_S = 0.2512 (0.824)

h_T = 0.9126 (2.994)

L = 5.6693 (18.6)

Masses: kg (slug)

m_B = 526.84 (36.1)

m_C = 16083 (1102)

m_S = 350.26 (24)

m_W = 1117.9 (76.6)

L_A = 5021.8 (344.1)

Moments of Inertia: kg m^2 (slug ft^2)

I_{BX} = 242.15 (178.6)

I_{BZ} = 242.15 (178.6)

I_{CX} = 17625 (13000)

I_{CZ} = 3.1726×10^5 (2.34×10^5)

I_{WY} = 71.993 (53.1)

I_{WZ} = 608.08 (448.5)

Suspension Parameters:

Stiffnesses:

k_B = 0 N m/rad (0 ft lb/rad)

k_{CP} = 67.79 N m/rad (50 lb ft/rad)

k_{PX} = 5.5457×10^8 N/m (38×10^6 lb/ft)

k_{PY} = 4.2323×10^8 N/m (29×10^6 lb/ft)

k_S = 0 N/m (0 lb/ft)

k_{SY} = 3.5026×10^5 N/m (24000 lb/ft)

k_{SZ} = 3.8791×10^6 N/m (2.658×10^5 lb/ft)

k_W = 5.4381×10^6 N m/rad (4.011×10^6 lb ft/rad)

Table 5.4. (continued)

Damping:	d_{CP} = 3266.1 N m s/rad	(2409 lb ft s/rad)
	$d_{PX} = d_{PY}$ = 0 N s/m	(0 lb s/ft)
	d_{SY} = 1.8454×10^5 N s/m	(12645 lb s/ft)
	d_{SZ} = 1.1072×10^5 N s/m	(7587 lb s/ft)
	d_W = 12633 N m s/rad	(9318 ft lb s/rad)
Creep Coefficients			
	f_{11} = 4.4245×10^6 N/wheel	(9.9946×10^5 lb/wheel)
	f_{12} = 6246.2 N m/wheel	(4,607 lb ft/wheel)
	f_{22} = 32.027 N m^2/wheel	(77.5 lb ft^2/wheel)
	f_{33} = 5.1288×10^6 N/wheel	(1.153×10^6 lb/wheel)

rails were used. The vehicle, initially centered on the track, moved along a tangent segment for 3.05 m (10 ft) before entering the spiral. Its critical speed was approximately 21.95 m/s (72 ft/s).

The lateral excursion of the wheelsets as the vehicle travels along the track is shown in Fig. 5.14. The outer wheel of the lead wheelset of the lead truck impacts the rail 1.15 seconds after entering the curve and continues to flange as the vehicle moves into the constant radius portion of the curve. The other wheelsets do not contact the flange. After the impact of the lead wheelset against the flange, oscillations in the wheelset angles of attack occur. These oscillations are more pronounced at the leading wheelset. The carbody and lead truck experience their peak lateral accelerations coincident with the initial flange impact. The lead truck also warps at this time.

Figure 5.15 shows the time history of lateral component of normal force on the left wheel for the four wheelsets. The impact of the lead wheelset's flange against the rail results in a large lateral component of normal force, termed the 'flange force'. This

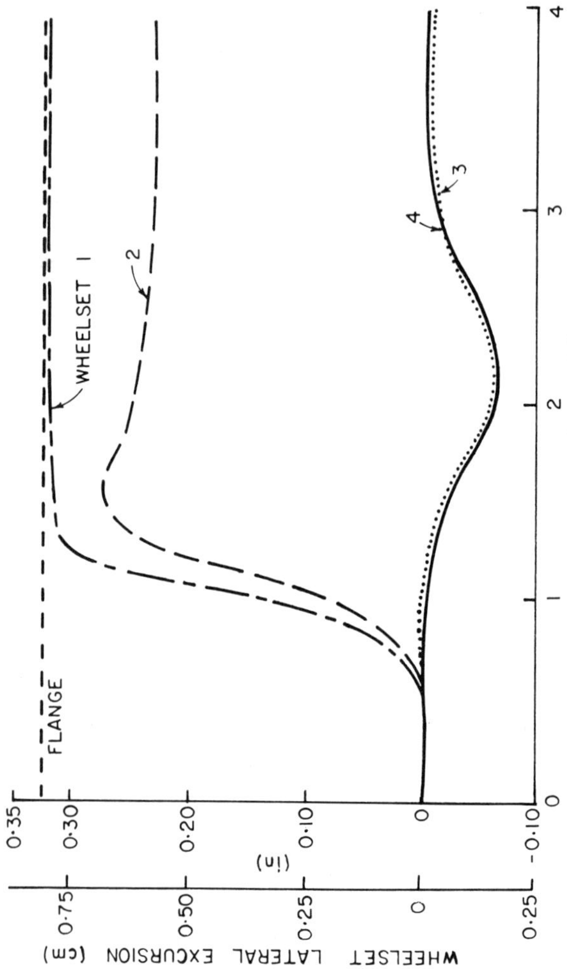

Fig. 5.14 Lateral excursion of the wheelsets of the baseline vehicle (redrawn from Krolewski [13])

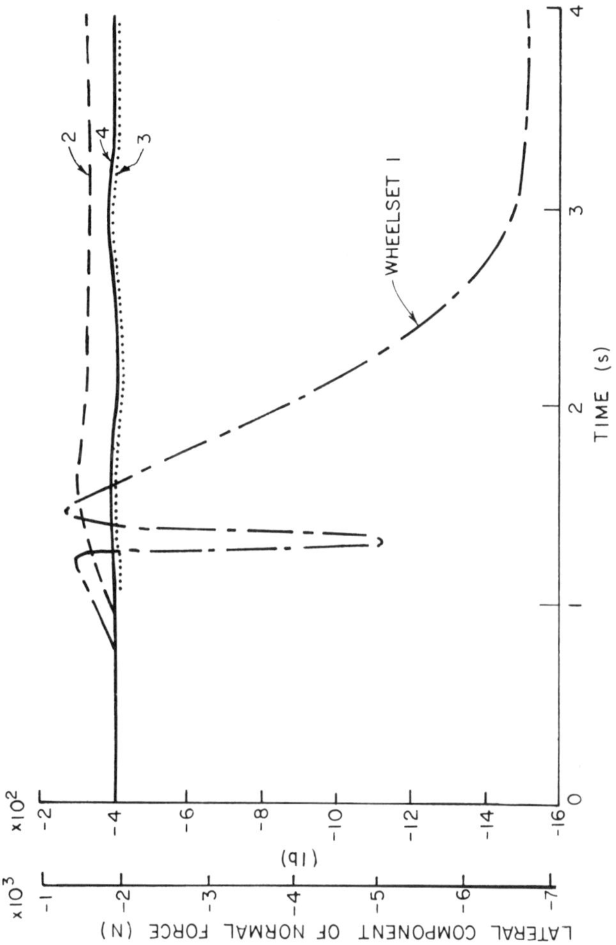

Fig. 5.15 Lateral component of normal force on left wheels of baseline vehicle (redrawn from Krolewski [13])

flange force is of short duration and creates an increase in work. The lateral force component and work of the trailing wheelsets do not exhibit this behaviour and are significantly lower in magnitude.

(b) Spiral Length: The length of the transition curve affects the peak lateral force of the lead wheelset. Reducing the spiral length increases the peak force resulting from the initial flange contact. For spirals up to a critical length, this peak force exceeds the steady state curving values. Beyond this length, the steady state force is of greater magnitude. For this model the critical length is approximately 45.72 m (150 ft). This effect was also observed by Law and Cooperrider [8].

(c) Degree of Curvature: Studies were conducted with the baseline vehicle running through several final degrees of curvature at balanced running speeds, keeping the length of the transition spiral constant in each case. Figure 5.16 shows the flange force of the lead wheelset for various final degrees of curvature. For tight curves, the wheel flange impacts the rail soon after curve entry and remains in contact for the duration of the curve and high flange forces are associated with this behaviour. As the radius of the curve increases, the wheelset oscillates laterally after initial flange contact and the corresponding lateral forces are lower. Further increase in radius of the curve results in low flange forces.

(d) Rail Flexibility: The lateral flexibility of the rail has an important effect on the vehicle response during curving. The effect of rail stiffness on the peak lateral force of a vehicle negotiating a five degree curve at balanced running speed was investigated. It was found that as the flexibility of the track was increased, while maintaining 31% damping, the net lateral flange force experienced by the vehicle decreased [2]. The net lateral force predicted by a model with rigid rails was significantly greater than with typical values of rail stiffness.

(e) Wheel/rail Profile: As wheels and rails wear in service, the geometry of the wheel/rail interface changes. The baseline

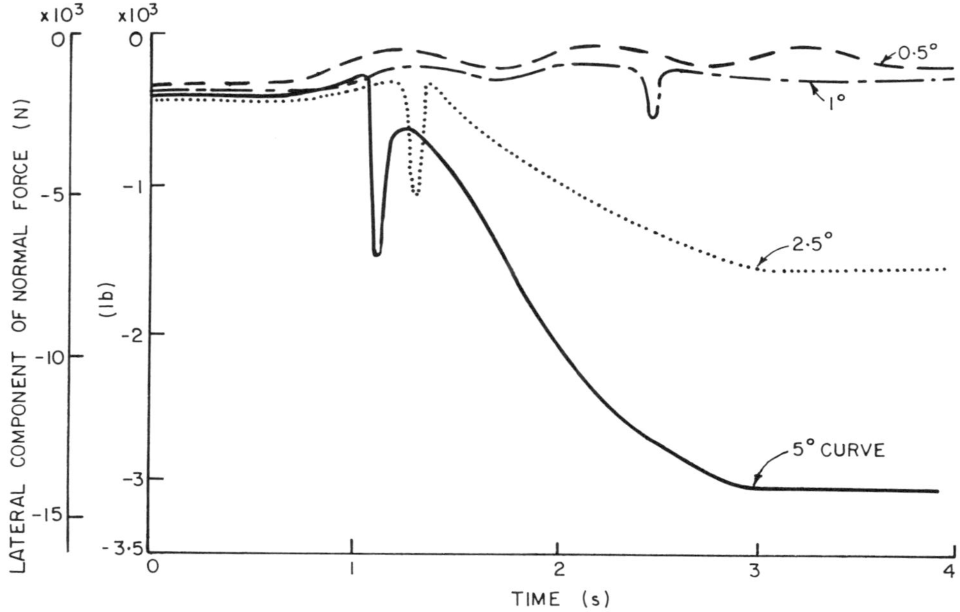

Fig. 5.16 Effect of final degree curve on the lateral component of normal force on the left lead wheel (redrawn from Krolewski [13])

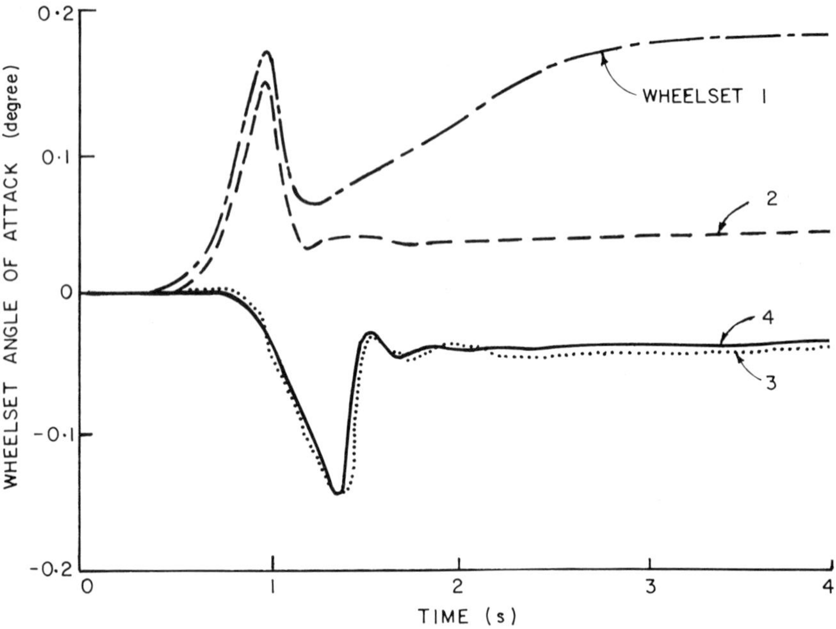

Fig. 5.17 Effect of worn wheel worn rail profile on wheelset angle of attack (redrawn from Krolewski [13])

operating conditions described earlier were repeated for the case of a worn wheel on worn rail profile and it was found that both lead wheelsets flange and the angles of attack are much greater (Fig. 5.17). The flange force resulting from the initial flange contact is nearly twice as large as with new wheelsets. The steady state flange force increases to a lesser extent. Therefore, the critical spiral length is greater than with new wheels. An explanation is that worn wheels produce a smaller steering moment than new wheels becuase their difference in rolling radius is less for a given gravitational stiffness.

(f) Wheelset Interconnection: Interconnecting the wheelsets of a conventional freight truck to produce a so-called 'radial truck' results in a significant improvement in the curving performance.

The angles of attack and flange forces of a conventional three-piece truck and a radial truck with varying shear stiffnesses were compared. In each case, travel through a 45.72 m (150 ft) spiral into a constant 2.5 degree curve was simulated. The mass and yaw moment of inertia of each wheelset of the radial truck were larger than the conventional truck by about 20% due to the addition of radial arms [9], the primary stiffnesses were smaller to allow the wheelsets to assume a radial position, and the bending stiffnesses of the trucks were equal. Figure 5.18 shows that the lead wheel-sets flange forces are smaller for the radial truck. This is expected since the wheelset's more radial position results in a smaller angle of attack. Increasing the shear stiffness of the radial truck by 100% had little effect on the angle of attack, work or flange force levels.

(g) Coulomb Friction: The linear viscous damping elements of the baseline vehicle were replaced by Coulomb friction elements, discussed in Chapter 3. Values for their breakout forces, breakout moments and spring constants are given in Table 5.5 [7]. Figure 5.19 shows the angles of attack obtained as the vehicle moved through the spiral. Whereas the angle of attack magnitudes are approximately the same as with the linear suspension, the flange force was found to be much more oscillatory.

5.5 LOCOMOTIVE RESPONSE TO TRACK IRREGULARITIES

This section presents Arslan's evaluation [3] of a 29 degree of freedom six-axle locomotive nonlinear model on tangent track by comparing the model's predictions with field test results from perturbed track tests (PTT) conducted at Pueblo, Colorado.

5.5.1 Physical Tests

The test train was comprised of two SDP-40F locomotives, one baggage car and a data acquisition vehicle. The three rear-truck axles of the second locomotive were instrumented to measure the vertical and lateral rail forces, and accelerometers were placed at various locations on the same locomotives carbody.

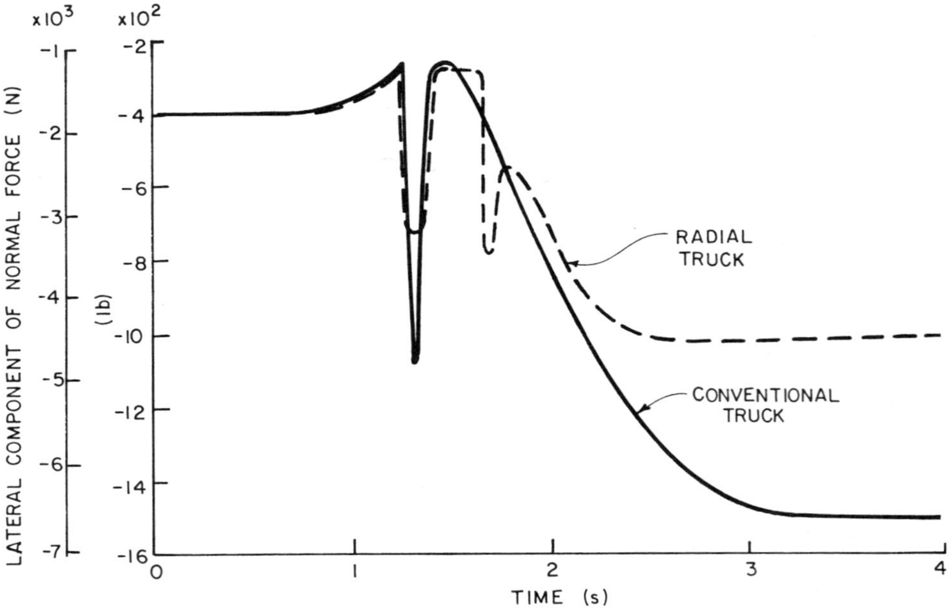

Fig. 5.18 Effect of wheelset interconnection on lateral component of the normal force on the left wheel (redrawn from Krolewski [13])

Table 5.5. Freight Car Nonlinear Suspension Parameters (from Krolewski [13]).

F_{Y0}	= 1.632×10^4 N (3669 lb)
F_{Z0}	= 15569 N (3500 lb)
k_{SY}	= 903835 N/m (61932 lb/ft)
k_{SZ}	= 3.7528×10^6 N/m (257150 lb/ft)
k_W	= 5.4421×10^7 N/m (3.729×10^6 lb/ft)
T_{CP0}	= 821.62 N m (606 ft lb)
T_{W0}	= 6354.63 N m (4687 ft lb)

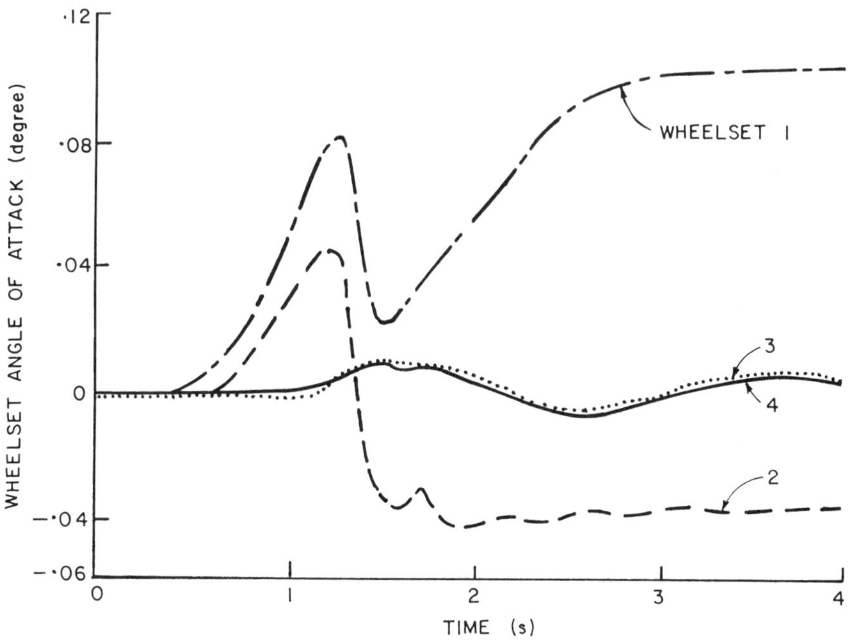

Fig. 5.19 Effect of Coulomb friction on wheelset angle of attack (redrawn from Krolewski [13])

A tangent track and a curved track zone were used in the perturbed track tests. However, only the tangent track test results were used in the present computer model validation. The tangent track zone of 77 rail lengths was 915.31 m (3003 ft) long. It consisted of four perturbed track sections separated by sections of high-quality track, of either 11 or 12 rail lengths, so that the vehicle's response to each section could be isolated. The perturbed track sections were fabricated with 11.89 m (39 ft) regular cut and 5.94 m (19.5 ft) long cut rails, for controlled vertical and lateral irregularities meeting Class 4 track standards, and each one contained five 23.77 m (78 ft) perturbation cycles. To provide an accurate reference between the measured onboard response data and specific track locations, Automatic Location Detectors (ALD) were installed at the rail joints. The perturbed track sections incorporated the following piecewise linear perturbations:

(1) Section 6: Profile perturbations;

(2) Section 7: Cross-level perturbations;

(3) Section 8: Alignment perturbations;

(4) Section 9: Combined cross-level and alignment perturbations.

After construction of the track, its geometry was measured at 15.24 cm (6 in) intervals by Track Geometry Vehicle T-6 [3].

5.5.2 Basis of Validation

Here, the basis is established for the process of evaluation, carried out in Section 5.5.2, by comparing the model's predictions of wheel/rail loads and carbody accelerations to those measured in track section 9 at eleven speeds in the range 56.3 - 127.1 km/h (35 - 79 mph). The test and simulation results were filtered by a low-pass filter with a cut-off frequency of 20 Hz. Time traces and LT40 values of both sets of results were compared. The time traces of lateral and vertical rail forces for the three instrumented axles of the rear truck, and the lateral and vertical accelerations measured at seven carbody locations, were compared at each speed. The LT40 value of a time trace was defined as the maximum value satisfying the constraint of trace continuity for the next 40 milliseconds. This value was used as a basis of comparison, rather than the actual minimum or maximum values, because of the noise in the time traces, even after filtering.

The assumptions pertinent to the model and the input data of Table 5.6 are as follows:

(1) Whereas the wheel/rail geometry was not measured during the tests, the wheels and track were new. Therefore, the geometry of a new 83.82 cm (33 in) diameter Association of American Railroads (AAR) freight car wheel on new rail was assumed for the study, but rolling radii were adjusted to 50.8 cm (20 in), to account for the difference in diameters between freight car and locomotive wheels. Some contact geometry data, such as rolling radii, contact angles, and roll angle for the new AAR freight car wheel on new rail, were prepared by means of the Heller and Cooperrider computer program [12].

(2) The suspension and mass parameter values listed in Table 5.6 were obtained from Reference [1].

(3) In Arslan's model [2], simple rail flexibility was assumed to modify the wheel/rail geometry tables used as input data to the computer program. Although the lateral rail stiffness is known to increase with vertical wheel/rail load increases, it was modeled by linear springs, for simplicity, and rail damping was neglected in both lateral and vertical directions. An approximate value of 140.1 kN/cm (80,000 lb/in) was used for lateral rail stiffness. This value is valid for a vertical load range of 111.2-200.2 kN (25,000 to 45,000 lb) [3]. A vertical track stiffness of 210.2 kN/cm (120,000 lb/in) was assumed.

(4) Track irregularity variations with speed and loading conditions were neglected and the data measured by the Track Geometry Vehicle T-6 at 80.46 km/h (50 mph) was used. This approximation was made in spite of the different axle loads: 18.144 tonnes (20 tons) for the T-6 vehicle; 29.937 tonnes (33 tons) for the locomotive. In fact, the larger loads would result in the different track perturbations at the locomotive than the ones measured by

Table 5.6. Input Data for the Nonlinear Model of a Six-Axle Locomotive on Tangent Track (from Arslan [1]).

Dimensional Data: cm (in)

a = 75.56 (29.75)
b_1 = 89.20 (35.12)
b_2 = 107.57 (42.35)
b_3 = 100.33 (39.50)
b_4 = 24.76 (9.75)
h_{ap} = 12.70 (5.00)
h_{cs} = 105.66 (41.60)
h_{tp} = 18.79 (7.40)
h_{ts} = 21.84 (8.60)
ℓ = 701.04 (276)
ℓ_1 = 223.90 (88.15)
ℓ_2 = −21.59 (−8.50)
ℓ_3 = 191.14 (75.25)
ℓ_4 = 191.14 (75.25)
ℓ_5 = −21.59 (−8.50)
ℓ_6 = 223.90 (88.15)
r_0 = 50.80 (20)

Clearances in Primary and Secondary Suspensions: cm (in)

d_{pxi} = 0.1524 (0.06)
d_{py1i} = 0.2540 (0.1)
d_{py2i} = 0.8890 (0.35) — $i = 1$ to 6
d_{sxi} = 0.079375 (0.03125)
d_{syi} = 2.8575 (1.125) — $i = 1$ to 4

Masses: N s^2/cm (lb s^2/in)

m_b = 28.89 (16.5)
m_c = 1341.5 (766)
m_t = 83.18 (47.5)
m_w = 47.98 (27.4)
W = 3.03×10^5 N (68161 lb)

Table 5.6. (continued)

<u>Moments of Inertia</u>:	N cm s^2 (lb in s^2)		
I_{bz}	$= 6.4\times10^5$	(5.6×10^4)	
I_{cx}	$= 1.35\times10^7$	(1.2×10^6)	
I_{cy}	$= 3.84\times10^8$	(3.4×10^7)	
I_{cz}	$= 4.06\times10^8$	(3.6×10^7)	
I_{tx}	$= 6.77\times10^5$	(6×10^4)	
I_{ty}	$= 3.55\times10^6$	(3.15×10^5)	
I_{tz}	$= 2.64\times10^6$	(2.34×10^5)	
I_{wx}	$= 1.64\times10^5$	(1.45×10^4)	
I_{wy}	$= 1.64\times10^5$	(1.45×10^4)	
<u>Friction and Creep Coefficients</u>			
μ	$= 0.3$		
f_{11}	$= 1.59\times10^7$ N/wheel	$(3.59\times10^6$ lb/wheel)	
f_{12}	$= 5.11\times10^4$ N m/wheel	$(4.52\times10^5$ lb in/wheel)	
f_{22}	$= 1.68\times10^2$ N m^2/wheel	$(5.88\times10^4$ lb in^2/wheel)	
f_{33}	$= 1.80\times10^7$ N/wheel	$(4.05\times10^6$ lb/wheel)	
<u>Primary and Secondary Suspension Stiffness and Damping</u>			
k_{pxi1}	$= 3.32\times10^3$ N/cm	(1,900 lb/in)	
k_{pxi2}	$= 1.75\times10^5$ N/cm	$(99.99\times10^3$ lb/in)	
k_{pzi}	$= 1.19\times10^4$ N/cm	(6840 lb/in)	
c_{pxi}	$= 10.5$ N s/cm	(6 lb s/in)	
c_{pzi}	$= 8.75$ N s/m	(50 lb s/in)	i = 1 to 1
k_{pyi1}	$= 6.65\times10^3$ N/cm	(3800 lb/in)	
k_{pyi2}	$= 3.83\times10^4$ N/cm	(21900 lb/in)	
k_{pyi3}	$= 1.75\times10^6$ N/cm	$(1\times10^6$ lb/in)	
c_{pyi}	$= 8.75\times10^2$ N s/cm	(500 lb s/in)	i = 1 to 6
$k_{sxi1}=k_{syi1}$	$= 1.75\times10^4$ N/cm	$(1\times10^4$ lb/in)	
$k_{sxi2}=k_{syi2}$	$= 1.75\times10^6$ N/cm	$(1\times10^6$ lb/in)	
c_{sxi}	$= 5.25\times10^2$ N s/cm	(300 lb s/in)	
c_{syi}	$= 1.22\times10^3$ N s/cm	(700 lb s/in)	i = 1 to 4

Table 5.6. (continued)

k_{szi}	$= 1.92\times10^5$ N/cm (11×10^4 lb/in)	
c_{szi}	$= 1.22\times10^3$ N s/cm (700 lb s/in)	i = 1 to 8
T_{cpi}	$= 1.12\times10^6$ N cm (1×10^5 lb in)	
c_{cpi}	$= 2.25\times10^6$ N cm s/rad (2×10^5 lb in s/rad)	i = 1 to 2
k_{RV}	$= 2.10\times10^5$ N/cm (12×10^4 lb/in)	
k_{RL}	$= 1.40\times10^5$ N/cm (8×10^4 lb/in)	
c_{RL}	= 0 N s/in. (0 lb s/in)	

the T-6 vehicle and these differences would be reflected in the creep and suspension forces.

(5) It was assumed that gauge variations due to the track construction and rail loads were negligible.

5.5.3 Validation for Combined Alignment and Cross-Level Perturbations

Using the above assumptions and the simplified wheel/rail load equations, the locomotive model validation for section 9 of the Perturbed Test Track is reported here. For other test track sections, the reader is referred to Arslan [2,3].

Section 9 of the Perturbed Test Track had combined alignment and cross-level perturbations and flanging occurred at all test speeds. Figures 5.20 to 5.22 indicate that the dominant mode shapes of the wheel/rail loads were predicted quite accurately by the model. Comparisons of the LT40 values at eleven speeds, showed that the model underestimated the maximum wheel/rail loads at high speeds. Figure 5.22 leads to the same observation regarding the amount of wheel unloading. Even though the amplitudes of the wheel/rail loads were not predicted accurately, the trends in LT40 values were similar for both the test and model results.

The lateral accelerations at location 6, 734.06 cm (289 in) in front of the locomotive carbody's center of gravity) and their LT40

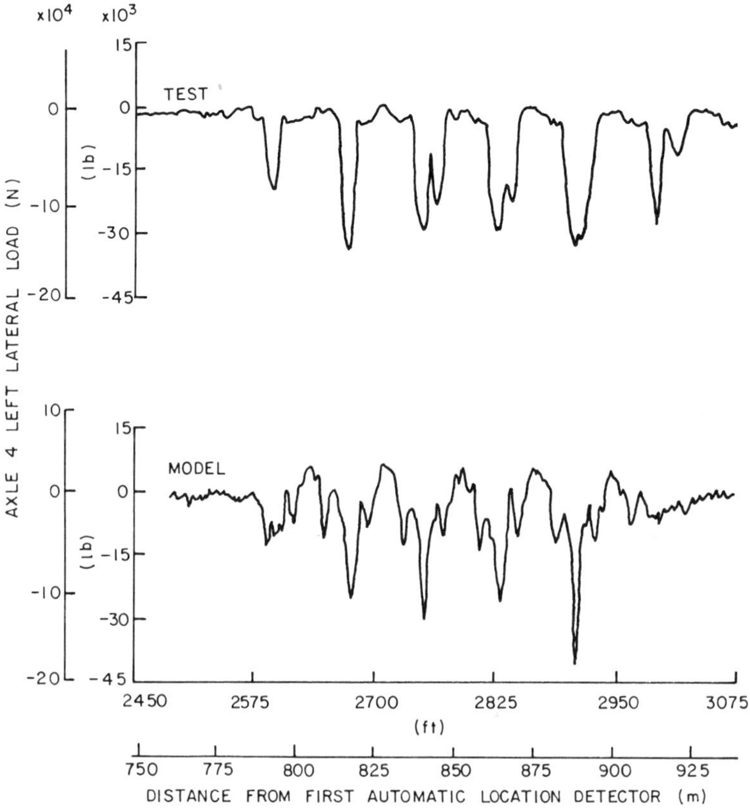

Fig. 5.20 Left lateral wheel/rail loads for axle no. 4 at 127.13 km/h (79 mph), from combined cross-level and alignment perturbations in test section 9 (from Arslan [1])

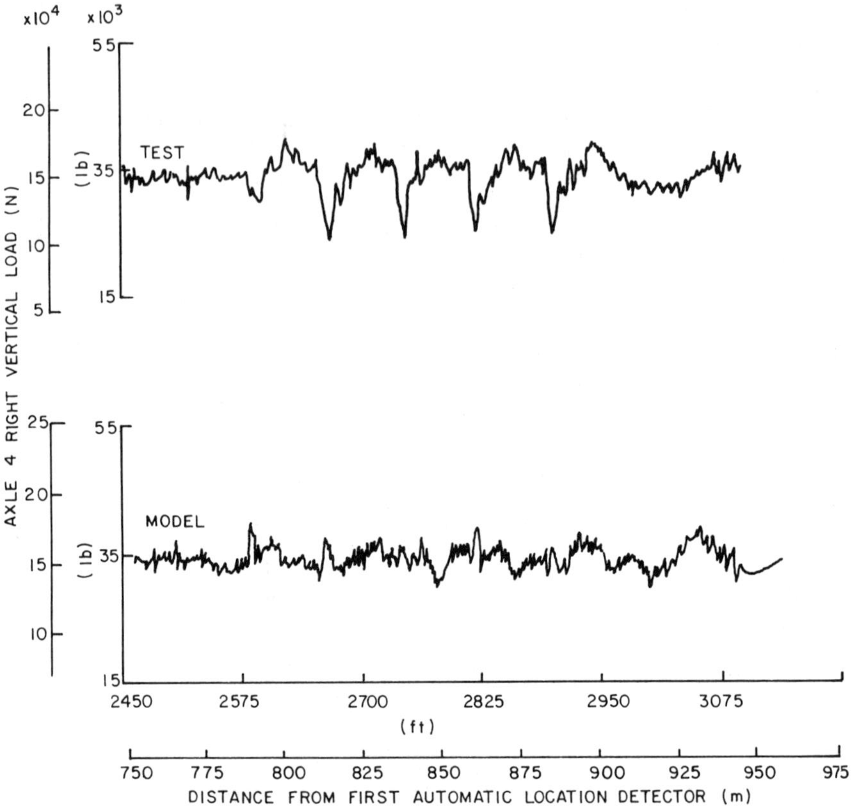

Fig. 5.21 Right vertical wheel/rail loads for axle no. 4 at 88.51 km/h (55 mph), from combined cross-level and alignment perturbations in test section 9 (from Arslan [1])

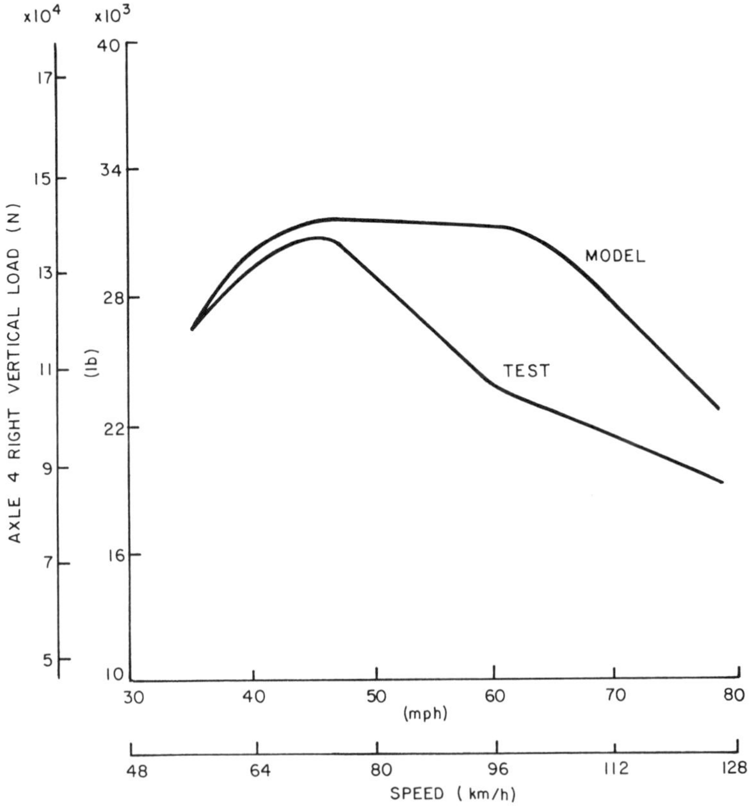

Fig. 5.22 LT40 values for right wheel unloading at axle no. 4 from combined cross-level and alignment perturbations in test section 9 (from Arslan [1])

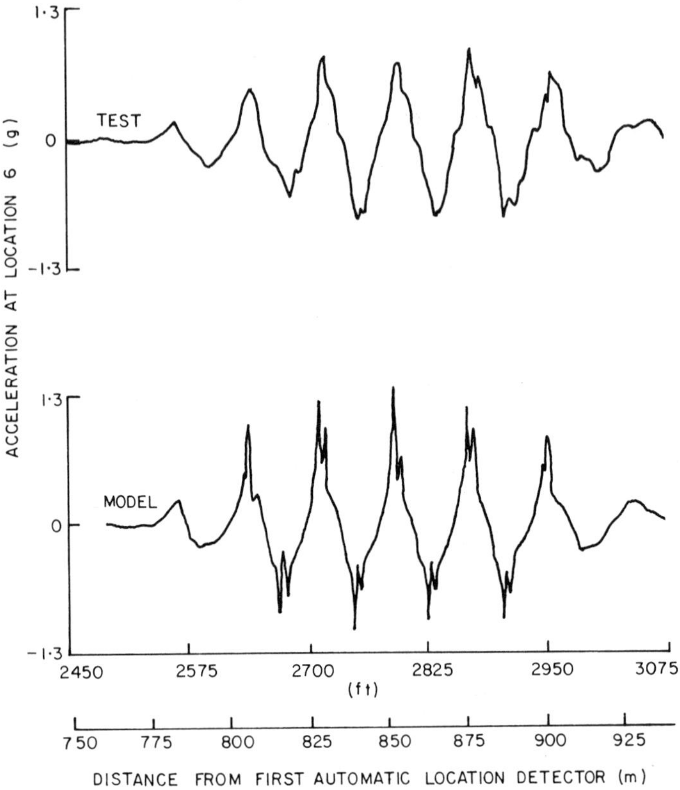

Fig. 5.23 Lateral carbody accelerations at location 6 for 127.13 km/h (79 mph) from combined cross-level and alignment perturbations in test section 9 (from Arslan [1])

values (Figs. 5.23 and 5.24) indicate that the acceleration predictions were quite accurate. Similar observations were valid for location 7, 792.48 cm (312 in) to the rear of the locomotive carbody's center of gravity). The extrema in Fig. 5.23 are associated with the rail joints.

Based on the above results pertaining to track section 9, it can be concluded that the locomotive and track model was validated qualitatively insofar as shapes and trends of responses were concerned. However, whereas numerical results were in the correct

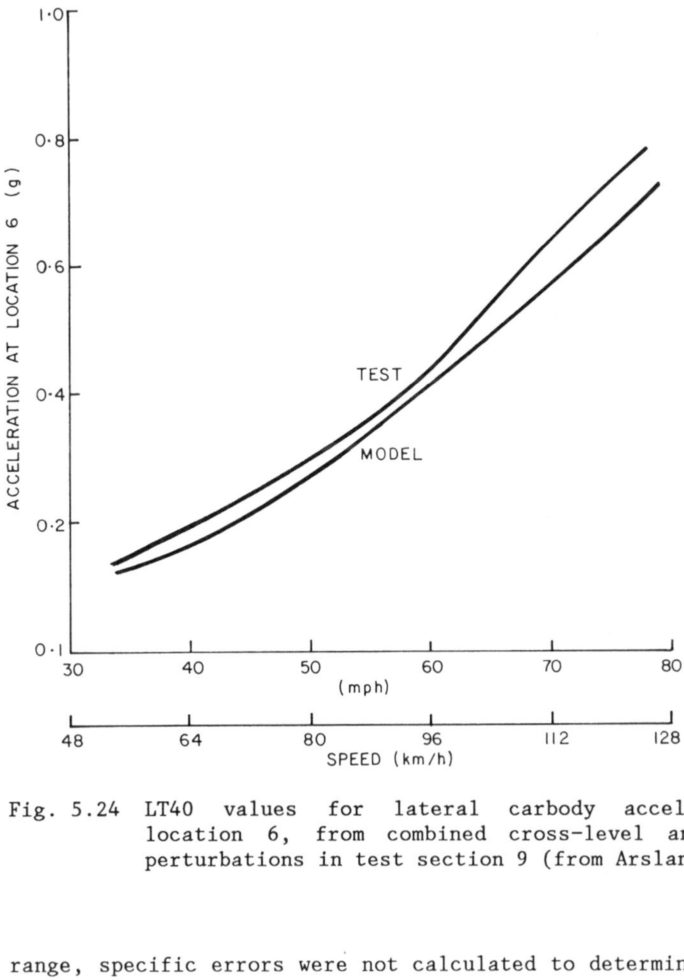

Fig. 5.24 LT40 values for lateral carbody accelerations at location 6, from combined cross-level and alignment perturbations in test section 9 (from Arslan [1])

range, specific errors were not calculated to determine the degree of quantitative validation.

5.6 LONGITUDINAL TRAIN ACTION

The Detailed Longitudinal Train Action Model (DTAM) developed by Martin and Tideman [16] for the simulation of longitudinal train dynamics was described in Section 3.7. Certain differences between various vehicles in a train can give rise to excessive longitudinal compressive or tensile forces. These coupler forces can cause failures of draft gears, jack-knifing of cars, or derailments due to wheel climb. The DTAM program can be used as an operational tool for the improvement of train make-up and handling practices to reduce longitudinal train action.

One of the primary functions of the program is to predict dynamic drawbar forces for various train handling tactics. Special Event 6 of the Steel Coil Train Test [16] was intended to test an emergency train braking operation in which high drawbar forces are developed. This validation study, extracted from References [15-17], compares simulation results with field test data for Special Event 6.

5.6.1 Field Test Description

The SP Steel Coil Train consisted of 4 SD 45T-2 locomotives, 52 cars with approximately equal loads of coiled steel and 4 other cars (Table 5.7). Event 6 of the SP Steel Coil Train Test was run

Table 5.7. SP Steel Coil Train Data (from Martin et al. [16].

Position	Car Number	Gross Weight tonnes	(tons)	Description
car 1	SP 595596	110.68	(122)	Instrument car
car 2	SP 595538	116.12	(128)	
car 3	SP 595559	115.21	(127)	
car 4	SP 595548	113.40	(125)	
car 5	SP 595526	110.68	(122)	
car 6	SP 595570	112.49	(124)	
car 7	SP 595503	112.49	(124)	
car 8	SP 595531	108.86	(120)	
car 9	SP 595593	117.93	(130)	
car 10	SP 595612	117.93	(130)	
car 11	SP 595573	112.49	(124)	Instrument car
car 12	SP 595563	111.58	(123)	
car 13	SP 595614	112.49	(124)	

Position	Car Number	Gross Weight tonnes (tons)	Description
car 14	SP 595529	111.58 (123)	
car 15	SP 595622	112.49 (124)	
car 16	SP 595615	117.93 (130)	
car 17	SP 595555	117.93 (130)	
car 18	SP 595519	113.40 (125)	
car 19	SP 595564	108.86 (120)	
car 20	SP 595510	115.21 (127)	
car 21	SP 595568	113.40 (125)	Instrument car
car 22	SP 595539	116.12 (128)	
car 23	SP 595572	114.30 (126)	
car 24	SP 595536	115.21 (127)	
car 25	SP 595556	113.40 (125)	
car 26	SP 595578	116.12 (128)	
car 27	SP 595511	108.86 (120)	
car 28	SP 595611	117.93 (130)	
car 29	SP 595605	117.03 (129)	
car 30	SP 595620	112.49 (124)	
car 31	SP 595610	111.58 (123)	Instrument car
car 32	SP 595619	112.49 (124)	
car 33	SP 595621	113.40 (125)	
car 34	SP 595582	117.03 (129)	
car 35	SP 595581	117.93 (130)	
car 36	SP 595522	112.49 (124)	
car 37	SP 595565	112.49 (124)	
car 38	SP 595590	112.49 (124)	
car 39	SP 595518	112.49 (124)	
car 40	SP 595534	112.49 (124)	
car 41	SP 595551	108.86 (120)	Instrumented Car
car 42	SP 595547	110.68 (122)	
car 43	SP 595617	115.21 (127)	
car 44	SP 595618	113.40 (125)	
car 45	SP 595583	116.12 (128)	
car 46	SP 595586	112.49 (124)	
car 47	SP 595537	116.12 (128)	

Table 5.7. (continued)

Position	Car Number	Gross Weight tonnes (tons)	Description
car 48	SP 595549	116.12 (128)	
car 49	SP 595523	115.21 (127)	
car 50	SP 595577	114.30 (126)	
car 51	SP 595516	115.21 (127)	Instrumented Car
car 52	SP 595512	115.21 (127)	
car 53	AAR 100	58.06 (64)	Research Car
car 54	SP 138	59.87 (66)	Crew Sleeping Car
car 55	SP 137	62.59 (69)	Kitchen and Support Car
car 56	SP 1741	23.58 (26)	Caboose

Miscellaneous Data:

Gross weight of load revenue cars - cars 1 to 52 = 5918.4 tonnes (6524 tons)

Gross weight of the 4 locomotives = 743.89 tonnes (820 tons)

Gross weight of the entire train = 6866.4 tonnes (7569 tons)

Train Length = 901.9 m (2959 ft)

on the main line between mile posts 801 and 802 near station STOVAL. This straight section of track had a slight grade (0.1-0.4%).

Upon entering the test section, the train was maintaining a steady speed of approximately 64 km/h (40 mph). Emergency brakes were applied soon after mile post 801 was passed and the locomotives were bailed off almost immediately to avoid a severe run-in from the rear. Then, in addition to the automatic (train) brakes, the independent brake was operated until the whole train came to a stop. The entire emergency braking operation occurred during a 43 second period and over a 471.83 m (1,548 ft) distance. Included in the measurements of interest were train speed, stopping time and distance, and the dynamic drawbar forces of the instrumented test vehicles.

5.6.2 Validation Procedure

The analog signals recorded on magnetic tape aboard the test train were converted to graphical plots by means of a computer program and the simulation results were plotted to the same scales. Thus, the two could be compared by overlaying the plots [15-18].

Included in the input data of the model was the starting distance of the train from a referenced mile post at which the emergency brakes were applied. This was deduced by analysing the data from the Automatic Location Detector (ALD) mounted aboard the train. Other input data, such as coupler slack, locomotive brake cylinder pressures, initial brake line pressure and brake operations, were also obtained directly from magnetic tape records (Figs. 5.25 and 5.26). Locomotive and car data were supplied by the Southern Pacific Transportation Company [16,17].

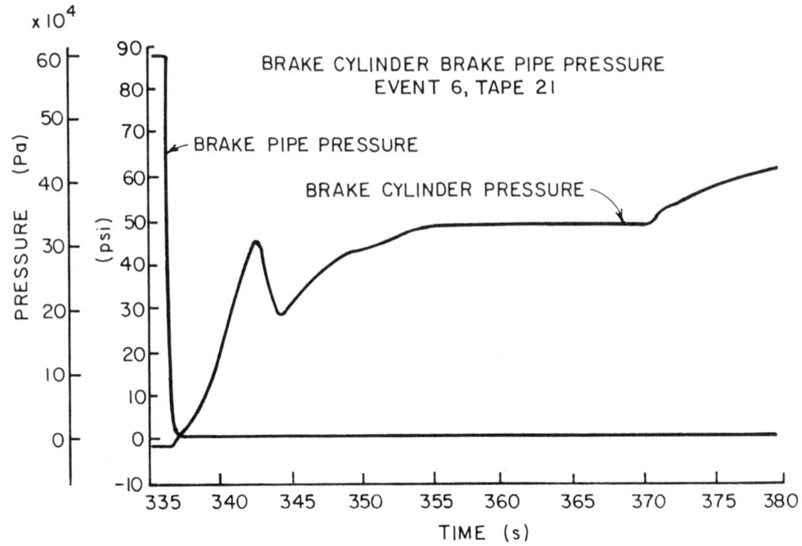

Fig. 5.25 Locomotive brake cylinder and brake pipe pressure (redrawn from Martin et al. [16])

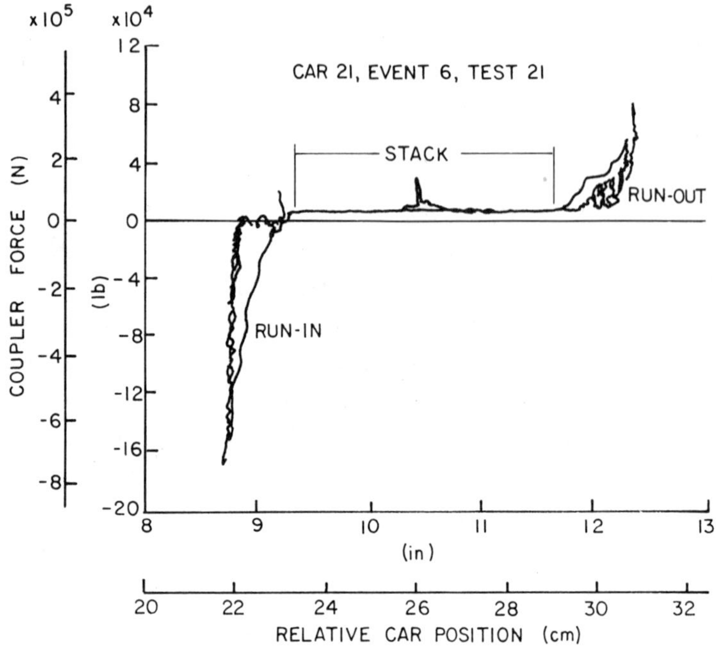

Fig. 5.26 Drawbar force vs. car relative position, car 21 Special Event 6, Steel Coil Train Test field data (redrawn from Martin et al. [16])

5.6.3 Comparison of Results

It can be seen from Fig. 5.27 that simulation and field test curves of train speed and distance are almost identical. The simulated stopping time and distances were too small by 1.16% and 0.97%, respectively (Table 5.8).

Coupler force comparisons were made for all six of the instrumented cars [16]. Simulation and test results confirmed that car 21 had the most severe run-ins among the six instrumented cars. Figure 5.28 shows a good agreement of results for car 21. The maximum simulated drawbar force in buff is 18.5% smaller than in the test, but its time of occurrence is correct to within 1.8% (Table 5.9).

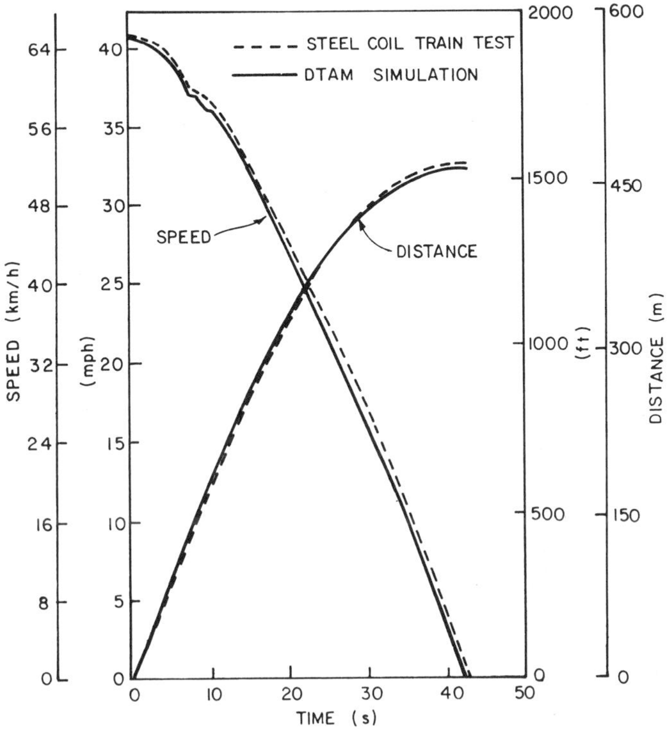

Fig. 5.27 Speed and distance, Special Event 6 - comparison of DTAM with field test data (redrawn from Martin et al. [16])

Table 5.8. Comparison of DTAM Model and SP Steel Coil Train Test Results for Train Stopping Time and Distance in Emergency Brake Application (from Martin et al. [16]).

Parameter	DTAM Model	SP Steel Coil Train Test
Time required to stop the train s	42.5	43.0
Train stopping distance m(ft)	467.26 (1533)	471.83 (1548)

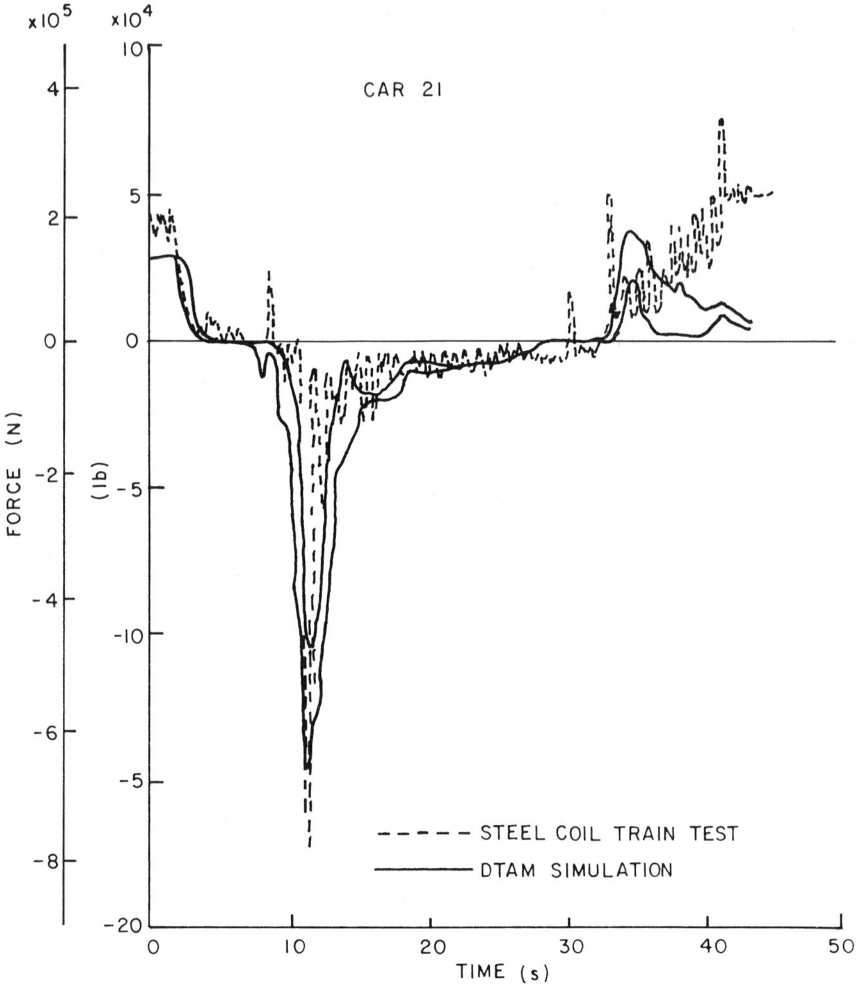

Fig. 5.28 Drawbar force of car 21, Special Event 6 (redrawn from Martin et al. [16])

Table 5.9. Comparison of DTAM Model and SP Steel Coil Train Test Results for Run-in Drawbar Forces During Emergency Braking (from Martin et al. [16]).

Car	Maximum Buff Force kN (kips)			Time of Occurrence s		
	DTAM	TEST	DTAM/TEST (%)	DTAM	TEST	DTAM/TEST (%)
1	333.62 (75)	364.75 (82)	91.5	11.0	11.8	93.2
11	569.37 (128)	569.37 (128)	100.0	11.5	10.8	106.5
21	627.20 (141)	769.54 (173)	81.5	11.0	11.2	98.2
31	498.20 (112)	556.03 (125)	89.6	12.3	12.0	102.5
41	427.03 (96)	311.37 (70)	137.1	12.0	15.3	78.4
51	231.31 (52)	- -	-	12.8	-	-

Table 5.10. Comparison of DTAM Model and SP Steel Coil Train Test Results for Run-out Drawbar Forces During Emergency Braking (from Martin et al. [16]).

Car	Maximum Draft Force kN (kips)			Time of Occurrence s		
	DTAM	TEST	DTAM/TEST (%)	DTAM	TEST	DTAM/TEST (%)
1	-	-	-	-	-	-
11	151.24 (34)	329.17 (74)	45.9	35	37.5	93.3
21	173.48 (39)	244.65 (55)	70.9	34.5	41.5	83.1
31	169.03 (38)	244.65 (55)	69.1	34.5	29.2	118.1
41	151.24 (34)	200.17 (45)	75.6	35.5	42.0	84.5
51	129.00 (29)	133.45 (30)	96.7	36.0	31.0	116.1

Tables 5.9 and 5.10 compare the maximum drawbar forces in the six instrumented cars and their times of occurrence during emergency braking from a velocity of 64.37 km/h (40 mph). The maximum simulated run-in drawbar forces vary between 81% and 137% of the test values and the times of occurrence vary in the range of 78 - 106%. The corresponding ranges for maximum run-out drawbar forces and their occurrence times are 46 - 97% and 83 - 118%, respectively.

5.7 SUMMARY

In this chapter, we have presented case studies illustrating the application of various rail vehicle and train mathematical models described in Chapter 3. These examples are an indication of how mathematical modeling and simulation can be used as an effective tool in the design and operation of railway vehicles and trains.

The results presented in Section 5.2 illustrated the dependence of the wheel/rail geometric constraint variables on the wheelset's lateral displacement. Since the constraint functions are highly nonlinear, the dynamic behaviour of a vehicle is a function of the amplitude of wheelset motion. The parametric study revealed that the geometric constraint variables are sensitive to wheel profile variations, and may be affected significantly also by rail wear variations in combination with certain wheel profiles.

The parametric vertical response study of a passenger vehicle presented in Section 5.3, showed that primary suspension variations did not affect the secondary stroke but caused reductions of nearly 25% in the rms acceleration of the carbody, at frequencies such that ride quality was only slightly improved. Variations in secondary suspension stiffness and damping indicated that smaller natural frequencies of the carbody rigid-body modes, improved ride comfort whereas damping ratio increases had the opposite effect.

The freight car dynamic curving model's usefulness in evaluating the effects of roadbed and wheel/rail geometry, rail stiffness and suspension design was illustrated in Section 5.4.

The locomotive response to tangent track irregularities model was evaluated, in Section 5.5, by comparing its predictions of wheel/rail loads and carbody accelerations with field test measurements. The dominant mode shapes of the time traces were accurately predicted, as were the trends in the plots of LT40 values (i.e., the maximum values satisfying the constraint of trace continuity for the next 40 milliseconds) versus speed.

The validation study of the Detailed Longitudinal Train Action Model presented in Section 5.6, showed sufficient agreement with field tests to justify its use as a tool in the design and operation of trains.

REFERENCES

1. ARSLAN A.V. (1981): Nonlinear Mathematical Model of a Six-Axle Locomotive on Tangent Track; *Association of American Railroads*, Report R-498, Chicago, U.S.A.
2. ARSLAN A.V. (1983): The Application of Statistical Linearization to Nonlinear Rail Vehicle Dynamics; *Ph.D. Thesis*, Massachusetts Institute of Technology, Cambridge, U.S.A.
3. ARSLAN A.V. (1983): Evaluation of a Nonlinear Locomotive Model Using Perturbed Track Test Data; *Association of American Railroads*, Report R-499, Chicago, U.S.A.
4. BELL C.E. (1981): Curving Mechanics of Rail Vehicles; *Ph.D. Thesis*, Massachusetts Institute of Technology, Cambridge, U.S.A.
5. COOPERRIDER N.K., LAW E.H., HULL R., KADALA P.S., and TUTEN J.M. (1975): Analytical and Experimental Determination of Nonlinear Wheel-Rail Geometric Constraint Relationships; *U.S. Department of Transportation*, Federal Railroad Administration, Report FRA-OR&D-76-244, Washington, D.C.
6. COOPERRIDER N.K. and LAW E.H. (1976): Wheel/Rail Geometry for Five Wheel Profiles and Three Rail Profiles; *U.S. Dept. of Transporation*, Report ERC-R-76009, Washington, D.C.

7. COOPERRIDER N.K. (1981): Freight Car Dynamics: Field Test Results and Comparison with Theory; *U.S. Department of Transportation*, Report FRA/ORD-81/46, Washington, D.C.

8. COOPERRIDER N.K. and LAW E.H. (1981): The Nonlinear Dynamics of Rail Vehicles in Curve Entry and Negotiation; *Proceedings of the 7th IAVSD-IUTAM Symposium on the Dynamics of Vehicles on Road and Tracks*, Cambridge, U.K., 7-11 Sept. 1981.

9. HADDEN J.A. (1976): The Effects of Truck Design and Component Flexibility on the Lateral Stability of Railway Freight Vehicles; *MS Thesis*, Clemson University, Clemson, U.S.A.

10. HEDRICK J.K., WORMLEY D.N., KAR A.K., MURRAY W., and BAUM W. (1979): Performance Limits of Rail Passenger Vehicles: Evaluation and Optimization; *U.S. Dept. of Transportation*, Federal Railroad Administration, Report DOT/RSPA/DPB-50/79/32, Washington, D.C.

11. HEDRICK J.K., WORMLEY D.N., KIM R.R., KAR A.K., and BAUM W. (1982): Performance Limits of Rail Passenger Vehicles: Conventional, Radial, and Innovative Trucks; *U.S. Dept. of Transportation*, Federal Railroad Administration, Report DOT/RSPA/DPB-50/81/28, Washington, D.C.

12. HELLER R. and COOPERRIDER N.K. (1977): Users' Manual for Asymmetric Wheel/Rail Contact Characterization Program; *U.S. Dept. of Transportation*, Federal Railroad Administration, Report FRA-ORD-78/05, Washington, D.C.

13. KROLEWSKI S.M. (1982): Model Development for Freight Car Dynamic Curving Simulation; *MS Thesis*, Massachusetts Institute of Technology, Cambridge, U.S.A.

14 LAW E.H., COOPERRIDER N.K., and TUTEN J.M. (1981): Measurement, Processing and Use of Wheel-Rail Geometric Constraint Functions; *Proceedings of the International Conference on Wheel-Rail Measurements*, Cambridge, U.S.A., Jan. 1981.

15. LOW E.M. and GARG V.K. (1978): Track Train Dynamics Report - Programmers Manual - Detailed Longitudinal Train Action Model;

Association of American Railroads, Report R-296, Chicago, U.S.A.

16. MARTIN G.C. and TIDEMAN H. (1977): Track Train Dynamics Report - Technical Documentation - Detailed Longitudinal Train Action Model; *Association of American Railroads*, Report R-221, Chicago, U.S.A.

17. MARTIN G.C., PLOUFFE W.E., AHMED S., ANTEZAK H., and TIDEMAN H. (1977): Track Train Dynamics Report - User's Manual - Detailed Longitudinal Train Action Model; *Association of American Railroads*, Report R-220, Chicago, U.S.A.

18. SHAW N. and D'SOUZA A.F. (1981): Interactive Simulation and Computer Graphics for the Detailed Longitudinal Train Action Model; *Association of American Railroads*, Report R-492, Chicago, U.S.A.

6

SOFTWARE SURVEY

6.1 INTRODUCTION

Computer simulation models can be valuable in planning physical tests, specifying operational procedures for trains, and assessing vehicle performance as a function of parametric variations under various controllable operating conditions. Fortunately, the number of available computer models in railway dynamics has increased dramatically in the past decade. Therefore, before making a committment to embark on a software development project, a first step in the acquisition of a computer model should be to determine if it already exists. This chapter presents the results of a comprehensive, though not exhaustive, survey to define the existence and sources of railway-dynamics computer models. Most can be purchased, or rented, and some may be borrowed; however, terms and conditions were excluded from the survey due to their volatility.

The choice of model depends on the user's perception of the capabilities and limitations of the software realtive to the physical problem under investigation. Accordingly, model class (linear versus nonlinear), complexity (degrees of freedom), documentation and validation information were part of the survey. Although most models were written in the FORTRAN programming language, for portability, some modifications are usually required when transporting a program from one computer to another; therefore, included in the survey tables, as available, is a list of the computers on which each model was installed. The author(s), organization(s) and pertinent reference article(s), as applicable, are also identified with each program.

A potential user would find it advantageous to obtain the pertinent references and documentation. The author(s) should be

Table 6.1. List of Program Tables

Table	Name
6.1	List of Program Tables
6.2	Track Structure Programs
6.3	Wheel-Rail Interaction Programs
6.4	Hunting Programs
6.5	Twist and Roll Programs
6.6	Pitch and Bounce Programs
6.7	Yaw and Sway Programs
6.8	Steady State Curving Programs
6.9	Quasi-Steady State Curving Programs
6.10	Spiral Negotiation Programs
6.11	Dynamic Curving Programs
6.12	Miscellaneous Programs
6.13	Longitudinal Train Dynamics Programs
6.14	Lateral Train Dynamics Programs
6.15	Vertical Train Dynamics Programs
6.16	Steady Buff and Draft Programs
6.17	Structural Mechanics Programs

contacted in order to ascertain whether or not a model and its validation have been improved. As the tables indicate, the models have not been validated over all operating ranges. Therefore, a user should determine the validation status of a model. Also, an author may be contacted to run his program on behalf of a user, presumably for a negotiated cost.

This search for computer models has uncovered an impressive array of programs dealing with most aspects of railway dynamics. They are tabulated in Section 6.7 and discussed in Sections 6.2 to 6.6, under the following application areas: track structure (Section 6.2), wheel-rail interaction (Section 6.3), vehicle dynamics (Section 6.4), train dynamics (Section 6.5) and structural mechanics (Section 6.6). The first group deals with the stresses and strains in track structures as functions of lateral and vertical wheel loads, whereas the last is applicable to the structural aspects of vehicle design. The others are concerned with the influence of various parameters on the interaction between wheels and rails and the dynamics of vehicles and trains.

The information pertaining to approximately 150 computer programs is presented in the form of 17 tables in Section 6.7 where, for conciseness, the names of source and user organizations are abbreviated. An alphabetical list of these abbreviations, together with the complete name and mailing address of each corresponding organization, can be found in Section 6.8.

6.2 TRACK STRUCTURE PROGRAMS

The track structure programs are presented in Table 6.2. They can predict stresses and deflections in track structures due to wheel loads, and therefore, may be used to: (a) assess the damage a new vehicle causes to a track, and (b) estimate whether or not maintenance-of-way costs would increase as a result of placing a particular vehicle in service. Some may serve as input data generators for several of the vehicle dynamics programs. As such, the long-term effect of variable track conditions, on a vehicle, can be studied. Conversely, via the dynamics of vehicles, it is possible to determine the influence of track irregularities on the overall track structure.

The infinite-beam on elastic-foundation theory was the basis of the earlier track analysis programs, predicting rail stresses and moments for simplified uniform conditions. Values of track modulus, defined as the stiffness per unit length of track, were obtained from field measurements. Since individual substructure layer properties were not characterized in these programs, the effects of substructure component properties could not be evaluated.

However, most programs in Table 6.2 belong to the more recent category of three-dimensional finite element models. They can give

Table 6.2. Track Structure Programs.

Source/Model	Author(s)	Usage	Hardware	Users' Manual/ Validation Description
1. BCL[54]/ Pyramid Model-1970	H.C. Meacham R.H. Prause D.R. Ahlbeck J.A. Kasuba	LTD	IBM 370/158 DEC	Y/Y (poor)/ Track stress and strain
2. CIGGT[65, 66]/ARTS Analysis of Rail Track	G.P. Raymond D.J. Turcke D.W. Siu	LTD		Y/P/Track stress and strain
3. UI[75,76]/ ILLI-TRACK 1976	S.D. Tayabji M.R. Thompson A.L. Robnett	LTD		Y/Y/Track stress and strain

4. AAR[71]/ PSA-BURMISTER Prismatic Solid Analysis	W. SO G.C. Martin B. Singh I.C. Chang E.H. Chang	LTD		Y/Y/Track Stress and Strain
5. BCL[63]/ MULTA Multi-layer Track Analysis 1978	R.H. Prause J.C. Kennedy	LTD		Y/Y/Track Stress and Strain
6. UM[8]/ GEOTRACK 1980	C.S. Chang C.W. Adegoke E.T. Selig	LTD	DEC 10	Y/Y/Track Stress
7. RDSO[14]/ ESMD	D. Deepak V.V. Singh	IR	IBM	Y/Y/Track Stiffness and damping

Table 6.2. (continued)

Source/Model	Author(s)	Usage	Hardware	Users' Manual/ Validation Description
8. CUED[23]/ TRKVERT	S.L. Grassie	CUED	IBM CDC	Y/P/Track Vertical force and receptance
9. CUED[24]/ TRKLAT	S.L. Grassie	CUED	IBM CDC	Y/P/Track Vertical force and receptance
10. CUED[25]/ TRKLONG	S.L. Grassie	CUED	IBM CDC	Y/N/Track longitudinal force and receptance

11. CUED-PANDROL [26]/TRKFLEX	S.J. Cox	CUED-PANDROL	IBM CDC	Y/P/Track vertical force and receptance
12. CUED-PANDROL/TRKHUNG	S.J. Cox	CUED-PANDROL	IBM CDC	Y/P/Track dynamic res-ponse with Hung sleepers
13. AAR/PHX	AAR	AAR	PC	Y/Y/Rail Fatigue Life
14. AAR/RPM	AAR	AAR	PC	Y/Y/Rail Performance
15. AAR/TIEREN	AAR	AAR	PC	Y/Y/Tie Renewal and Failures
16. BR/TRACK	BR	BR	NI	Y/Y/3-D Rail Stress

Table 6.2. (continued)

Source/Model	Author(s)	Usage	Hardware	Users' Manual/ Validation Description
17. BR/TRACK	BR	BR	NI	Y/Y/Track Deterioration
18. BR/RAIL JOINT	BR	BR	NI	Y/Y/Dipped Rail Joint Analysis
19. BR/TRACK DYNAMICS	BR	BR	NI	Y/Y/Track Dynamic Analysis

the response of railroad tracks represented as multilayer elastic structures. Their outputs include railset load reactions, tie/ballast reactions, bulk and principal stresses, deflections, and bending moments. Some of the parameters are: (1) single- or dual-axle loading conditions, (2) size and stiffness of rails, (3) size, stiffness, bending and elimination of ties, (4) strength, stiffness and thickness of ballast and subballast, and (5) strength and stiffness of subgrade. These are determined from basic information on boundary conditions, triaxial test results, rail/tie system geometry and material properties, such as Young's modulus and Poisson's ratio.

6.3 WHEEL-RAIL INTERACTION PROGRAMS

Table 6.3 conains a list of wheel-rail interaction programs. These programs are concerned with the geometry of contact, contact stresses, and the phenomenon of creep. Some programs are concerned with the creep forces that develop at the contact region of two rolling bodies as well as the shape and size of the contact area between two elastic bodies in contact. One such program 'CONFORM' (Conformal Contact of Two Elastic Bodies) in Table 6.3 is a FORTRAN program for the analysis of contact stresses between two elastic bodies in conformal contact. It is used to determine the pressure distribution between the two bodies, the contact patch boundary, and the total load corresponding to a given depth of penetration.

This program is a generalization of a previous program (COUNTACT) which was restricted to the case of counterformal contact. The program CONFORM is applicable to counterformal as well as conformal cases. Built into the program are specialized subroutines which enable the user to conveniently specify the rail and wheel profiles. The user supplies an initial estimate of the contact patch boundary and the program corrects it by an iterative method. Although CONFORM may be used for counterformal or conformal contact, program COUNTACT is more efficient for the latter case.

Table 6.3. Wheel-Rail Interface Programs.

Source/Model	Author(s)	Usage	Hardware	Model Class/Users' Manual/Validation
1. UOP[58]/ CONFORM	B. Paul J. Hashemi	S	Univac 90/70	NI/Y/Y
2. UOP[56]/ COUNTACT	B. Paul J. Hashemi	S	IBM 370/168 Univac 90/70	NI/Y/Y
3. CU[22]/ EXACT	J.G. Goree	S,A	IBM 360/165 IBM 360/3165-II	NL/Y/P
4. CU[21]/ SIMPLIFIED	J.G. Goree E.H. Law	S,A	IBM 360/65 IBM 370/3165-II	NL/Y/P
5. DU[36,37]/ DUVOROL	J.J. Kalker	S	IBM 370/158	NL/Y/Y

6. ASU[12]/ WHRAIL	N.K. Cooperrider E.H. Law	S	DEC 10 Univac	NL/Y/P
7. ASU[35]/ WHRAILA	N.K. Cooperrider R. Heller	E	DEC 10 Univac	NL/Y/P
8. PU[73]/ NI	L.M. Sweet A. Karamel	PU	IBM CDC	NL/Y/L
9. DU[38,39]/ FASTSIM	J.J. Kalker	E	IBM 370/158	NL/Y/P
10. DU[36,38]/ CONTACT	J.J. Kalker	E	IBM 370/158	NL/Y/P
11. DU[38,39]/ SIMPLE 2	J.J. Kalker	E	IBM 370/158	NL/Y/P

Table 6.3. (continued)

Source/Model	Author(s)	Usage	Hardware	Model Class/Users' Manual/Validation
12. DU[38,39]/ LINE CON 3	J.J. Kalker	S	IBM 370/158	NL/Y/P
13. DU[39]/ LINE CON 4	J.J. Kalker	S	IBM 370/158	NL/Y/P
14. UNSW[11]/ NI	A.K. Hellier	LTD	DEC VAX 11/780	FE/Y/Y
15. CUED[11]/ DISSIP	S.L. Grassie	LTD	IBM 370/158	NI/Y/P
16. CIGGT[53]/ RWCM	E.R. McIlveen A.M. Birk J.D. Young R. Glaser	E	PDP 11-03 IBM 370/158	NI/Y/Y
17. BPA/CONWHEEL	B. Paul S. Singh	UOP	UNIVAC 1100 VAX	NL/NA/P
18. SJ/KYTA	SJ	SJ	IBM	L/Y/Y

Input

(a) Modulus of elasticity and Poisson's ratio for both bodies.

(b) Rigid body "approach" of the two bodies.

(c) Information on the (x,y) coordinates of estimated contact boundary, and the desired meshwork of rectangular cells which this estimated contact patch is to be divided into.

(d) An initial estimated contact patch which the program will refine by an iterative procedure. To establish this initial contact boundary, for the case of wheel and rail, two sub-programs of CONFORM, called MIDSEP and INTERPEN, are used.

(e) The initial separation (before deformation occurs) of neighboring points on each of the two surfaces must be calculated. When using the standard version, the user needs only to supply numerical data on the sets of straight lines and circular arcs which describe the profiles of arbitrary (unworn) wheels and rails. The user must also indicate that point on the wheel and that point on the railhead which make initial contact as loading begins.

Output

The output includes a reprint of all input data. Then, the following output data is printed for each iteration:

(a) Coordinates of a set of points along the boundary of the current contact boundary.

(b) A table of the contact pressures at a set of points within the contact patch.

(c) The resultant force (and moment, if any) of the contact pressure if it is positive throughout the contact patch.

Software Operation

The program is suitable for batch or time-sharing.

Hardware

The program was tested on UNIVAC 90/70 (using BG-4 compiler); requires 120K storage.

Capabilities

Program CONFORM determines the shape of the contact region, the distribution of contact pressure, and the resultant force, on the surface of the two elastic bodies in conformal contact, with a specified approach (rigid body displacement).

6.4 VEHICLE DYNAMICS PROGRAMS

An all-inclusive vehicle dynamics program would be impractical. The complexity of such a program would make it too difficult for wide usage and its development cost would be prohibitive. Consequently, Tables 6.4 to 6.11 present special-purpose programs developed to model specific degrees of freedom and operating conditions, as discussed in (a) to (h) below. Other miscellaneous programs have been grouped in Table 6.12.

The 'classes' of dynamic models, referred to in the tables, can be summarized as shown in Fig. 6.1. Thus, a model can be linear, quasi-linear or nonlinear. In each case, except the latter, it can be a frequency-domain or time-domain model. Nonlinear models are not amenable to frequency-domain analysis.

Several types of frequency domain techniques can be applied to rail vehicle dynamics. They are based on linear assumptions, but can be adapted for use on nonlinear systems by applying quasilinearization techniques such as describing functions [1]. Eigenvalue analyses are used to determine critical speeds, natural frequencies, damping ratios and mode shapes. For validation purposes, critical speeds can be compared with the speeds at which hunting becomes apparent in tests. Transfer functions are often used to calculate vehicle response spectra and rms acceleration levels of interest in ride quality and lading damage studies.

Time Domain Solutions

The majority of vehicle dynamics models use time-domain solution techniques. For linear systems, the solution can be computed by linear algebra, using the state transition matrix. In general, linear or nonlinear systems can be solved by a variety of numerical integration methods. In any case, time histories of response and auxiliary variables are available.

The 'quasi-static' (algebraic) analysis may be applied to steady-state models to predict the dynamic equilibrium performance of vehicles. These models are relatively simple to validate

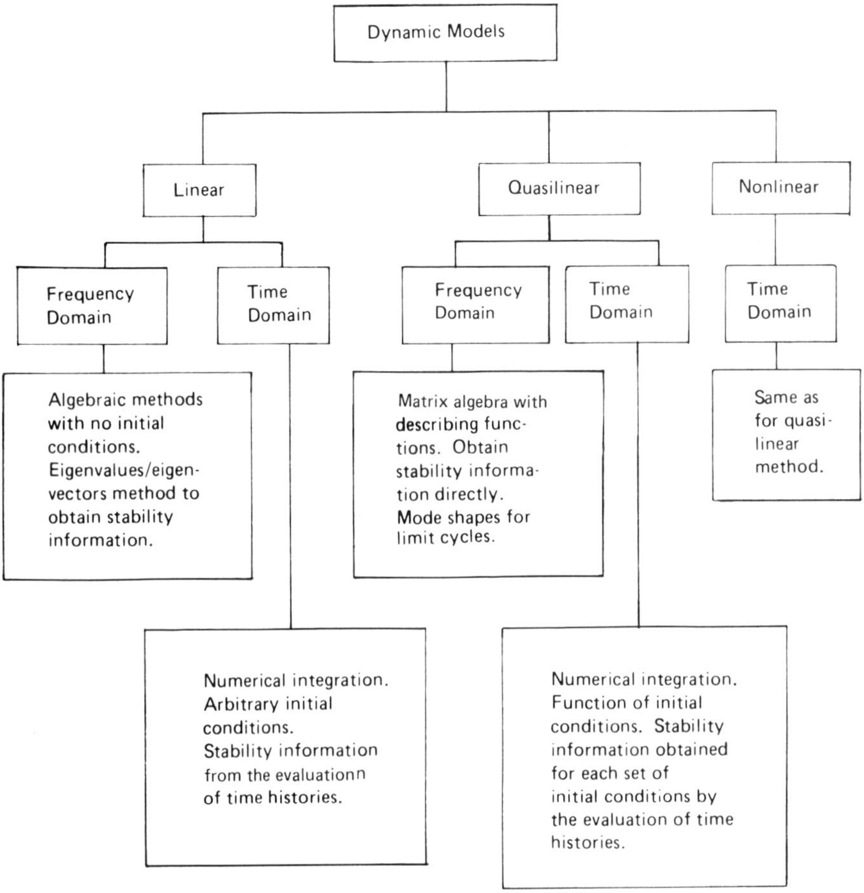

Fig. 6.1 Classes of Dynamic Models

because each output is a single number which can be compared with test results. An example is a model which calculates equilibrium forces and displacements in steady curving, where the speed and curvature are constant.

Let us now consider the special-purpose groups of vehicle dynamics programs:

(a) Hunting (Table 6.4): This is a form of instability manifested by self-excited oscillations of a wheelset, truck or carbody. It can arise on a perfect track and is self-sustaining.

Table 6.4. Hunting Programs.

Source/Model	Author(s)	Usage	Hardware	Vehicle/Model Class/ Degrees of Freedom/ Users' Manual/ Validation
1. CU-ASU[41]/ Freight Car Lateral Stability Models	E.H. Law N.K. Cooperrider J.A. Hadden	DOT-FRA A	IBM 370/265 UNIVAC 1110	F/L,FD/5-23/NI/LTD
2. ASU-MIT[31]/ NI	N.K. Cooperrider E.H. Law J.K. Hedrick	TSC-FRA A	UNIVAC 1110 DEC 10	F/QL/9/NI/N

3.	ASU-MIT[32]/ NI	J.K. Hedrick A.V. Arslan	TSC-FRA A	UNIVAC 1110 DEC 10	F/NL/39/NI/N
4.	AAR[10]/ Freight Car Hunting Model	T.H.W. Cheung V.K. Garg G.C. Martin	AAR	IBM 370/158	F/L,FD/25/Y/Y
5.	SPTC[3]/ Freq. Domain Model (FDM)	Technical Research & Development Group	TDOP	IBM 370/158	F/L,FD/13/Y/Y
6.	BCL[3]/ TRKVPSD (Mod IIB)	D.R. Ahlbeck G.R. Doyle	A	CDC Cyber 70	F/L/19/Y/P
7.	BCL[3]/ GENTRK	J.A. Hadden	A	CDC 6400	F/L/11/N/P

Table 6.4. (continued)

Source/Model	Author(s)	Usage	Hardware	Vehicle/Model Class/ Degrees of Freedom/ Users' Manual/ Validation
8. BCL[3]/ GENCAR	J.A. Hadden	A	CDC 6400	F/L/23/N/P
9. BCL/ TRKHNTT	G.R. Doyle P.H. Prause	A	CDC 6400	F/L,FD/9/N/N
10. BCL/ CARHNT II	G.R. Doyle R.H. Prause	A	CDC 6400	PA,LO/L,FD/17/N/N
11. AAR[19]/ Locomotive Truck Hunting Model	V.K. Garg P.W. Hartmann G.C. Martin	AAR	IBM 370/158	LO/L/2-21/Y/LTD

12. MIT[34]/ NI	J.K. Hedrick	MIT-AAR	DEC 10 VAX	LO/QL/11/Y/N
13. MIT[33]/ NI	J.K. Hedrick	MIT-AAR	DEC 10 VAX	LO/NL/21/Y/N
14. MIT[33,79]/ NI	J.K. Hedrick	MIT-PU-PN	DEC 10 VAX	PA/L/6-15/Y/Y
15. RDI/ Truck Hunting	V.T. Hawthorne	RDI	TRS80 TI59	F/L/8/N/Y
16. DOFASCO/ Hunting	W.H. Elmaraghy	DOFASCO	IBM 370/158 CDC 6400	F,PA,LO/L/19/Y/P
17. DOFASCO/ Dynamic Stability	W.H. Elmaraghy	DOFASCO	IBM 380/158 CDC 6400	LO/L/21/Y/P
18. BCL/ NINEDOF	NI	BCL	CDC 6400	F/L/9/N/P

Table 6.4. (continued)

Source/Model	Author(s)	Usage	Hardware	Vehicle/Model Class/ Degrees of Freedom/ Users' Manual/ Validation
19. BCL/ 19 DOF	NI	BCL	CDC 6400	F/L/19/N/P
20. BUDD/ Stability	NI	BUDD	DG Eclipse	PA/L/NI/Y/Y
21. BI/ SLV	F.B. Blader R. Dumoulin	BI	CDC 6400	TC/L/9/N/Y
22. BI/ SLVA	F.B. Blader R. Dumoulin	BI	CDC 6400	TC/L/27/N/Y
23. BI/ SLRC	F.B. Blader R. Dumoulin	BI	CDC 6400	PA/L/19/N/Y

24. RDSO/ MULRAW	D. Deepak V.V. Singh	IR	IBM 370/158 CDC 6400	F/L/NI/Y/Y
25. ASF/ HU7	D.S. Garivaltis	S	CYBER 175	F/L/7/Y/Y
26. ASF/ HU9	D.S. Garivaltis	S	CYBER 175	F/L/9/Y/Y
27. ASF/ HU11	D.S. Garivaltis	S	CYBER 175	F/L/11/Y/Y
28. ASF/ HU21	D.S. Garivaltis	S	CYBER 175	F/L/21/Y/Y
29. ASF/ HU23	D.S. Garivaltis	S	CYBER 175	F/L/23/Y/Y
30. DSL/ DSL-1	P. Marcotte R. Cheng	S	IBM 370/158 CDC 6400	LO/NL/5/N/N

Table 6.4. (continued)

Source/Model	Author(s)	Usage	Hardware	Vehicle/Model Class/ Degrees of Freedom/ Users' Manual/ Validation
31. CP/ DTM	NI	S	IBM 370/158 CDC 6400	T/L/17/N/N
32. TA/ HUNT	F.B. Blader	S	PC	T,RV,F,PA/L/9/N/Y
33. IGUTKY[72]/ VCR4	Y. Suda	IGUTKY	HITAC M280/M680	PA/LF/4/N/N

34. IGUTKY[72]/ VCR4	Y. Suda	IGUTKY	HITAC M280/M680	PA/LF/6/N/N
35. IGUTKY[72]/ VCR9	Y. Suda	IGUTKY	HITAC M280/M680	PA/LF/6/N/P
36. IGUTKY[72] VCR17	Y. Suda	IGUTKY	HITAC M280/M680	PA/LF/17/N/N
37. IGUTKY[72] VCR19	Y. Suda	IGUTKY	HITAC M280/M680	PAP/LF/19/N/N
38. AAR/NUCARS	F.B. Blader J.A. Elkins	AAR	PC	G/NL/V/N/P
39. AAR/FCLH	AAR	AAR	PC	F/L,FD/25/Y/Y
40. AAR/LTHM	AAR	AAR	PC	LO/L/2-21/Y/LTD

Hunting programs are used for studying the hunting motions of carbody, truck and wheelsets of a vehicle. The programs predict the critical speed at which a vehicle becomes unstable and the hunting frequency. Typical input to these programs consists of vehicle dimensional data, spring stiffness and damping data, mass and inertia data, and creep coefficients. The output may include the reflection of all the input data, mass, stiffness and damping matrices, velocity, degrees of freedom, eigenvalues and eigenvectors, critical speed and hunting frequency.

(b) Twist and Roll (Table 6.5): These are low-speed, externally excited, resonance-type motions in which a carbody oscillates about an axis parallel to the train. Twist refers to torsional bending; roll corresponds to rotational motion about a longitudinal axis. Twist and roll is often associated with high center-of-gravity cars whose truck spacings lie in a small range of distances. Operation can be on bolted-rail, staggered-joint tracks with "dipped" joints; or, on newly-installed, continuously-welded rail, with "dips" induced dynamically by car-roll motions. Although creep guidance does play a role in twist and roll, it is subordinate to periodic track inputs and is customarily excluded from these models. These programs simulate a railroad freight vehicle operated on staggered rails, either half-staggered or zero-staggered. As predictors of rock and roll and bouncing behaviour, they can be used as freight car design tools and in derailment investigations.

The inputs to the programs include mass and rotational moment of inertia of carbody and truck, geometrical data of car and truck, stiffness and damping data, and simulation data (rail profile, vehicle velocity and length of simulation). Typical outputs include translational and rotational acceleration, velocity and displacement of carbody, bolsters and wheelsets, rail forcing functions, centerplate and sidebearing loads, rail loads and deflections, damping and spring forces, and suspension spring deflections.

Table 6.5. Twist and Roll Programs.

Source/Model	Author(s)	Usage	Hardware	Vehicle/Model Class/ Degrees of Freedom/ Users' Manual/ Validation
1. MITRE/ FRATE	G. Kachadourian N.E. Sussman J.R. Anderes	S	CDC 7600	F/NL/33-39/Y/LTD
2. IIT[69] NI	K.L. Shum T. Willis	S	UNIVAC 1108	F/NL/27/N/LTD
3. BCL/ PERTRK II	D.R. Ahlbeck	A	CDC 6400 CYBER 74	F/NL/17/N/P
4. BCL/ SPICOM	G.R. Doyle	A	CDC 6400	PA/NL/6/N/N
5. BCL/ TRAKVEH	G.R. Doyle	A	CDC 6400	F/NL/19/N/N
6. MIT[62]/ NI	B.E. Platin	MIT-AAR-TSC	IBM 370/158 DEC 10	F/NL/19/N/N

Table 6.5. (continued)

Source/Model	Author(s)	Usage	Hardware	Vehicle/Model Class/ Degrees of Freedom/ Users' Manual/ Validation
7. MIT[5]/ NI	J.J. Beaman	MIT	PDP-11	F/QL/6/N/N
8. AAR[78]/ NI	Y.H. Tse G.C. Martin	NI	NI	F/NL/22/Y/Y
9. WYLE[30]/ NI	J.J. Healy	NI	NI	F/NL/11/N/N
10. WYLE[43]/ NI	A.A. Liepens	NI	NI	F/L/8/N/N

11. RDI/ Carbody roll	S.A. Schwam A.M. Whitman	RD	TRS 80	F/NL/1/N/Y
12. ASF/ ROCK	D.S. Garivaltis	ASF	CYBER 175	F/L/2/Y/Y
13. ASF/ ROC3	D.S. Garivaltis	ASF	CYBER 175	F/L/3/Y/Y
14 THRALL/ NI	NI	THRALL	DEC 20	F/L/2/Y/Y
15. TASC/ SIMCAR	F.B. Blader N. Acharya	TSC-TASC	DEC 10	F/NL/14-16/N/N
16. AAR/NUCARS	F.B. Blader J.A. Elkins	AAR	PC	G/NL/V/N/P
17. AAR/FBRFC	AAR	AAR	PC	F/NL/22/Y/Y

Table 6.6. Pitch and Bounce Programs.

Source/Model	Author(s)	Usage	Hardware	Vehicle/Model Class/ Degrees of Freedom/ Users' Manual/ Validation
1. AAR[6]/ NI	E.H. Chang V.K. Garg C.H. Goodspeed	NI	NI	LO/NL/38/N/N
2. TSC[57]/ NI	A.B. Perlman F.P. Dimasi	TSC	DEC 10	F/L/6/Y/N

3. TSC[57]/ NI	A.B. Perlman F.P. Dimasi	TSC	DEC 20	PA/L/8/Y/N
4. THRALL/ NI	NI	THRALL	DEC 20	F/L/3/Y/Y
5. ASF/ PDBNC	D.S. Garivaltis	ASF	CYBER 175	F/L/3/Y/Y
6. RDI/ NI	S.A. Schwam A.M. Whitman	RD	TRS 80 TI 59	F/L/1/Y/Y
7. TASC/ SIMCAR	F.B. Blader N. Acharya	TSC-TASC	DEC 10	F/NL/14-16/N/N
8. AAR/NUCARS	F.B. Blader J.A. Elkins	AAR	PC	G/NL/V/N/P

(c) Pitch and Bounce (Table 6.6): These motions are excited by track geometry variations, especially symmetrical irregularities. Pitch is rotational motion of a carbody about a lateral axis whereas bounce is vertical motion. Pitch and bounce occasionally contribute to derailments by causing wheel-lifts and carbody-truck separations, but their greatest impact is on human discomfort and lading damage.

The wheelset guidance mechanism is negligible; so, unless track dynamics are included, pitch and bounce models tend to be especially simple.

(d) Yaw and Sway (Table 6.7): Yaw is rotational motion of a carbody about a vertical axis whereas sway refers to lateral motion. These usually-oscillatory motions are self-induced when critical hunting speeds are exceeded; but, at subcritical speeds, they are externally-excited by track irregularities. The large lateral wheel-rail forces which can be generated, increase the likelihood of derailment. These programs are used to study the dynamic response of a vehicle to rail irregularities. They are an effective tool in designing and selecting the optimum suspension system for a vehicle.

Typical inputs include vehicle, truck and carbody geometrical and dimensional data, mass and inertia data, spring stiffness and damping data, creep coefficients, wheel conicity, and track vertical and lateral irregularities data. The output includes reflection of input data, lateral and vertical displacements, velocities, accelerations and forces of wheelsets, trucks and carbody.

(e) Steady-State Curving (Table 6.8): Large steady-state lateral forces are possible at unbalanced speeds. They may be increased by large-wheelbase trucks and inadequate maintenance of sidebearings and centerplates, tending to cause binding. These programs calculate the lateral wheel rail forces associated with the negotiation of a section of curved track by a truck or vehicle for a variety of operating conditions. The may be used as as truck design tools.

Table 6.7. Yaw and Sway Programs.

Source/Model	Author(s)	Usage	Hardware	Vehicle/Model Class/ Degrees of Freedom/ Users' Manual/ Validation
1. AAR[7]/ Locomotive Response Model	E.H. Chang V.K. Garg P.W. Hartman	AAR	IBM 370/158 DEC 10	LO/L/21/Y/N
2. AAR[6]/ NI	E.H. Chang V.K. Garg C.H. Goodspeed	AAR	IBM 370/158 DEC 10	LO/NL/38/Y/N
3. MIT[34]/ NI	J.K. Hedrick	MIT-AAR	DEC 10 VAX	LO/QL/11/Y/N
4. MIT[34]/ NI	J.K. Hedrick	MIT-AAR	DEC 10 VAX	LO/NL/21/Y/N

Table 6.7. (continued)

Source/Model	Author(s)	Usage	Hardware	Vehicle/Model Class/ Degrees of Freedom/ Users' Manual/ Validation
5. BCL/ TRKVSPD	G.R. Doyle	BCL	CDC 6400 CYBER 70	LO,PA/L/14-15/Y/P
6. MIT[79]/ NI	D.N. Wormeley	MIT-PU-PN	DEC 10 VAX	PA/L/15/Y/N
7. TSC[57]/ NI	A.B. Perlman F.P. Dimasi	TSC	DEC 10	PA/L/15/Y/N

8. MIT[79]/ NI	D.N. Wormeley	MIT-PU-PN	DEC 10 VAX	PA/NL/15/Y/P
9. BCL/ TRAKVEH	G.R. Doyle	BCL	CDC 6400 CYBER 70	LO,PA/NL/19/Y/Y
10. BCL/ TRKVPSD II	G.R. Doyle	BCL	CDC 6400 CYBER 70	LO,PA/L/18/Y/Y
11. BCL/ PERTRK II	G.R. Doyle	BCL	CDC 6400 CYBER 70	LO,PA/NL/17/Y/Y
12. AAR/NUCARS	F.B. Blader J.A. Elkins	AAR	PC	G/NL/V/N/P
13. AAR/LRM	AAR	AAR	IBM370/158 DEC-20/60	LO/L/21/Y/Y

Table 6.8. Steady State Curving Programs.

Source/Model	Author(s)	Usage	Hardware	Vehicle/Model Class/ Degrees of Freedom/ Users' Manual/ Validation
1. BCL/ SSCUR2	G.R. Doyle D.R. Ahlbeck	TSC-BCL	CDC CYBER 73	PA,LO/NL/7/N/P
2. BCL/ SSCUR2 II	G.R. Doyle D.R. Ahlbeck	TSC-BCL	CDC CYBER 73	PA,LO/NL/7/Y/P
3. BCL/ SSCUR 3	G.R. Doyle D.R. Ahlbeck	TSC-BCL	CDC CYBER 73	PA,LO/NL/9/Y/N
4. CU-ASU[42]/ Steady Curve Negotiation	E.H. Law N.K. Cooperrider	CU	IBM 370/158	ITA/NL/7/N/N

5. CU-ASU[42,61]/ Steady Curving	E.H. Law N.K. Cooperrider	CU	IBM 370/158	F/NL/9/N/N
6. BCL[42/ NI	NI	BCL	CDC CYBER 73	PA/NL/11/N/N
7. TASC/ SIMCAR	NI	TSC-TASC	DEC 10 IBM 370/158	F/NL/14-16/N/N
8. CU-ASU[61]/ NI	E.H. Law N.K. Cooperrider	CU	IBM 370/158	PA/NL/11/N/N
9. CU-ASU[61] NI	E.H. Law N.K. Cooperrider	CU	IBM 370/158	F/NL/17/N/N
10. BCL/ SSCUR3	G.R. Doyle	BCL-TSC-TASC	CDC CYBER 73	LO/NL/8/N/N

Table 6.8. (continued)

Source/Model	Author(s)	Usage	Hardware	Vehicle/Model Class/ Degrees of Freedom/ Users' Manual/ Validation
11. AAR[70]/ 2,3,4 Axle Truck Curving	K.R. Smith	GM	IBM 370/158	LO/QS/NI/Y/Y
12. MIT[33,34]/ NI	J.K. Hedrick	MIT-AAR	DEC 10 VAX	PA/NL/17-21/Y/N
13. TA/ CURE	F.B. Blader	S	CDC 6400 IBM 370/158 PC	PA,F/NL/8/N/Y
14. BI/ CORE2	F.B. Blader R. Dumoulin N. Acharya	S	CDC 6400 IBM 370/158	PAT/NL/8/N/Y

15. DSL[50,51]/ DSL-3	P. Marcotte R. Mathewson	S	CDC 6400 IBM 370/158	F/NL/14/Y/Y
16. TASC/ SIMCAR	F.B. Blader N. Acharya	TSC-TASC	DEC 10 VAX	F/NL/14-16/N/N
17. IGUTKY[72] CUR4	Y. Suda	IGUTKY	HITAC M280/M680	PA/LF/4/N/N
18. BN-UCL[4,9, 67]/ Autodyn	Th. Chatelle P. Maes J.C. Samin G. Schmitz	BN UCL	IBM 4381 SUN	G/NL/V/P/N
19. AAR/NUCARS	F.B. Blader J.A. Elkins	AAR	PC	G/NL/V/N/P
20. AAR/RTCNM	AAR	AAR	PC	LO/QS/NI/Y/Y
21. AAR/SSCRV	AAR	AAR	PC	F/NL/NI/Y/P
22. AAR/SSLTCM	AAR	AAR	PC	LO/NL/NI/Y/P

Typical inputs: Number of axles or wheelsets per truck, tractive and braking effort data, buff force on truck, truck speed, clearance between axle and truck frame, lateral displacement allowed on suspension, wheel diameter, wheel loads on axle, wheel base, and coefficient of friction.
Typical outputs: Reflection of input data, lateral centerplate load, lateral flange forces on wheels, lateral friction forces on wheels, and distance of instantaneous center of rotation from axle or truck centerline.

(f) Quasi-Steady State Curving (Table 6.9): A general curve is negotiated, starting from arbitrary initial conditions; however, acceleration is ignored. The response of a vehicle is determined at discrete locations along a curved track by applying the kinematic constraints. The typical inputs and outputs of these programs are as for the steady-state curving programs described above.

(g) Spiral Negotiation (Table 6.10): A spiral segment can be regarded as a voluntary warp designed into a track. If it is short relative to the length of a torsionally-stiff vehicle, or truck, the vertical load at a wheel-rail contact may be reduced considerably while the lateral load is increased due to centrifugal forces. These programs are similar to the curving programs described in (e) and (f) above.

(h) Dynamic Curving (Table 6.11): Dynamic curving programs can usually represent the dynamics of curve entry and exit. Some may even include the effects of track irregularities and speed variations in curves. These programs predict the transient and steady-state wheel-rail interaction forces in curve negotiation by a truck or vehicle. Coupler forces may also be computed.
Typical inputs: Truck and vehicle geometric data, conicity, truck and carbody mass, roll and inertia data, suspension stiffness and damping, rail stiffness and damping, track data, wheel diameter, creep coefficients and whell-rail coefficient of friction. The output generally includes reflection of all input data, lateral,

Table 6.9. Quasi-Steady State Curving Programs.

Source/Model	Author(s)	Usage	Hardware	Vehicle/Model Class/ Degrees of Freedom/ Users' Manual/ Validation
1. TSC/ NI	A.B. Perlman H. Weinstock	TSC	DEC 10	I/NL/4/Y/N
2. AAR[61]/ RTCN	TTD Steering Committee	TTD-AAR	IBM 370/158	LO/NL/10/Y/Y
3. AAR/NUCARS	F.B. Blader J.A. Elkins	AAR	PC	G/NL/V/N/P
4. BR/QUASI-CURVING	BR	BR	NI	G/NL/V/Y/P
5. SJ/KVASIM	SJ	SJ	IBM	F/NL/15/N/P

Table 6.10. Spiral Neogtiation Programs.

Source/Model	Author(s)	Usage	Hardware	Vehicle/Model Class/ Degrees of Freedom/ Users' Manual/ Validation
1. BCL/ CURVENT	G.R. Doyle	BCL	CDC 6400 IBM 370/158 UNIVAC	PA/NL/9/N/N
2. BCL/ SPICOM	G.R. Doyle	BCL	CDC 6400	PA/NL/6/N/P
3. BCL/ PERTCRV	G.R. Doyle	BCL	CDC 6400	PA/NL/17/P/N
4. IGUTKY[72]/ PEN4	Y. Suda	IGUTKY	HITAC M280/M680 S810	PA/NL/19/N/N
5. AAR/NUCARS	F.B. Blader J.A. Elkins	AAR	PC	G/NL/V/N/P

Table 6.11. Dynamic Curving Programs.

Source/Model	Author(s)	Usage	Hardware	Vehicle/Model Class/ Degrees of Freedom/ Users' Manual/ Validation
1. BCL/ SPICOM	G.R. Doyle	BCL	CDC 6400	PA/NL/6/N/N
2. CU-ASU[61]/ Half-car Curve Entry	E.H. Law N.K. Cooperrider	CU-ASU	IBM 370/158	PA/NL/9/N/N
3. AAR/ Freight Car Curving	V.K. Garg	TTD-AAR	IBM 370/158	F/NL/43/N/N
4. AAR/ Locomotive Curving	V.K. Garg	TTD-AAR	IBM 370/158	LO/NL/59/N/N

Table 6.11. (continued)

Source/Model	Author(s)	Usage	Hardware	Vehicle/Model Class/ Degrees of Freedom/ Users' Manual/ Validation
5. CU-ASU [15,61]/ CURVLOCO	E.H. Law N.K. Cooperrider	TSC-TASC A	IBM 370/158	LO/NL/27/Y/N
6. TASC[61]/ RVDCADET 2	NI	TASC	DEC 10 IBM 370/3031	F/NL/14/N/N
7. MIT/ NI	J.K. Hedrick	MIT	DEC 10 VAX	F/NL/23/Y/N

8. TASC/ SIMCAR	F.B. Blader N. Acharya	TSC-TASC	DEC 10 VAX	F/NL/14-16/N/N
9. BCL/ CURVENT	G.R. Doyle	A	IBM 370/158 CDC UNIVAC	PA/NL/9/N/N
10. BCL/ PERTCRV	G.R. Doyle	A	CDC IBM UNIVAC	PA/NL/17/N/N
11. DSL[50,51, 81]/DSL-3	P. Marcotte R. Mathewson	S	IBM CDC	F/NL/14/Y/Y
12. TA/SIMFRT	F.B. Blader	A,S	PC	F/NL/16/N/P
13. SBB/KURVES	J.C. Peng	SBB	IBM 370	PA,F,L/NL/16/N/P
14. TUW/DYN.CURV	K.O. Endlicher	TUW	CDC 180-860	F,PA/NL/14/N/N
15. AAR/NUCARS	F.B. Blader J.A. Elkins	AAR	PC	G/NL/V/N/P

vertical and yaw response of axles, sideframes, bolsters and carbody, wheel tread friction forces and flange forces.

(i) Miscellaneous Programs (Table 6.12): These programs are either special-purpose or general-purpose. Many digital simulation programs have evolved for the analysis of complex mechanical systems. Formalisms for automatic generation of the equations of motion for multibody systems play a central role. Since most applications originated from tracked ground transportation systems, where only small deviations from a nominal motion are of interest, linear formalisms with several applications were also undertaken. As a result of the activities, many programs emerged. One such program, FADYNA (Table 6.12), is a comprehensive program with a linear multibody formalism including linear constraints, elastic bodies and a variety of linear and nonlinear, passive and active coupling elements (e.g. suspensions). The program also enables the simulation of vehicles riding over elevated elastic guideways. Time and frequency response and vehicle ride comfort can be obtained. An important application of this program is in vehicle-guideway interaction studies.

A successor of FADYNA, characterized by a completely novel data processing concept with interactive and integrated features, is program MEDUSA. Its linear formalism is more rigorous taking into consideration the effects of rotors, gravitational forces, and nominal loads. Other highlights include finding the static equilibrium, improved modeling of elastic bodies and a pre-processor for interface with finite element programs. One main application area is in the design of advanced rail vehicles.

6.5 TRAIN DYNAMICS PROGRAMS

The railroad and supply industries are faced with the need for analysis tools which can shed insight into the many problems of railroad operations. Computer modeling of train dynamics can be used to study the interactions of train handling, train make-up, and equipment and track designs for a more profitable rail-service operation.

Differential velocities between vehicles are caused by accelerating, braking, and grade and curvature variations. They generate dynamic forces which, when high enough, can result in train separation, lading and equipment damage, derailments, and personal injuries. In addition to the longitudinal buff and draft motions, the lateral dynamics of trains on curves, turn-outs and cross-overs are of concern.

Railway organizations are interested in train performance comparisons. For example, when re-coupling in hump yard operations, freight cars are subjected to impact forces. Lading damage due to vertical vibrations may be encountered in along-track travel. The amount of lading and equipment damage might be reduced by a better selection of suspensions and shock absorbing devices for controlling the bounce and pitch motions and impact forces, respectively.

As for vehicle dynamics, specialized train dynamics programs have been developed instead of general programs. In Tables 6.13 to 6.15 and the corresponding discussions (a) to (c) below, they are grouped into longitudinal, lateral and vertical dynamics programs, respectively. The non-dynamical models of Table 6.16, discussed in (d) below, deal with steady buff and draft conditions. A more detailed description of each group follows:

(a) Longitudinal Train Dynamics (Table 6.13): A long train traversing undulating terrain may have some portions of it descending a gradient while other portions are ascending another. The descending portions are pulled forward by gravity while the ascending portions are pulled backward. In this manner, severe longitudinal oscillations of the train may develop, which can only partially be controlled by the train operator through the use of throttle and brakes. The longitudinal oscillation generates high buff and draft forces, which can result in derailment or broken couplers. When a coupler breaks, the separation of cars severs the airbrake line, resulting in automatic application of the emergency brakes and the possibility of either a derailment or a collision between the separated parts of the train.

Longitudinal train dynamics programs are intended to predict coupler forces as well as speed, distance and time relationships.

Table 6.12. Miscellaneous Programs.

Source/Model	Author(s)	Usage	Hardware	Vehicle/Model Class/ Degrees of Freedom/ Users' Manual/ Validation
1. JWWC[29]/ DYNALYST II	T.K. Hasselman A. Bronswicki	DOT-TSC A	CDC 6600	PA/L/8,14/Y/N
2. DSL[52]/ DSL-2	P. Marcotte R. Mathewson	SR	IBM CDC	LO/NL/15/N/Y
3. DSL[52]/ DSL-1	P. Marcotte R. Mathewson	SR	IBM CDC	LO/L/5/N/Y
4. BUDD/ Wayside Vibration	NI	SR	DG Eclipse 2000	PA/NI/NI/Y/Y

5. BUDD/ Ride quality	NI	SR	DG Eclipse 2000	PA/NI/74/Y/P
6. BUDD/ DYNSIM	NI	SR	DG Eclipse 2000	PA/NL/60+/Y/P
7. BUDD/ Airspring Simulation	NI	SR	DG Eclipse 2000 and Analog	NI/NI/NI/Y/Y
8. REA/ DYNSIM	G.F. List H.A. List	S	CDC 6400 Univac 90/70 DG Eclipse C-350	PA,F/48/Y/Y
9. DFVLR[40]/ MEDUSA	O. Wallrapp et al.	S	IBM CDC	RV/NL/NI/Y/Y

Table 6.12. (continued)

Source/Model	Author(s)	Usage	Hardware	Vehicle/Model Class/ Degrees of Freedom/ Users' Manual/ Validation
10. TU[40]/ LINDA	K. Knothe et al.	S	IBM CDC	RV/L/NI/Y/Y
11. DFVLR[40]/ FADYNA	W. Duffek O. Wallrapp	S	IBM CDC	WR/NL/NI/Y/Y
12. RIDL/MLLZ	P. Sixta L. Rus	CKD	ICL 1900	LO/L/6.N/AC/Y
13. RIDL/MLVA	P. Sixta J. Zahradka	CKD	ICL 1900	LO/L/100/AC/Y

14. RIDL/SADKO	P. Sixta J. Kolar	CKD	ICL 1900	LO/L/NI/AC/Y
15. AAR/NUCARS	F.B. Blader J.A. Elkins	AAR	PC	G/NL/V/N/P
16. AAR/TEM	AAR	AAR	PC	T/NL/NI/Y/P
17. AAR/VHCNP	AAR	AAR	PC	G/NL/NI/Y/N
18. AAR/PCAP1	AAR	AAR	PC	F/NI/NI/Y/Y
19. AAR/PCAP2	AAR	AAR	PC	F/NI/NI/Y/Y
20. AAR/PCAP3	AAR	AAR	PC	G/NI/NI/Y/Y
21. UCL/AUTODYN	J.C. Samin	UCL	IBM	G/NL/V/Y/Y
22. UCL/ROBOTRAN	J.C. Samin	UCL	IBM	G/NL/V/Y/P
23. DFVLR/MEDYNA	DFVLR	S	IBM, DEC SIEMENS	G/NL/V/Y/Y

Table 6.13. Longitudinal Train Dynamics Programs.

Source/Model	Author(s)	Usage	Hardware	Train/Model Class/ Degrees of Freedom/ Users' Manual/ Validation
1. AAR[44,46]/ TOS Train Operations Simulator	N.W. Luttrell R.K. Gupta E.M. Low G.C. Martin V.K. Garg	E	IBM 370/158	F/NL/1PB/Y/P
2. AAR[20,45, 48,49,68]/ DTAM Detailed Longitudinal Train Action	G.C. Martin W.E. Plouffe S. Ahmed H. Antczak	LTD	IBM 370/158	F/NL/1PC/Y/P
3. BN-UCL[4,9, 67]/Autodyn	Th. Chatelle P. Maes J.C. Samin G. Schmitz	BN UCL	IBM 4381 SUN	G/NL/V/P/N

The application areas of these programs include train makeup, train handling, derailment investigations, power selection, brake system design and cushioning devices design. As an example, the program 'Train Operations Simulator' (TOS) developed by the AAR [44,46] (Table 6.13) and used extensively by several major railroads in the USA, is briefly described here (Pilkey [61]).

Capability

This program can simulate the performance of a freight train powered by diesel-electric locomotives over a length of railroad track with any combination of curvature and grades. Any combination of throttle and various modes of braking manipulations are possible. It has been applied in derailment investigations, train make-up, and power selection studies, train operation training, and signal spacing and track layout design.

Method

The net force on the train is calculated from the available locomotive tractive effort and the total train resistance. The latter is determined from the Davis Resistance Formula, the dynamic braking characteristics, and the air-brake force characteristics. The net force determines whether the train will accelerate, decelerate, or remain at a constant speed. Train velocity and distance are computed by numerical integration. The relative velocities between adjacent vehicles, together with the amount of coupler slack, govern the buff and draft actions in the train and the magnitude of draw bar forces calculated according to draft gear characteristics.

Limitations and Restrictions

The drawbar forces and L/V ratios are quasidynamic but are good approximations. Only ABD valves are correctly handled during brake release. The program shows much quicker than actual release on AB valves and does not provide for ABDW brake valves. Simulation of cyclic braking involving complete release is not accurate and brake pipe leakage is assumed to be evenly distributed throughout the train.

Input

Track Data: Curvature data includes degree of curve, length of spiral and curve, and superelevation. Profile data includes mileage and elevation.

Train Data: Data required are for locomotive and freight cars. If different from those in the Standard Library (which comes with the program) the user must supply various geometric parameters as well as the brake valve type, brake shoe type, and forces. For locomotives not in the Standard Library, data such as tractive effort and dynamic braking characteristics are required.

Train Operations Commands: These include commands such as speed, throttle, brake releases, and independent and dynamic brake applications.

Title Information: This is the title that a user wishes to appear in the output.

Output

Train Data: The number of locomotives, number of cars, train length, tonnages, and a list of primary characteristics of each vehicle.

Power Summary: The tractive effort and dynamic braking characteristics, including power on rail and transition speeds.

Performance Print-out: (1) Time, station and mile post location of front end of train; (2) Speed limit and actual train speed; (3) Locomotive throttle setting and amperage; (4) Train brake setting and pressures in brake pipe and cylinder; (5) Maximum values and locations of drawbar forces and L/V ratios; (6) Fore and aft drawbar forces and L/V ratios on vehicles selected by user for each print interval; (7) A summary of maximum buff, draft, and L/V, with corresponding location in train; and (8) An indication of changes in operation commands, grade, and/or curvatures.

Software Operation

The program is available for both batch and time-sharing modes.

Hardware

The program was developed on an IBM 370/158 machine and requires 216 K bytes of core storage. With minor modifications, it can be run on other computers.

Running Time

A typical stopping simulation for a freight train at 80.5 - 96.6 km/h (50-60 mph) requires about 1 minute of IBM 360/168 CPU time. The automatic mode requires approximately twice as much time as the manual mode.

(b) Lateral Train Dynamics (Table 6.14): These program combine wheel-rail interaction forces, centrifugal forces and forces arising from track superelevation with the lateral components of coupler forces. Intended primarily for train stability studies, they can also provide information on gauge widening, rail roll-over and wheel climbing. The following information on program QLTS [61] is presented here as a typical example of lateral train dynamics programs.

Capability

Program QLTS simulates the lateral stability of a train consist in curve negotiation under buff or draft drawbar forces. It calculates derailment tendencies through L/V ratios and coupler lateral angles. The program can be used for evaluation of performance with long-car, short car, loaded and empty combinations of vehicles in a train. It has been used to determine minimum tangent lengths between reverse curves, the effect of spiral lengths in curve negotiation, and guidelines for train make-up. It has also been applied as a design tool for freight car geometry.

Method

Based on the geometric interaction of a track and a train, the amount of buff or draft forces, the train speed and the amount of superelevation in a curve, this program computes the equilibrium condition. The resultant lateral load at bolster centers is used to calculate L/V ratios and, hence, the propensity for rail roll-over and wheel climb.

Table 6.14. Lateral Train Dynamics Program.

Source/Model	Author(s)	Usage	Hardware	Train/Model Class/ Degrees of Freedom/ Users' Manual/ Validation
1. AAR[1]/ DLTS Detailed Lateral Train Stability	TTD Steering Committee	LTD	DEC 10 IBM 370/158	F/NL/5PC/Y/P
2. AAR[17,77] QLTS Quasi-Static Lateral Train Stability	L.R. Thomas R.D. MacMillan G.C. Martin	LTD	IBM 370/158	F/NL/2PC/Y/N
3. RIDL/MNXF	L. Rus M. Herold	CKD	ICL 2958	LO/L/34/AC/Y

4. RIDL/MNXM	L. Rus M. Herold	CKD	ICL 2958	LO/L/34/AC/Y
5. RIDL/MNXH	L. Rus	CKD	ICL 2958	LO/L/34/AC/Y
6. BN-VEL[4]/ SP15B	Th. Chatelle P. Hecq J.C. Samin	BN	IBM 370 IBM 4381	PA/L/15/P/N
7. BN-UCL[4]/ SP27B	Th. Chatelle P. Hecq J.C. Samin	BN	IBM 370 IBM 4381	PA/L/27/P/N
8. BN-UCL[4,9, 67]/Autodyn	Th. Chatelle P. Maes J.C. Samin G. Schmitz	BN UCL	IBM 4381 SUN	G/NL/V/P/N
9. AAR/QSLTS	AAR	AAR	PC	F/NL/2PC/Y/Y
10. SJ/KVASIM	SJ	SJ	IBM	F/NL/15/N/P
11. ASEA/PSL	ASEA	SJ	IBM	PA/NL/37/Y/P

Limitations and Restrictions

This quasi-static model does not provide for transient inertial force considerations. A maximum of 762 m (2,500 ft) of track and 100 vehicles can be simulated. A limited selection of couplers is available in the model.

The maximum allowable drawbar force is 2,002 kN (450,000 lbf). There is no lateral coupler swing limit and the user must manually check each calculation of the coupler angle against the maximum allowable for the equipment. The L/V ratios calculated do not take into account the coupler arms' contacting strikers.

Input

Track Data: Requires lengths of tangent, lead and trail spirals, length and degree of curve, and superelevation.

Operation Data: Includes train speed, drawbar force, start and stop times, and interval of simulation.

Train Data: Includes bolster (truck) center distance, length between bolster center and coupler pin, coupler length, maximum lateral offset of bolster center from track center line, type of alignment control, vehicle weight, and the net lateral load at the leading outer wheel of each vehicle.

Output

Track Generator: The track generator gives an echo of the input data and generates the x-y coordinates of points along the track centerline and the angle to the centerline origin. It shows the radii of curvature and superelevation of every chord length defined by the user.

Results at Equilibrium: Location of vehicle on track, bolster lateral displacements and reactions, coupler lateral angles and pin reactions, centrifugal and superelevation forces, alignment control moments, drawbar forces and L/V ratios.

Software Operation

Both batch and time-sharing modes are available. This program was developed on the IBM 370/158 machine and requires 292 K bytes of core for execution. Only minor modifications, if any, are required to operate on other computer systems.

Running Time

The running time on an IBM 370/158 machine varies from a few CPU seconds to over a minute, depending on the train's length and speed and the calculation interval specified by the user.

(c) Vertical Train Dynamics (Table 6.15): These programs are used to study train stability in the vertical plane. They can be used to predict coupler disengagements and carbody separations from trucks. As a typical example, program VTS is described briefly, based on Pilkey [61].

Capability

Program VTS can be used as an analysis tool for the vertical forces and motion due to longitudinal in-train dynamic forces which, when they become large, can cause vertical coupler disengagement and/or carbody separation from trucks. Various train configurations can be evaluated for vertical stability.

Method

This is a two-dimensional model developed to study the vertical and pitch motions and forces of lading and trucks under the action of in-train longitudinal forces. Carbody, trucks, and lading are assumed to be rigid masses interconnected by nonlinear spring and damper systems. The equations of motion are integrated numerically using a modified Euler's method to obtain velocities and displacements.

Limitations and Restrictions

The current version of the program does not have a train operation simulation capability. Drawbar forces may be obtained by means of the Train Operations Simulator (TOS), or the Detailed Train Action Model (DTAM), or tests.

Input

Train Data: Includes the number of cars, first and last car numbers, and train speed.

Car Data: Includes geometry, weight, stiffness, snubber damping forces, coupler slacks, pitch moment of inertia, and coefficient of friction.

Coupler Forces: Horizontal component of coupler forces acting at front and rear end of train segments as functions of time.

Simulation data: Includes time duration of simulation, integration time step, and plot option.

Output

Echo: Print-out of input data.

Carbody: Acceleration, velocity and displacement.

Couplers: Horizontal and vertical forces and vertical slippage.

Trucks: Horizontal and vertical reactions and vertical displacements.

Lading: Includes acceleration, velocity and displacement.

Draft gears: Force-displacement characteristics.

Software Operation

Operation is in batch mode only and no pre- or post-processors are available.

Hardware

The program was developed on an IBM 370/158 computer and requires 156 K bytes of core for execution.

Running Time

A typical one second run-in simulation on a seven-car train segment uses about 20 seconds of IBM 370/158 CPU time.

Longitudinal Impact

Whereas the longitudinal train action problem involves oscillations of several cars, the longitudinal impact problem concerns the behaviour of one car (or locomotive or caboose) under the influence of a single impulsive buff force encountered during car coupling operations or, possibly, derailment scenarios. The programs in Table 6.15 may also be used to study such problems.

(d) Steady Buff and Draft (Table 6.16): When locomotive brakes are applied, high compressive or "buff" forces can develop in the train. These buff forces cause an accordian-like buckling motion of the train during which cars may yaw (rotate about a vertical axis) or be pushed sideways, resulting in large lateral forces between wheels and rails. With light (empty) cars, the lateral to

Table 6.15. Vertical Train Dynamics Programs.

Source/Model	Author(s)	Usage	Hardware	Train/Model Class/ Degrees of Freedom/ Users' Manual/ Validation
1. WU[59,80]/ VTA Vertical Train Action	S.K. Yin	LTD	IBM 360/65	F/NL/3PC/Y/N
2. AAR[64]/ VTS Vertical Train Stability	J.R. Raidt K.L. Shum G.C. Martin V.K. Garg	LTD	IBM 360/158	F/NL/4PC/Y/P
3. BN-UCL[4]/ SP12A	Ph. Chatelle P. Hecq J.C. Samin	BN	IBM 370 IBM 4381	PA/L/12/P/Y

Table 6.15. (continued)

Source/Model	Author(s)	Usage	Hardware	Train/Model Class/ Degrees of Freedom/ Users' Manual/ Validation
4. BN-UCL[4]/ SP15A	Ph. Chatelle P. Hecq J.C. Samin	BN	IBM 370 IBM 4381	PA/L/15/P/Y
5. BN-UCL[4]/ SP7A	Ph. Chatelle P. Hecq J.C. Samin	BN	IBM 370 IBM 4381	PA/NL/7/P/N
6. BN-UCL[4,9, 67]/Autodyn	Ph. Chatelle P. Maes J.C. Samin G. Schmitz16v	BN UCL	IBM 4381 SUN	G/NL/V/P/P
7. ASEA/SIMFOV	ASEA	SJ	IBM	PA/NL/6/Y/P

Table 6.16. Steady Buff and Draft Programs.

Source/Model	Author(s)	Usage	Hardware	Train/Model Class/ Degrees of Freedom/ Users' Manual/ Validation
1. AAR[77]/ QLTS Quasi-Static Lateral Train Stability	L.R. Thomas R.D. MacMillan G.C. Martin	LTD	IBM 370/150	F/NL/2PC/Y/N
2. TSC[74]/ TSA Train Stability Analysis	J. Tanne R. Brantman	TSC	IBM 370/150 DEC 10	F/NL/2PC/Y/N
3. AAR[2]/ DLTS Detailed Lateral Train Stability	TTD Steering Committee	S	IBM 360/158 DEC 10 CDC 6400	F/NL/5PC/Y/P

vertical force ratios may become large enough to cause derailment. Although this phenomenon occurs on tangent and curved track, its severity is greater on the latter. Alternatively, when a locomotive is operating at high tractive effort while negotiating a curve (for example, when climbing a gradient at low speeds), high tensile or "draft" forces can develop in the train. These forces tend to straighten the train, thus creating a tendency for cars to be derailed by being pulled to the inside of a curve.

6.6 STRUCTURAL MECHANICS PROGRAMS

General-purpose structural mechanics programs can sometimes be used to analyze rail, roadbed and vehicle structures for stresses and strains. Pilkey and Pilkey [60] produced a survey report on such programs. Some of the ones which have been used by the railway industry are listed in Table 6.17.

An integrated package, SAFEM, has been developed by the AAR. It includes three programs: (1) SPAR, a finite element program, (2) RAINFLOW, a peak-and-valley counting program for a time history, and (3) FLAP, a fatigue life calculation program. Given a track input spectrum, SAFEM can perform the fatigue life calculations for freight car structural components.

WHEEL is a three-dimensional stress analysis program used by the AAR to calculate stresses, strains and displacements, at points throughout a wheel, due to mechanical loads and temperature variations. However, since the input data requirements are very bulky and time consuming for three-dimensional analysis, an interface program, GIFWHL, has been developed. This program uses the graphics and mesh generation features of a commercial program, GIFTS, for which minimal input information is needed to generate the two-dimensional mesh and element connectivity. Program GIFWHL uses the GIFTS output to prepare the WHEEL program input. The time saving is very substantial and an element of human error in data preparation is also eliminated. The following details regarding program WHEELS are supplied to give some insight into the nature of the structural mechanics programs of Table 6.17.

Table 6.17. Structural Mechanics Programs.

Source(s)	Program	Author(s)
1. UG, UAN, MSC	NASTRAN	NASA
2. ICES, MCD	STRUDL	ICES
3. UA, UCC, SMSC	GIFTS	H.A. Kamel
4. SAS, RA	ANSYS	J.A. Swanson
5. SDRC, RA	SUPERB	SDRC
6. CDC	STARDYNE	R. Rosen (MRI)
7. USC	SAP IV	K.J. Bathe E.L. Wilson F.E. Peterson (UCB)
8. NASA(L)	SPAR	NASA
9. IBM	STRESS	IBM
10. AAR	SAFEM	AAR
11. AAR	WHEEL	AAR

Table 6.17. (continued)

Source(s)	Program	Author(s)
12. AAR	RAINFLOW	AAR
13. AAR	GIFWHL	AAR
14. BR	NEWPAC	BR

Capabilities

- Finite difference analysis of transient and steady-state temperature distributions.
- Finite element analysis of stesses, strains and displacements in arbitrarily shaped solids.
- Interpolation of nodal temperatures of two-dimensional finite element models.
- Generation of three-dimensional temperature information.

Input

- Geometric dimensional data of a wheel.
- Boundary conditions and initial conditions for both thermal analysis and stress analysis.
- Rates, periods and locations of heat generating sources.
- Mechanical loads.
- Material properties, including Young's modulus of elasticity, Poisson's ratio and linear thermal expansion coefficient.

Output

- Solid element stresses and strains corresponding to each element number and node number.

Restrictions

- The thermal analysis is axisymmetric.
- The material is restricted to a linear elastic stress-strain relationship.

6.7 TABULATIONS

The software survey results are summarized in eighteen tables. The first is a list of these tables for easy reference, whereas the others contain the descriptive information about the computer programs encountered in the survey. Abbreviations, defined in Section 6.8, are used liberally for the sake of conciseness. The various table headings are interpreted as follows:

(a) Source/Model: The 'source' is the name of the organization from which a program can be obtained. Often, the source is followed by numbers in square brackets, denoting pertinent items in the list of references. The 'model' is the name, acronym or description of a computer model (program). The slash '/' is a delimiter used to separate 'source' from 'model' information. It is also used in the fifth column of the tables to separate information regarding items (e) to (j) below.

(b) Author(s): The author(s) responsible for the development of a program are listed.

(c) Usage: The extent of use and/or the names of organizations and individuals using a program are indicated.

(d) Hardware: The computer systems on which a program has been installed are listed.

(e) Vehicle: The type of vehicle (e.g., passenger, freight, locomotive, transit car), to which a program applies, is identified.

(f) Train: The type of train (e.g., passenger, freight), to which a program applies, is specified.

(g) Model Class: The class of model (e.g., linear, nonlinear, quasi-linear, frequency domain, finite element) is indicated. A breakdown of classes of dynamic models is shown in Fig. 6.1.

(h) Degrees of Freedom: The number of degrees of freedom is specified as an indication of the size and complexity of a program.

(i) Users' Manual: The availability of a users' manual is stated.

(j) Validation: The existence or extent of validation of a model against practical test results is indicated.

6.8 ABBREVIATIONS

The following is a list of abbreviations used for conciseness in the software survey tables. The complete mailing addresses of organizations are given wherever possible.

A Author(s)

AAR Association of American Railroads, Technical Center, 3140 South Federal Street, Chicago, Illinois, 60616, U.S.A.

AC	Available in Czech Language.
ASEA	ASEA Traction AB, S-721 73 Västeras, Sweden.
ASF	American Steel Foundaries, 3600 Prudential Plaza, Chicago, Illinois, 60601, U.S.A.
ASU	Arizona State University, Department of Mechanical Engineering, Tempe, Arizona, 85281, U.S.A.
BCL	Battelle Columbus Laboratories, Applied Dynamics and Acoustics Section, 505 King Avenue, Columbus, Ohio, 43201, U.S.A.
BI	Bombardier Inc., Rail and Diesel Power Products Division, 1505 Dickson Street, Montreal, Quebec, Canada, H1N 2H7, Canada
BN	Constructions Ferroviaires et Métalliques S.D. (Brugeoise et Nivelles), rue Montoyer, 1040 Bruselles, Belgium
BPA	Burton Paul Associates, 204 Dodds Lane, Princeton, N.J., 08540 U.S.A.
BR	British Rail.
BUDD	The Budd Company Technical Center, 375 Commerce Drive, Fort Washington, Pennsylvania, 19034, U.S.A.
CDC	Control Data Corporation
CIGGT	Canadian Institute of Guided Ground Transport, Queens' University, Kingston, Ontario, K7L 3N6, Canada
CKD	CKD Lokomotivka - Sokolovo, Ceskomoravska 205, 190 02 Praha 9, CSSR.
CP	Canadian Pacific, Department of Research, Windsor Station, Montreal, Quebec, H3C 3E4, Canada
CU	Clemson University, Department of Mechanical Engineering, Clemson, South Carolina, 29632, U.S.A.
CUED	Cambridge University Engineering Department, Trumpington Street, Cambridge, CB2 1PZ, England
DEC	Ditial Equipment Corporation
DFVLR	Deutsche Forschungs-und Versuchsanstalt für Luft-und Raumfahrt, Multibody Dynamics Group, Institute for Flight Systems Dynamics, German Aerospace Research Establishment, D-8031 Wessling, West Germany.

DG	Data General
DOFASCO	Dofasco Inc., P.O. Box 460, Hamilton, Ontario, L8N 3J5, Canada
DOT	U.S. Department of Transportation, Nassif Building, 400 Seventh Street South West, Washington, D.C., 20590, U.S.A.
DSL	Dynamic Sciences Limited, 359 St. Croix Blvd., Suite 200, St. Laurent, Quebec, H4N 2L3, Canada
DU	Delft University of Technology, Department of Mathematics and Informatics, P.O. Box 356, 2600 AJ Delft, The Netherlands
E	Extensive
F	Freight
FD	Frequency Domain
FE	Finite Element
FRA	Federal Railroad Administration,
G	General
GM	General Motors Corporation, Electro-Motive Division, 9301 55th Street, LaGrange, Illinois, 60525, U.S.A.
I	Idealized
IBM	International Business Machines
ICES	ICES Users Group Inc., P.O. Box 8243, Cranston, Rhode Island, 02920, U.S.A.
IGUTKY	Iguchi Laboratory, Department of Mechanical Engineering, Faculty of Engineering, University of Tokyo, 7-3-1 Hongo, Bunkyo-ku, Tokyo 113, Japan.
IIT	Illinois Institute of Technology, Department of Mechanics, Mechanical and Aerospace Engineering, 3110 South State Street, Chicago, Illinois, 60616, U.S.A.
IR	Indian Railways, Ministry of Railways, LUCKNOW, U.P., 226011, India
ITA	Idealized two-axle vehicle
JWWC	J.W. Wiggins Co., 1650 South Pacific Coast Highway, Redondo Beach, California, 90277, U.S.A.
L	Linear

LF	Linear Frequency Domain
LO	Locomotive
LTD	Limited
MCD	McDonnell Douglas Automation Co., Box 516, St. Louis, Missouri, 63166, U.S.A.
MIT	Massachusetts Institute of Technology, Department of Mechanical Engineering, Cambridge, Massachusetts, 02139, U.S.A.
MITRE	The Mitre Corporation, 1820 Dolly Madison Blvd., McLean, Virginia, 22101, U.S.A.
MRI	Mechanics Research Inc., 9841 Airport Blvd., Los Angeles, California, 90045, U.S.A.
MSC	MacNeal Schwindler Corp., 7442 Figueroa Street, Los Angeles, California, 90041, U.S.A.
N	No
NASA	National Aeronautics and Space Administration, Lewis Research Center, Cleveland, Ohio, 44135, U.S.A.
NASA(L)	NASA Langley Research Center, Hampton, Virginia, 23665, U.S.A.
NI	No information obtained
NL	Nonlinear
P	Partial
PA	Passenger
PAA	Passenger Pendulum Car
PANDROL	Pandrol Limited, 1 Vincent Square, London, SW1P 2PN, United Kingdom
PAT	Passenger truck
PC	Personal Computer
PN	Pullman Standard, Champ Carry Technical Center, 1414 Field Street, Indiana, 46320, U.S.A.
PU	Princeton University, Department of Mechanical and Aerospace Engineering, Princeton New Jersey, 08540, U.S.A.
QL	Quasi-linear
QS	Quasi-static
RA	Remote Access from several commercial computer services

RDI	Railroad Dynamics Inc., 10 East Athens Avenue, P.O. Box 550, Ardmore, Pennsylvania, 19003, U.S.A.
RDSO	Research Designs and Standards Organization, Government of India, Ministry of Railways, Lucknow, U.P., India, 226011, India
REA	Railway Engineering Associates, Transportation and Distribution Associates, 1818 Market Street, Philadelphia, Pennsylvania, 19103, U.S.A.
RIDL	Research Institute for Diesel Locomotives, Ceskomoravska 205, 190 02 Praha 9, CSSR
RV	Rail vehicle
S	Several
SAS	Swanson Analysis Systems, 870 Pineview Drive, Elizabeth, Pennsylvania, 15037, U.S.A.
SBB	Schweizerische Bundesbahnen (Swiss Federal Railways), Bahnhofplatz 10B, 3030 Bern.
SDRC	Structureal Dynamics Research Corporation, 5729 Dragon Way, Cincinnati, Ohio, 45227, U.S.A.
SJ	Statens Järnvägar, Huvudkontoret, Maskinavdelningen, S-105 50 Stockholm, Sweden
SMSS	Structural Mechanics Software Series, University Press of Virginia, Box 3068, University Station, Charlottesville, Virginia, 22903, U.S.A.
SPTC	Southern Pacific Transportation Company, Technical Research and Development Group, One Market Plaza, San Francisco, California, 94105, U.S.A.
SR	Safety research
T	Truck
TA	Transangle, 97 Minot Road, Concord, Massachusetts, 01742, U.S.A.
TASC	The Analytic Sciences Corporation, 6 Jacob Way, Reading, Massachusetts, 01867, U.S.A.
TC	Transit car
TDOP	Truck Design Optimization Project, U.S. Department of

Transportation, Federal Railroad Administration, 400 Seventh Street South West, Washington, D.C., 20590, U.S.A.

THRALL Thrall Car Manufacturing Company, 26th and State Street, Chicago Heights, Illinois, 60411, U.S.A.

TR Train

TSC Transportation Systems Center, U.S. Department of Transportation, Kendall Square, Cambridge, Massachusetts, 02142, U.S.A.

TTD Track Train Dynamics, Association of American Railroads, Technical Center, 3140 South Federal Street, Chicago, Illinois, 60616, U.S.A.

TU Technische Universität Berlin, Institut für Luft and Raumfahrt, Salznfer 17-19, D1000, Berlin 10, West Germany.

TUW Technische Universität Wien, Institut für Mechanik, Karisplatz 13/325, A-1040 Wien.

UA University of Arizona, Tucson, Arizona, 85721, U.S.A.

UAN Universal Analytics

UCB University of California, Berkely, California, U.S.A.

UCC United Computing Systems, 3130 Broadway, Kansas City, Missouri, 64111, U.S.A.

UCL Université Catholique de Louvain, Laboratoire d'automatique, de dynamique et d'analyse des systèmes, Bâtiment Maxwell, Place du Levant, 3 1348 Louvain-La-Neuve, Belgium.

UG University of Georgia, Computer Software Management and Information Center, 112 Barrow Hall, Athens, Georgia, 30602, U.S.A.

UI University of Illinois at Urbana-Champaign, Department of Civil Engineering, Urbana-Champaign, Illinois, U.S.A.

UM University of Massachusetts, Department of Mechanical Engineering, Amherst, Massachusetts, U.S.A.

UNSW	University of New South Wales, School of Metallurgy, Kensington, Sydney, 2033, Australia.
UOP	University of Pensylvania, Deptartment of Mechanical Engineering and Applied Mechanics, 224-D3 Towne Building, Philadelphia, Pennsylvania, 19104, U.S.A.
USC	University of Southern California, Los Angels, California, 90007, U.S.A.
V	Variable
WR	Wheel-rail interaction
WU	Washington University, School of Engineering and Applied Science, St. Louis, Missouri, 63130, U.S.A.
WYLE	Wyle Laboratories, 4620 Edison Street, Colorado Springs, Colorado, 80915, U.S.A.
Y	Yes
1PB	1 degree of freedom per nodal block
1PC	1 degree of freedom per car
2PC	2 degrees of freedom per car
3PC	3 degrees of freedom per car
4PC	4 degrees of freeom per car
5PC	5 degrees of freedom per car
2P5C	2 degrees of freedom per 5 cars
6·N	6·Number of nodes

6.9 SUMMARY

The extensive list of programs presented in this chapter provides a good foundation for a more thorough and up-to-date software survey on simulation in railway dynamics. The addresses of source organizations should make it possible to obtain information on terms and conditions pertaining to the use, or acquisition, of a specific program. The amount of description and reference to the literature are hopefully sufficient to allow for the choice of an appropriate set of programs to be examined further for a given application. Such an investigation should be included in the make-or-buy decision process regarding software acquisition.

REFERENCES

1. ANON. (1976): Technical Documentation - Lateral Train Stability Model; *Association of American Railroads*, Research Report, Chicago.
2. ANON. (1976): Technical Documentation - Lateral Train Stability Model; *Association of American Railroads*, Chicago, U.S.A.
3. ANON. (1978): TDop-Phase I Frequency Domain Model; *U.S. Dept. of Transportation*, Federal Railroad Administration, Report FRA/ORD-78-12, Vol III, Washington, D.C.
4. ANON. (1986): Final report on *Etude du comportement dynamique de véhicules ferroviaires.* Research Program Supported by IRSIA under contract no. 3952, 1983-1987, Brussel, Belgium.
5. BEAMAN J.J. (1976): Freight Car Operations Computational Methods and Engineering Data for Response to Cross-level Variations; *U.S. Department of Transportation*, Cambridge, U.S.A.
6. CHANG E.H., GARG V.K. and GOODSPEED C.H. (1979): Comparative Study on the Linear and Nonlinear Response of Locomotive; *J DYN SYST MEAS CONTROL, TRANS ASME*, Volume 101, No. 3, 263-271.
7. CHANG E.H., GARG V.K. and HARTMAN P.W. (1978): Track Train Dynamics Report - Technical Documentation-Locomotive Response Model; *Association of American Railroads*, Report R-295, Chicago, U.S.A.
8. CHANG C.S., ADEQOKE C.W. and SELIG S.T. (1980): GEOTRACK Model for Railroad Track Performance; *Journal of the Geotechnical Engineering Division*, November 1980.
9. CHATELLE, Th., DUPONCKEEL, J. and SAMIN, J.C. (1984): "Investigation on Non-conventional Railway Systems through a Generalized Multi-Body Approach", *Proc. of 8th IASVD Symposium*, Cambridge, Massachusetts, August 15-19, 1983, 43-57.

10. CHEUNG T.H.W., GARG V.K. and MARTIN G.C. (1977): Track Train Dynamics Report - Users' Manual for Freight Car Hunting Model; *Association of American Railroads*, Report R-251, Chicago, U.S.A.

11. CHIPPERFIELD C.G., GROENHOUT R. and HELLIER A.K. (1983): Analysis of Rail Head Stresses in the Heavy Haul Regime; *Proceedings of Railway Engineering Symposium*, Melbourne, Australia, August 1983, 42-48.

12. COOPERRIDER N.K., LAW E.H., HULL R., KADALA P.D. and TUTEN J.M. (1975): Analytical and Experimental Determination of Wheel-Rail Constraint Relationships; *U.S. Department of Transportation*, Federal Railroad Administration, Report FRA-OR&D-76/244, Washington, D.C.

13. COX S.J. and GRASSIE S.L. (1984): The Dynamic Response of Railway Track with 'Hung' Sleepers to High Frequency Vertical Excitation; *Pandrol Limited*, London.

14. DEEPAK D. and SINGH V.V. (1984): Half Space Theory for Evaluation of Track Stiffness and Track Damping; *Rail International*, 1984.

15. DENOVCHEK J.H. and COOPERRIDER N.K. (1978): Locomotive Dynamic Curving Analysis Program - Users' Manual; *U.S. Dept. of Transportation*, Federal Railroad Administration, Interim Report, Washington, D.C.

16. EHRENBECK R. and POLCARI S. (Eds.) (1984): Vehicle/Track Interaction Assessment Techniques; *U.S. Department of Transportation*, Federal Railroad Administration, Report DOT/FRA/ORD-84/01.2, Washington, D.C.

17. ESTEN S. and BANTMAN R. (1976): Track Train Dynamics Report - Users' Manual - Modified Quasi-Static Lateral Train Stability Program; *U.S. Dept. of Transportation*, Transportation Systems Center, Report AAR-R-207, Cambridge, U.S.A.

18. FALLON W.J., COOPERRIDER N.K. and LAW E.H. (1978): An Investigation of techniques for validation of Railcar Dynamic Analyses; *U.S. Dept. of Transportation*, Federal Railroad Administration, Report FRA-OR&D-78-19, Washington, D.C.

19. GARG V.K., MARTIN G.C., HARTMANN P.W. and TOLOMEI J.G. (1976): Track Train Dynamics Report - Technical Documentation-Locomotive Truck Hunting Model; *Association of American Railroads*, Report R-219, Chicago, U.S.A.

20. GOGINENI B.R. and GARG V.K. (1978): A Dynamic Model for Longitudinal Train Action; *Association of American Railroads*, Report R-297, Chicago, U.S.A.

21. GOREE J. and LAW E.H. (1977): Users' Manual for Kalker's Simplified Nonlinear Creep Theory; *U.S. Department of Transportation*, Federal Railroad Administration, Report FRA-OR&D-78/06, Washington, D.C.

22. GOREE J. (1978): Users' Manual on Kalker's Exact Nonlinear Creep Theory; *U.S. Department of Transportation*, Federal Railroad Administration, Report FRA/ORD-78/50, Washington, D.C.

23. GRASSIE S.L., GREGORY R.W., HARRISON D. and JOHNSON K.L. (1982): The Dynamic Response of Railway Track to High-Frequency Vertical Excitation; *Journal of Mechanical Engineering Science*, (24), No. 2, 77-90.

24. GRASSIE S.L., GREGORY R.W., and JOHNSON K.L. (1982): The Dynamic Response of Railway Track to High-Frequency Lateral Excitation; *Journal of Mechanical Engineering Science*, (24), No. 2, 91-95.

25. GRASSIE S.L., GREGORY R.W., and JOHNSON K.L. (1982): Dynamic Response of Railway Track to High-Frequency Longitudinal Excitation; *Journal of Mechanical Engineering Science*, (24), No. 2, 97-102.

26. GRASSIE S.L. and COX S.J. (1984): The Dynamic Response of Railway Track with Flexible Sleepers to High Frequency Vertical Excitation; *Proc. of Institution of Mechanical Engineers*, Part D.

27. HARTMANN P.W. and GARG V.K. (1977): Track Train Dynamics Report - Programmers' Guide - Locomotive Truck Hunting Model; *Association of American Railroads*, Report R-278, Chicago, U.S.A.

28. GRASSIE S.L. and JOHNSON K.L. (1984): Periodic Micro-Slip Between a Rolling Wheel and a Corrugated Rail; *Cambridge University Engineering Department*, Cambridge, England.

29. HASSELMAN T.K., BRONOWICKI A. and HART G.C. (1975): DYNALIST-II, A Computer Program for Stability and Dynamic Response Analysis of Rail Vehicle Systems; *U.S. Department of Transportation*, Federal Railroad Administration, Report FRA-OR&D-75-22-I, Washington, D.C.

30. HEALY J.J. (1976): A Computer Method for Calculating Dynamic Responses on Nonlinear Flexible Rail Vehicles; *ASME-IEEE Joint Railroad Conference*, Paper 76-RT-5, Chicago, U.S.A., April 1976.

31. HEDRICK J.K., COOPERRIDER N.K. and LAW E.H. (1978): Application of Quasi-Linearization to Rail Vehicle Dynamic Analyses; *U.S. Dept. of Transportation*, Federal Railroad Administration, Report FRA/OR&D-78-56, Washington, D.C.

32. HEDRICK J.K. and ARSLAN A.V. (1978): An Analytical Method for Nonlinear Rail Vehicle Design; *ASME Winter Annual Meeting*, San Fransisco, U.S.A.

33. HEDRICK J.K., WORMLEY D.N., KAR A.K., MURRAY W. and BAUM W. (1979): Performance Limits of Rail Passenger Vehicles - Evaluation and Optimization; *U.S. Dept. of Transportation*, Office of University Research, Report DOT-SPA-DPB-50-79-32, Washington, D.C.

34. HEDRICK J.K. WORMLEY D.N., ARSLAN A.V. and CHIN R. (1980): Nonlinear Analysis and Design Tools for Rail Vehicles; *Association of American Railroads*, Report R-463 Chicago, U.S.A.

35. HELLER R. and COOPERRIDER N.K. (1977): Users' Manual for Asymmetric Wheel-Rail Contact Characterization Program; *U.S. Dept. of Transportation*, Federal Railroad Administration, Report FRA-ORD-78/05, Washington, D.C.

36. KALKER J.J. (1979): Variational Principles of Contact Elastostatics; *J. Inst. Maths Applics*, Vol. 20, 199-219.

37. KALKER J.J. (1979): Computation of Three Dimensional Rolling

Contact with Dry Friction; *Int. J. Num. Meth. Eng.*, Vol. 14, No. 9, 1293-1307.

38. KALKER J.J. (1982): The Contact Between Wheel and Rail; *Delft University of Technology*, Dept. of Mathematics, Report 82-87, Delft, The Netherlands.

39. KALKER J.J. (1982): A Fast Algorithm for the Simplified Theory of Rolling Contact: *Vehicle System Dynamics*, Vol. 11, No. 1, 1-13.

40. KORTUM W. (1982): Modeling and Simulation of Actively Controlled Mechanical Systems; *Proceedings of the 10th IMACS on World Congress on System Simulation and Scientific Computations*, North-Holland Publishing Company, 91-101.

41. LAW E.H., HADDEN J.A. and COOPERRIDER N.K. (1977): General Models for Lateral Stability Analyses of Railway Freight Vehicles; *U.S. Dept. of Transportation*, Federal Railroad Administration, Report FRA-OR&D-77-36, Washington, D.C.

42. LAW E.H. and COOPERRIDER N.K. (1978): Nonlinear Dynamic and Steady-State Curving of Rail Vehicles; *ASME Winter Annual Meeting*, San Francisco, U.S.A., December 1978.

43. LIEPENS A.A. (1965): Digital Computer Simulation of Railroad Freight Cars; *ASME Winter Annual Meeting*, Paper 65-WA/RR-8, August 1965.

44. LOW E.M. and GARG V.K. (1977): Technical Documentation, Train Operations Simulator; *Association of American Railroads*, Report R-269, Chicago, U.S.A.

45. LOW E.M. and GARG V.K. (1977): Detailed Longitudinal Train Action Model - Programming Manual; *Association of American Rialroads*, Report R-296, Chicago, U.S.A.

46. LOW E.M. and GARG V.K. (1978): Train Operations Simulator-Validation; *Association of American Railroads*, Report R-335, Chicago, U.S.A.

47. LUTTRELL N.W., GUPTA R.K., LOW E.M. and MARTIN G.C. (1977): Users' Manual - Train Operations Simulator; *Association of American Railroads*, Report R-198, Chicago, U.S.A.

48. MARTIN G.C. and TIDEMAN H. (1977): Track Train Dynamics

Report - Detailed Longitudinal Train Action Model - Technical Documentation; *Association of American Railroads*, Report R-221, Chicago, U.S.A.

49. MARTIN G.C., PLOUFFE W.E., AHMED S., ANTCZAK H. and TIDEMAN H. (1976): Track Train Dynamics Report - Detailed Longitudinal Train Action Model - Users' Manual; *Association of American Railroads*, Report R-220, Chicago, U.S.A.

50. MARCOTTE P.P., MATHEWSON J.J.R. and YOUNG R. (1981): A Practical Model of Rail Vehicle Curve Negotiation; *Proc. of the 7th Symposium of the International Association for Vehicle System Dynamics*, Vol. 10, Cambridge, England, September 1981.

51. MARCOTTE P.P., CALDWELL W.N. and LIST H.A. (1982): Performance Analysis and Testing of a Conventional Three-Piece Freight Car Truck Retrofitted to Provide Axle Steering; *ASME Journal of Dynamic Systems, Measurement, and Control*, Vol. 104, No. 1, 93-99.

52. MARCOTTE P.P. and MATHEWSON K.J.R. (1982): Track Train Dynamics Analysis and Test Program - Methodology Development for the Derailment Safety Analysis of Six-Axle Locomotives; *Martin-Marietta Areospace Corporation*, Report NASA-CR-162026, Denver, U.S.A.

53. MCILVEEN E.R., BIRK A.M., YOUNG J.D. and GLASER R. (1981): Rail Wear Cost Model - Users' Manual; *Transport Canada*, Report TP-3727E, Montreal, Canada.

54. MEACHAM H.C., PRAUSE R.H., AHLBECK D.R. and KASUBA J.A. (1970): Studies for Rail Vehicle Track Structures; *Battelle Memorial Institute*, Report FRA-RT-71-45, Columbus, Ohio, U.S.A.

55. NAYAK L. and JOHNSON K.L. (1979): Pressure Between Elastic Bodies Having a Slender Area of Contact and Arbitrary Profiles; *Int. J. Mech. Sci.*, Vol. 21, No. 4, 237-247.

56. PAUL B. and HASHEMI J. (1977): Users' Manual for Program COUNTACT (Counterformal Contact Stress Problems); *U.S. Dept.*

of Transportation, Federal Railroad Administration, Report FRA/ORD-78/27, Washington, D.C.

57. PERLMAN A.B. and DIMASI F.P. (1975): Frequency Domain Computer Programs for Prediction and Analysis of Rail Vehicle Dynamics - Vol. 1, technical Report; *U.S. Department of Transportation*, Federal Railroad Administration, Report FRA-OR&D-76-135-I, Washington, D.C.
58. PAUL B. and HASHEMI J. (1978): Users' Manual for Program CONFORM (Conformal Contact Stresses Between Wheels and Rails); *U.S. Department of Transportation*, Federal Railroad Administration, Report FRA/ORD-78/40, Washington, D.C.
59. PETERS D.A. and YIN S.K. (1976): Non-Destructive Impact Between Railroad Cars - Experimental and Analytical Study; *U.S. Dept. of Transportation*, Federal Railroad Administration, Report FRA-OR&D-76/247, Washington, D.C.
60. PILKEY W.D. and PILKEY B.F. (1979): Directory-Structural Mechanics Programs; *Association of American Railroads*, Track Train Dynamics Report R-366, Chicago, U.S.A.
61. PILKEY W.D. (1981): Review and Summary of Computer Programs for Railway Vehicle Dynamics; *U.S. Department of Transportation*, Federal Railroad Administration, Report FRA/OR&D-81/17, Washington, D.C.
62. PLATIN B.E. (1978): Dynamic Analysis and Stabilization Control of Freight Car Rocking; *ScD Dissertation*, Massachusetts Institute of Technology, Cambridge, U.S.A.
63. PRAUSE R.H. and KENNEDY J.C. (1977): Parametric Study of Track Response; *U.S. Department of Transporation*, Federal Railroad Administration, Report FRA-ORD-77-75, Washington, D.C.
64. RAIDT J.B., SHUM K.L., MARTIN G.C. and GARG V.K. (1977): Track Train Dynamics Report, Technical Documentation - Detailed Vertical Stability Models; *Association of American Railroads*, Report R-261, Chicago, U.S.A.
65. RAYMOND G.P. and TURCKE D.J. (1978): Analysis of Rail Track

Structures (ARTS) Users' Manual; *Canadian Institute of Guided Ground Transportation*, Kingston, Ontario, Canada.

66. RAYMOND G.P., TURCKE D.J. and SIU D.W. (1979): Analysis of Rail Track Structures (ARTS) Users' Manual; *Canadian Institute of Guided Ground Transportation*, Report TP-2603, Kingston, Ontario, Canada.
67. SAMIN, J.C., WILLEMS, P.Y. (1985): "Multibody Formalism Applied to Non-Conventional Railway Systems; *Proc. of IUTAM/IFToMM Symposium*, Udine, 237-248.
68. SHAW N. and D'SOUZA A.F. (1981): Interactive Simulation and Computer Graphics for the Detailed Longitudinal Train Action Model; *Association of American Railroads*, Report R-492, Chicago, U.S.A.
69. SHUM K.L. and WILLIS T. (1977): A Mathematical-Computer Simulation of the Dynamics of a Freight Element in a Railroad Freight Car; *U.S. Department of Transportation*, Report FRA/OR&D-77/28, Washington, D.C.
70. SMITH K.R., MACMILLAN L.D. and MARTIN G.C. (1976): Track Train Dynamics Report, Technical Documentation - 2, 3, and 4 Axle Rigid Truck Curve Negotiation Model; *Association of American Railroads*, Report R-206, Chicago, U.S.A.
71. SO W., MARTIN G.C., SINGH B., CHANG I.C. and CHANG E.H. (1977): Mathematical Models for Track Structures; *Association of American Railroads*, Chicago, U.S.A.
72. SUDA, Y., FUJIOKA, T., IGUCHI, M. (1986): "Dynamic Stability and Curving Performance of Railway Vehicles", *Bulletin of JSME*, Vol. 29, No. 256, October, 1986.
73. SWEET L.M. and KARMEL A. (1981): Analytical and Experimental Studies of Derailment Process of Railway Vehicles; *Doctoral Thesis*, Princeton University, New Jersey, U.S.A.
74. TANNE J. and BRANTMAN R. (1976): Users' Manual - Train Stability Analysis Program; *U.S. Dept. of Transportation*, Transportation Systems Centre, Cambridge, Massachusetts, U.S.A.

75. TAYABJI S.D., THOMPSON M.R. and ROBNETT A.L. (1976): Program Illi-Track - A Finite Element Analysis of Conventional Railway Track Support Systems-Users' Manual & Program Listing; *University of Illinois*, Dept. of Civil Engineering, Illinois, U.S.A.

76. TAYABJI S.D., THOMPSON M.R. and ROBNETT A.L. (1976): Track Support Systems Parameters Study-Ballast & Foundation Materials Research Program; *University of Illinois*, Dept. of Civil Engineering, Illinois, U.S.A.

77. THOMAS L.R., MACMILLAN R.D. and MARTIN G.C. (1976): Track Train Dynamics Report, Users' Manual - Quasi-Static Lateral Train Stability Model; *Association of American Railroads*, Report R-207, Chicago, U.S.A.

78. TSE Y.H. and MARTIN G.C. (1976): Users' Manual-Flexible Body Railroad Freight Car; *Association of American Railroads*, Report R-200, Chicago, U.S.A.

79. WORMLEY D.N. (1978): Computational Performance Models for Passenger Truck Design; *U.S. Dept. of Transportation*, Transportation Systems Centre, Report on contract DOT-TSC-1471, Washington, D.C.

80. YIN S.K. (1980): Theoretical Manual and Users' Guide: Longitudinal - Vertical Train Action Model; *U.S. Dept. of Transportation*, Federal Railroad Administration, Report FRA/ORD-76/278, Washington, D.C.

81. YOUNG R. and MARCOTTE P.P. (1981): Curving Performance Analysis of Freight Car Trucks; *Transport Canada*, Report TP-3211, Montreal, Canada.

7

TRACK SIMULATORS

7.1 INTRODUCTION

The purpose of this chapter is to: (1) introduce physical track simulators through the medium of a specific case, namely, a curved-track simulator (CTS) under development at the National Research Council of Canada (NRCC), and (2) present a survey of existing and planned simulators.

The CTS is designed to simulate the dynamical behaviour of single railway vehicles traveling along prescribed test track layouts in accordance with velocity profiles specified as functions of along-track vehicle position. Acceptable test track layouts can include straight (tangent) segments, simple curves and transition curves with superelevation but are restricted to zero grade. It replaces a test track by rollers which simulate the receding speed of rails relative to the vehicle which remains stationary except for motions associated with guidance. Each wheelset is supported by one roller stand with three variables in need of control: left roller speed, right roller speed, and yaw angular position of rollers. We define left and right from the viewpoint of an observer sitting in the vehicle and facing in the intended travel direction. Also, inertial moments and forces such as centrifugal and Coriolis forces, which occur naturally on curved track, need to be applied to the vehicle masses in the CTS and controlled according to calculations based on vehicle masses, inertias, velocity and track layout. The hardware includes roller drive motors and actuators to control roller yaw angles and apply inertial forces and moments. The motors and actuators will be controlled in real time by an on-line computer(s).

Matsudaira et al. [8] have reported on tangent track simulator studies of the Japanese National Railways. They concluded that full-scale tangent track simulator tests may be utilized as a substitute for field tests. In a discussion of their paper, Wickens [8] stated that the British Railway Research Department had found agreement, to within about 10-15 percent, in the critical speed results of tangent track simulator and field tests. Therefore, there is some degree of validation of full-scale tangent track simulators which, hopefully, will be extended to the broader context of curved track simulators.

The above authors indicated that, due to similarity problems, it is difficult to predict the critical speed of an actual vehicle from sub-scale simulator tests. Nevertheless, the 1/5-scale and 1/10-scale model tests of the Japanese National Railways were deemed effective for elaborate observations of hunting behaviour, investigation into the basic nature of hunting, and search for methods of prevention.

The same usefulness applies to full-scale simulator tests, which are likewise easier and safer to perform than high-speed field tests. Compared with purely computational simulation, they have the advantage of not requiring a mathematical model of the tested vehicle. Potential applications of a validated CTS include testing for: (a) dynamic performance of vehicles, (b) derailment proneness, and (c) vibration damage to vehicles and cargo.

We begin by describing the CTS architecture (Section 7.2). A general analytical foundation of curved track simulators is presented in Section 7.3. Control, instrumentation and validation aspects of the CTS are discussed in Sections 7.4 and 7.5. A survey of existing and planned physical track simulators is given in Section 7.6, and a convenient list of symbol definitions can be found in Section 7.7.

7.2 ARCHITECTURE

The full-vehicle CTS would include all four roller stands. Figure 7.1 illustrates a side elevation of a vehicle held longitud-

Fig. 7.1 Full-vehicle CTS (Adapted from Smith and Dukkipati [10]).

inally on four stands by a single end-of-car restraint. Otherwise, the vehicle is free to move laterally and vertically, and to roll, pitch and yaw within the kinematic constraints of wheel-roller contact. In case of derailment, a harness arrangement would prevent the vehicle from falling.

Some aspects of vehicle dynamics may be amenable to study by means of a half-vehicle CTS in which one truck is supported by a track structure (Fig. 7.2) or a crane (Fig. 7.3). For example, it may be possible to study steady-state curving performance, derailment proneness, and critical speed of hunting of a truck. However, in the interest of presenting the general principles of track simulators, we will concentrate on the full-vehicle CTS.

The inertial forces and moments, such as centrifugal forces, which occur naturally in the field due to track curvature, must be applied artificially by means of actuators in the simulator. Aerodynamic forces and gravitational forces, if the rollers are not

Fig. 7.2 Half-vehicle CTS with one truck supported by a track structure (Adapted from Smith and Dukkipati [10].

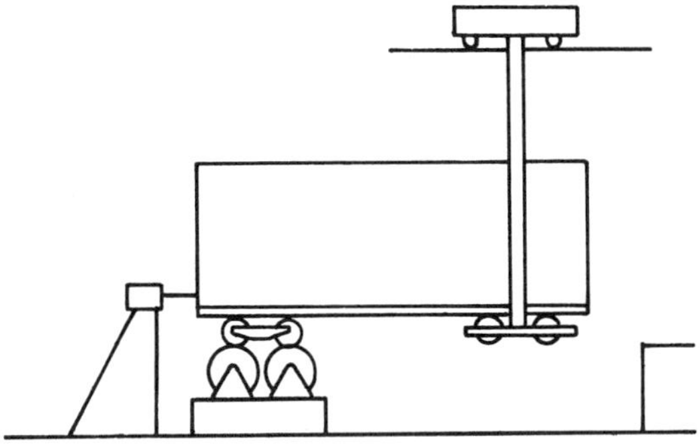

Fig. 7.3 Half-vehicle CTS with one truck supported by a crane (Adapted from Smith and Dukkipati [10]).

superelevated in accordance with track superelevation, may also be involved. Figure 7.4 demonstrates the application of correction forces F_B, F_T, F_{W1} and F_{W2} to the carbody, truck and two wheelsets, respectively.

7.2.1 Roller Stands

A perspective diagram of a pair of CTS roller stands appears in Fig. 7.5. The entire weight of rollers, axles, gear boxes and motors is supported by floating arm castings with hydro-static bearings which slide easily on smooth granite surfaces. A lever system allows the axle assemblies to be controlled for curving track simulations. The longitudinal and lateral track forces are transmitted to the building walls by means of a transverse *earth* beam.

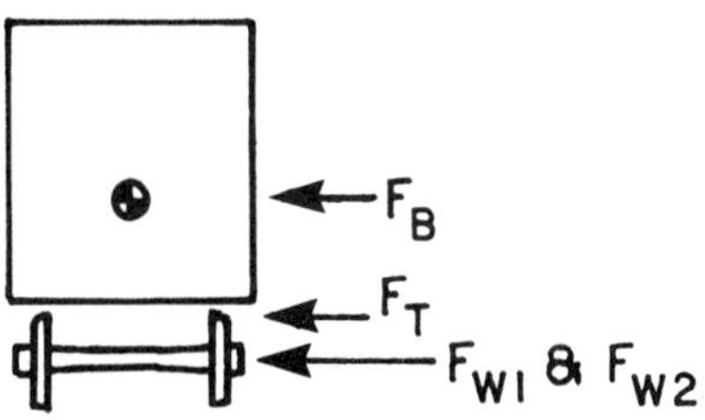

Fig. 7.4 Application of CTS correction forces (Adapted from Smith and Dukkipati [10]).

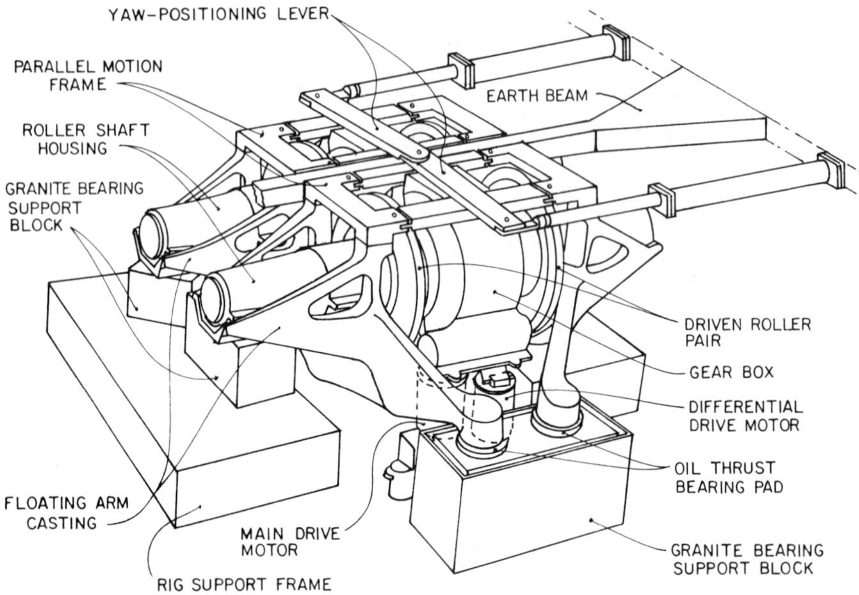

Fig. 7.5 Pair of CTS roller stands (Adapted from Smith and Dukkipati [10]).

A roller-to-roller torsional stiffness of approximately ten times that of a wheelset is achieved by means of a differential gear system and two constant velocity couplers (Figs. 7.6 and 7.7). The couplers are required for two major reasons: (a) In tangent track simulations, the roller cant angles must be matched with the wheel conicities. This is accomplished by tilting each axle stub about its constant velocity coupler and positioning the roller rims such that the contact points are centered on the wheel treads. Plans are to accommodate wheel conicities of 0, 1/40 and 1/20. (b) Additionally, in curving track simulations and to allow for various truck types, it is necessary to be able to move each roller longitudinally in order to keep it centered under its corresponding wheel. A range of 8 cm (3 in.) is provided.

Attached to the lower surface of each differential gear box is a 112 kW (150 HP) main drive motor and a 37 kW (50 HP) differential

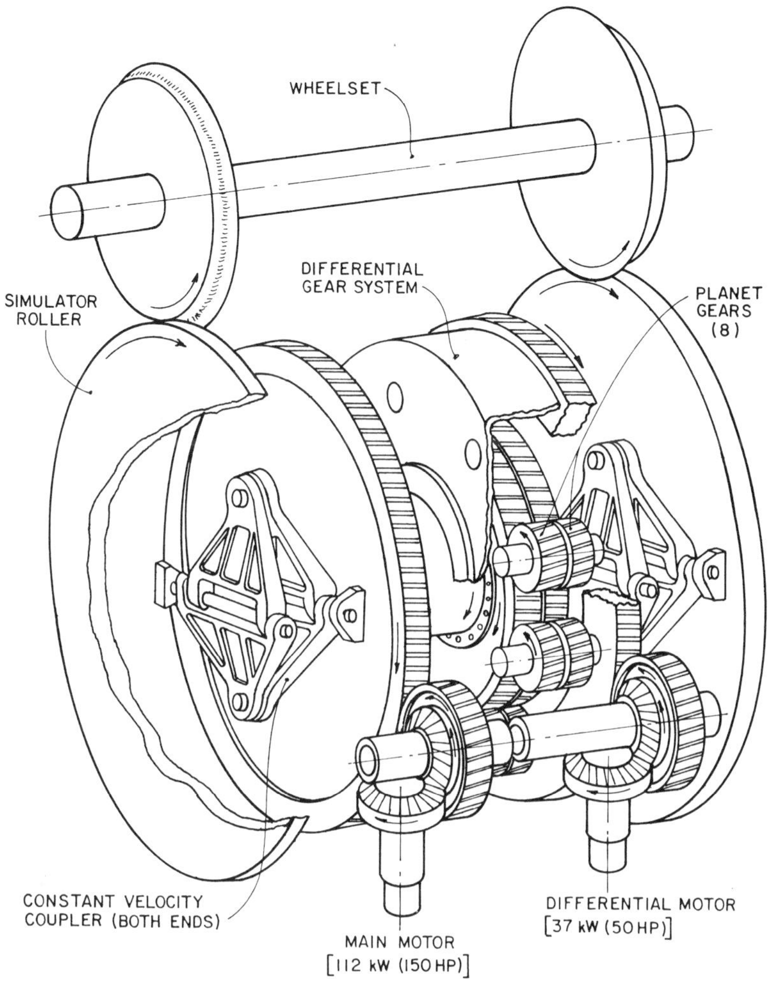

Fig. 7.6 Cut-away sketch of CTS drive train (Adapted from Smith and Bowler [9]).

drive motor. The speed difference between rollers in curving track simulations is obtained by driving the differential gear cage with the differential motor at speeds shown in Fig. 7.8 as percentages of main motor speeds. For tangent track, the differential/main motor speed ratio is 100%; the differential and main motor speeds

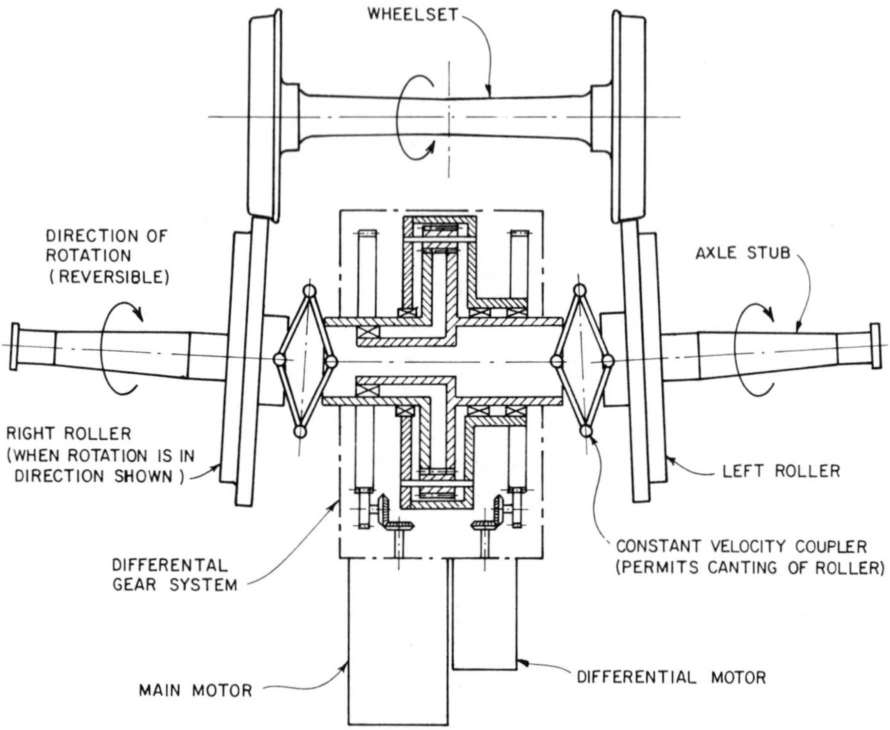

Fig. 7.7 Schematic diagram of CTS drive train (Adapted from Smith and Dukkipati [10]).

are equal. Due to roller stand lever system limits, the maximum right or left curvatures is 14 degrees per 30.48 m (100 ft) chord. The differential motor speeds are respectively 53% and 146% of the main motor speed for the corresponding right and left curves. Due to differential motor speed limits, the corresponding maximum track (i.e., vehicle) speeds which can be simulated are 240 km/h (150 mph) and 164 km/h (103 mph); the minimum track speeds are 9 km/h (5.6 mph) and 20 km/h (12.5 mph), respectively. These lower limits are due to the minimum main motor speed which can be achieved, taking the differential motor loading effect into account. The CTS operational limits defined by Fig. 7.8 show that the right and left curvatures which can be simulated lie between 0 and 14 degrees per

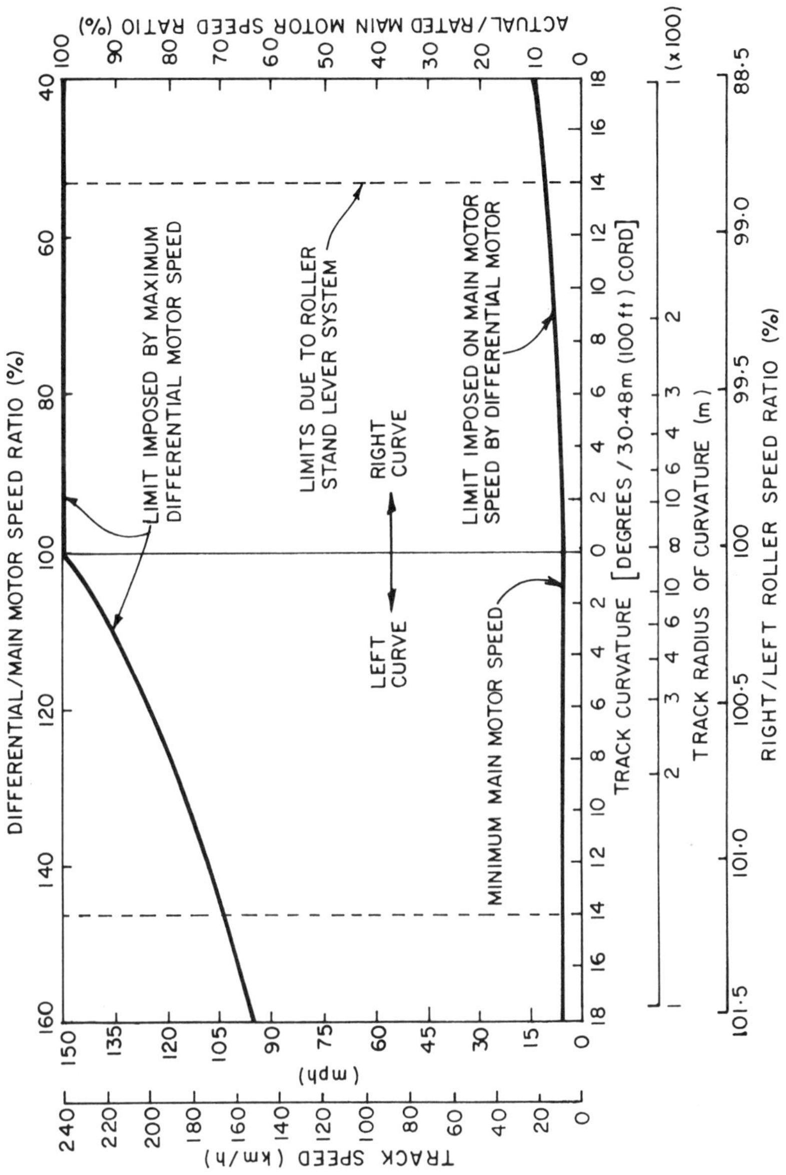

Fig. 7.8 CTS operational limits (Adapted from Smith and Dukkipati [10]).

30.48 m (100 ft) chord; track speeds can vary between 9 km/h (5.6 mph) and 240 km/h (150 mph), depending on the differential and main motor speed limits which, in turn, are functions of track curvature.

7.2.2 Control Systems

The four main motors of a fully-develped CTS are controlled to run at the same speed by means of a digital tachometer signal fed back from each right axle and compared with a desired reference speed signal to produce an error signal which is used to adjust the speed of the corresponding dc motor. On the other hand, depending on track curvature, the four differential motors are not necessarily controlled to run at the same speed. Each one is controlled by an error signal which is the difference between a digital tachometer signal from the corresponding left roller and a desired reference speed signal derived from Fig. 7.8. Thus, track speed and curvature are set independently by means of the desired reference speed signals applied to the main and differential motor control systems, respectively.

An overview diagram of the overall CTS control system is shown in Fig. 7.9 which highlights the front roller stand. A computer is required to compute the desired values of variables acted upon by controllers of the various motors and actuators. The computer performs these computations continuously as the vehicle moves hypothetically along the track. It computes vehicle velocity and actually-achieved track curvature from main and differential motor speed measurements and integrates the velocity to determine along-track position. The position and velocity are used in conjunction with the prescribed velocity mission, test track layout and vehicle data to determine desired values of: (1) main motor speed, (2) ratio of differential to main motor speed, (3) roller stand yaw angle, and (4) inertial moments and forces. These desired values are then acted upon by actuators and motors and their controllers.

In addition, data can be generated to produce two-dimensional displays of: (a) desired and actually-achieved test track layouts,

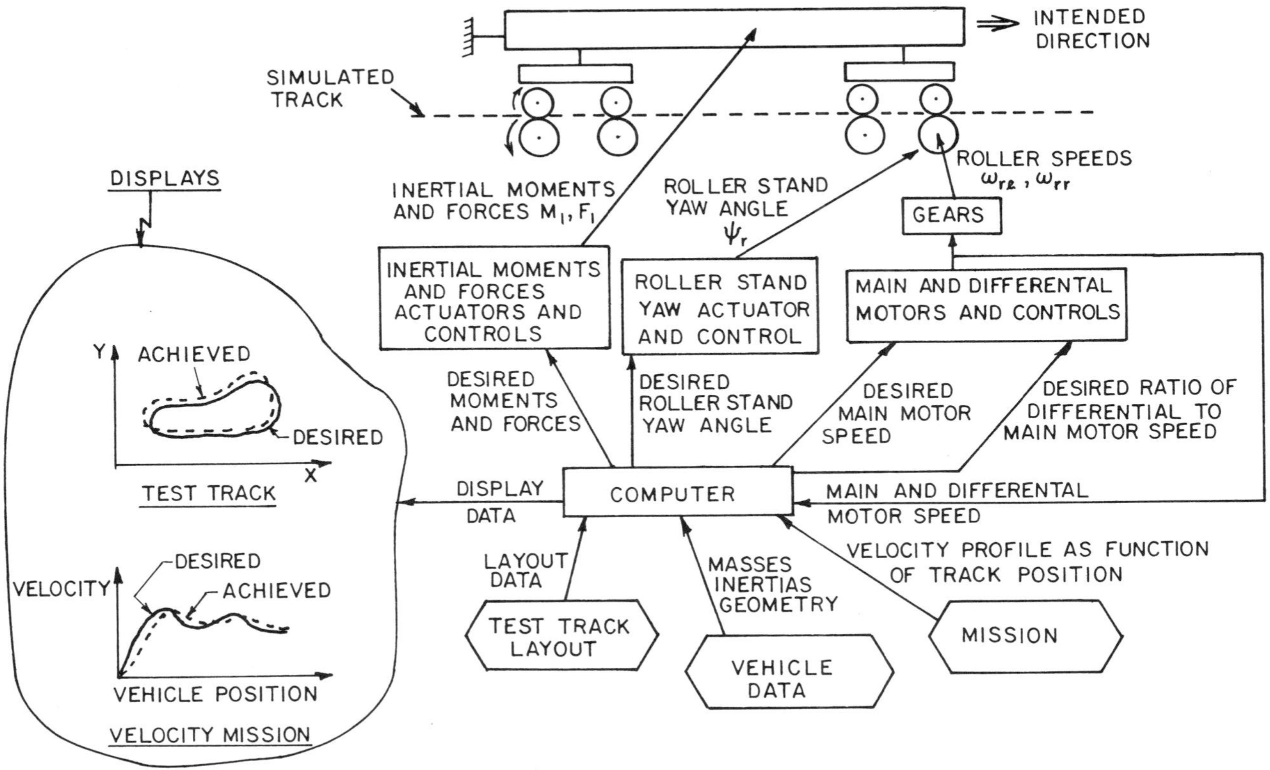

Fig. 7.9 Overview diagram of CTS control (Adapted from Amyot [5]).

and (b) desired and actually-achieved velocity missions. Various dynamical display options are conceivable. For example, a desired track layout and velocity mission could be pre-plotted to serve as visual reference for the actually-achieved loci which would then be plotted dynamically in real time as the vehicle would move hypothetically along the track.

7.3 ANALYTICAL FOUNDATION

Simulated track geometry and velocity mission data are the bases of control action in roller-type curved track simulators.

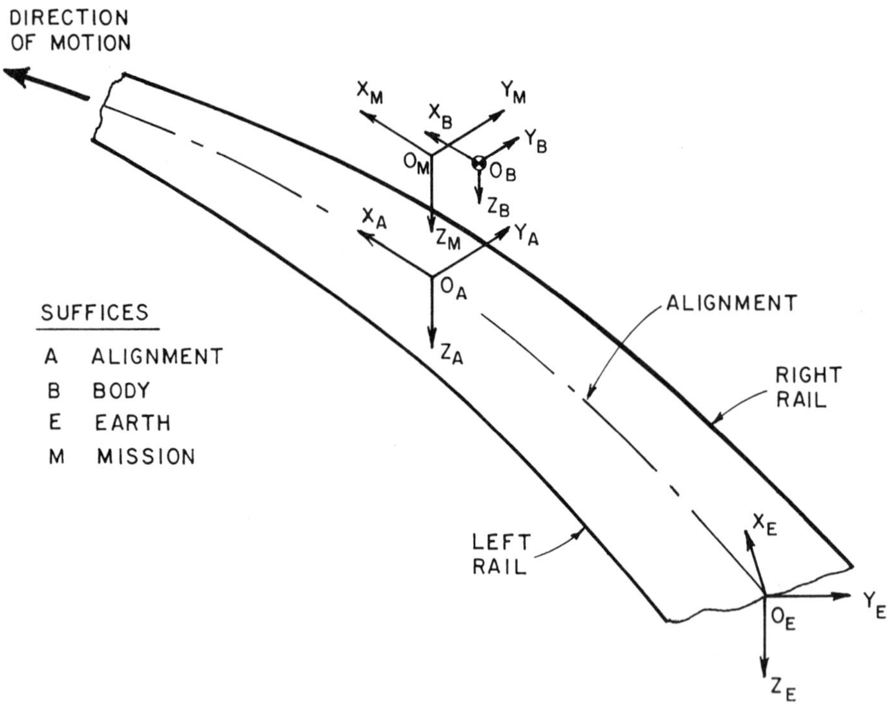

Fig. 7.10 Illustration of coordinate systems (Adapted from Ho et al. [7]).

The required speed and orientation of rollers are derived directly from these data and the dimensions of wheels and rollers. The required actuator forces are derived indirectly through vehicle dynamics.

7.3.1 Coordinate Systems

The terms *track centre-line* and *track alignment* are used interchangeably to define the locus of points midway between the rails on the set of straight lines intersecting both rail surface center lines at right angles. The set of coordinate systems illustrated in Fig. 7.10 is established as reference for the track geometry and vehicle dynamics equations. Thus, four right-handed Cartesian coordinate systems with their Z axes pointing downwards are defined (Figs. 7.11 and 7.12).

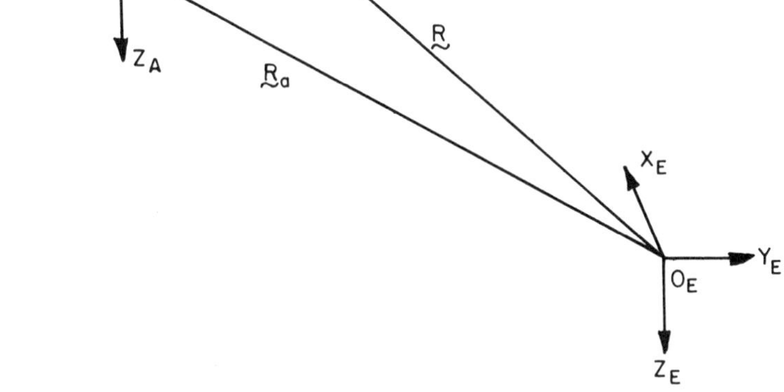

Fig. 7.11 Relative positions of alignment and mission axes (Adapted from Ho et al. [7]).

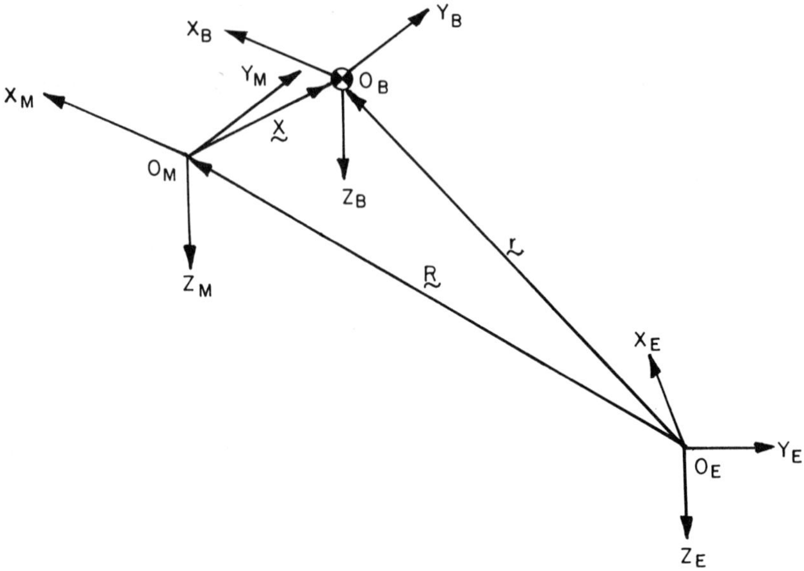

Fig. 7.12 Relative positions of mission and body axes (Adapted from Ho et al. [7]).

(1) Earth axes: Neglecting the earth's motion, they are fixed in space with origin 0_E preferably at a fixed point on the track center-line; the X_E axis is tangent to the center-line, pointing in an assumed direction of travel, and Y_E is parallel to the ties.

(2) Alignment axes: Alignment axes move with the vehicle such that their origin 0_A is the point where a plane through the body's geometric center meets the alignment normally; the X_A axis is tangent to the alignment, pointing in the direction of motion, and Y_A is parallel to the ties. The track can be completely specified by the translational and rotational position of the alignment axes relative to the earth axes as a function of along-track center-line distance, ℓ.

(3) Mission axes: A given mission reference frame (mission axes) is parallel to the alignment reference frame (alignment axes) and its origin is fixed relative to it. In practice, a distinct mission frame may be defined for each rigid body such that its

origin coincides with the body's center of mass when the vehicle is at rest on tangent track. However, it may be more convenient to choose only one mission frame for a given subset of vehicle rigid bodies. Dynamical equations of translational motion are written relative to mission axes so that displacements are relative to nearby nominal positions.

(4) Body axes: They are principal axes whose origins 0_B coincide with the rigid bodies' centers of mass. Dynamical equations of rotational motion are written relative to body axes.

In tangent track simulations, the alignment reference frame is fixed relative to the CTS building, with X_A parallel to the walls and floor, Y_A parallel to the floor, and 0_A directly below the carbody's geometric center. Then, the vehicle's motions in the CTS environment are equivalent to its local motions relative to alignment axes in the field.

In curved track simulations, the effects of global motions relative to earth axes are accounted for by the application of inertial forces and moments using actuators. Furthermore, since a long vehicle defines a chord on a curve, any resulting deviations of alignment axes from carbody axes must be considered in the interpretation of CTS results. The CTS building and carbody reference frames are nominally parallel and any motion of the latter relative to the former is limited to small values by the wheel-roller contact constraints.

7.3.2 Track Geometry

It will be seen that the translational acceleration (Eq. (7.20)) and angular velocity and acceleration of mission axes (Eqs. (7.21) and (7.22)) required in the dynamical equations of motion of a given vehicle rigid body (Eqs. (7.15) to (7.19)) are expressed in terms of four three-dimensional vector functions $\utilde{E}$, $\utilde{F}$, $\utilde{G}$ and $\utilde{H}$ which are, in turn, functions of $\utilde{x}_{am}$ and $\utilde{\alpha}_{me}$, $\utilde{\alpha}'_{me}$, $\utilde{\alpha}''_{me}$ either directly (Eqs. (7.23) to (7.26)) or indirectly via $\utilde{R}_a^{(e)}$, $\utilde{R}_a^{(e)''}$ (Eqs. (7.14) and (7.27)) and transformations $[T_{me}]$, $[A_{me}]$,

$[A'_{me}]$ (Eqs. (7.28) to (7.30)). The angular position of mission axes relative to earth axes

$$\underset{\sim}{\alpha}_{me} = [\phi_a \ \theta_a \ \psi_a]^T \tag{7.1}$$

and its distance derivatives

$$\underset{\sim}{\alpha}'_{me} = [\phi'_a \ \theta'_a \ \psi'_a]^T \tag{7.2}$$

$$\underset{\sim}{\alpha}''_{me} = [\phi''_a \ \theta''_a \ \psi''_a]^T \tag{7.3}$$

are uniquely defined by the track as functions of along-track distance ℓ. This Euler-angle vector representation assumes a yaw, pitch and roll order of rotation in conformity with civil engineering practice. A track is designed by a three-step procedure: (1) projection, which defines the yaw angle ψ_a as a function of horizontal distance L, thereby determining the horizontal layout, (2) development, which defines the pitch angle θ_a as a function of along-track distance ℓ, thereby determining the vertical layout, and (3) superelevation, which defines the bank angle ϕ_a as a function of horizontal distance L.

Thus, a track may be defined to consist of a sequence of vertical segments (grade or vertical curve) superimposed on a sequence of horizontal segments (tangent, curve, tangent-to-curve spiral, curve-to-tangent spiral, or curve-to-curve spiral). Then, the above angular position and its derivatives are given by the expressions in Tables 7.1 and 7.2.

As an example of the derivation of these expressions and others, let us consider the case of Fig. 7.13, i.e., a (horizontal) tangent-to-curve spiral for the projection combined with a (vertical) curve for the development. For simplicity, we shall neglect subscript 'a' in the derivation. Since

$$dL = R_H d\theta \tag{7.4}$$

and

$$R_H = R_s L_s / L \tag{7.5}$$

for a spiral segment (Fig. 7.13(b)), we have

Table 7.1. Vertical layout expressions.

Variable	Vertical Feature	
	Grade (G)	Vertical Curve (VC)
L	$\ell\cos\theta$	$R_V(\sin\theta-\sin\theta_o)$
ℓ	$L/\cos\theta$	$R_V\{\sin^{-1}[(\ell/R_V)+\sin\theta_o]-\theta_o\}$
θ	θ_o	$\theta_o+\ell/R_V$
θ'	0	R_V^{-1}
θ''	0	0

$$\psi = \psi_o + \int_o^L \frac{L}{R_s L_s}\, dL$$

$$= \psi_o + \frac{L^2}{2R_s L_s} \tag{7.6}$$

Also, since

$$d\ell = R_V d\theta \tag{7.7}$$

and R_V is constant for a vertical curve, it follows that

$$\theta = \theta_o + \int_o^\ell (1/R_V)d\ell$$

$$= \theta_o + \ell/R_V \tag{7.8}$$

Furthermore,

$$dL = \cos\theta\ d\ell \tag{7.9}$$

as seen from Fig. 7.13(c). Therefore,

Table 7.2(a). Horizontal layout expressions for the yaw angle and its distance derivatives.

Combination of Features	Variables		
	ψ	ψ'	ψ''
T and G	ψ_o	0	0
T→C and G	$\psi_o+L^2/(2R_sL_s)$	$L\cos\theta/(R_sL_s)$	$\cos^2\theta/(R_sL_s)$
C→T and G	$\psi_1-(L_s-L)^2$ $.(2R_sL_s)^{-1}$	$(L_s-L)\cos\theta/(R_sL_s)$	$-\cos^2\theta/R_sL_s)$
C→C and G	$\psi_o+(L/R_{so})$ $+(R_{so}-R_{s1})$ $.L^2/2R_{so}R_{s1}L_s)$	$\cos\theta/R_{so}$ $+(R_{so}-R_{s1})$ $.L\cos\theta/(R_{so}R_{s1}L_s)$	$(R_{so}-R_{s1})$ $.\cos^2\theta/(R_{so}R_{s1}L_s)$
C and G	$\psi_o+(L/R_c)$	$\cos\theta/R_c$	0
T and VC	ψ_o	0	0
T→C and VC	$\psi_o+L^2/(2R_sL_s)$	$L\cos\theta/(R_sL_s)$	$\cos^2\theta/(R_sL_s)$ $-L\sin\theta/(R_vR_sL_s)$
C→T and VC	$\psi_1-(L_s-L)^2$ $.(2R_sL_s)^{-1}$	$(L_s-L)\cos\theta/(R_sL_s)$	$-\cos^2\theta/(R_sL_s)$ $-(L_s-L)\sin\theta/(R_vR_sL_s)$
C→C and VC	$\psi_o+(L/R_{so})$ $+(R_{so}-R_{s1})$ $.L^2/(2R_{so}R_{s1}L_s)$	$\cos\theta/R_{so}$ $+(R_{so}-R_{s1})$ $.L\cos\theta/(R_{so}R_{s1}L_s)$	$-\sin\theta/(R_vR_{so})$ $+(R_{so}-R_{s1})$ $.\cos^2\theta/(R_{so}R_{s1}L_s)$ $-(R_{so}-R_{s1})L$ $.\sin\theta/(R_vR_{so}R_{s1}L_s)$
C and VC	$\psi_o+(L/R_c)$	$\cos\theta/R_c$	$-\sin\theta/(R_cR_v)$

Table 7.2(b). Horizontal layout expressions for the bank angle and its distance derivatives.

Combination of Features	Variables		
	ψ	ψ'	ψ''
T and G	0	0	0
T→C and G	$\phi_1 L/L_s$	$\phi_1 \cos\theta/L_s$	0
C→T and G	$\phi_o-(\phi_o L/L_s)$	$-\phi_o \cos\theta/L_s$	0
C→C and G	$\phi_o+(\phi_1-\phi_o)$ $.L/L_s$	$(\phi_1-\phi_o)$ $.\cos\theta/L_s$	0
C and G	ϕ_o	0	0
T and VC	0	0	0
T→C and VC	$\phi_1 L/L_s$	$\phi_1 \cos\theta/L_s$	$-\phi_1 \sin\theta/(R_v L_s)$
C→T and VC	$\phi_o-(\phi_o L/L_s)$	$-\phi_o \cos\theta/L_s$	$\phi_o \sin\theta/(R_v L_s)$
C→C and VC	$\phi_o+(\phi_1-\phi_o)L/L_s$	$(\phi_1-\phi_o)\cos\theta/L_s$	$(\phi_o-\phi_1)$ $.\sin\theta/(R_v L_s)$
C and VC	ϕ_o	0	0

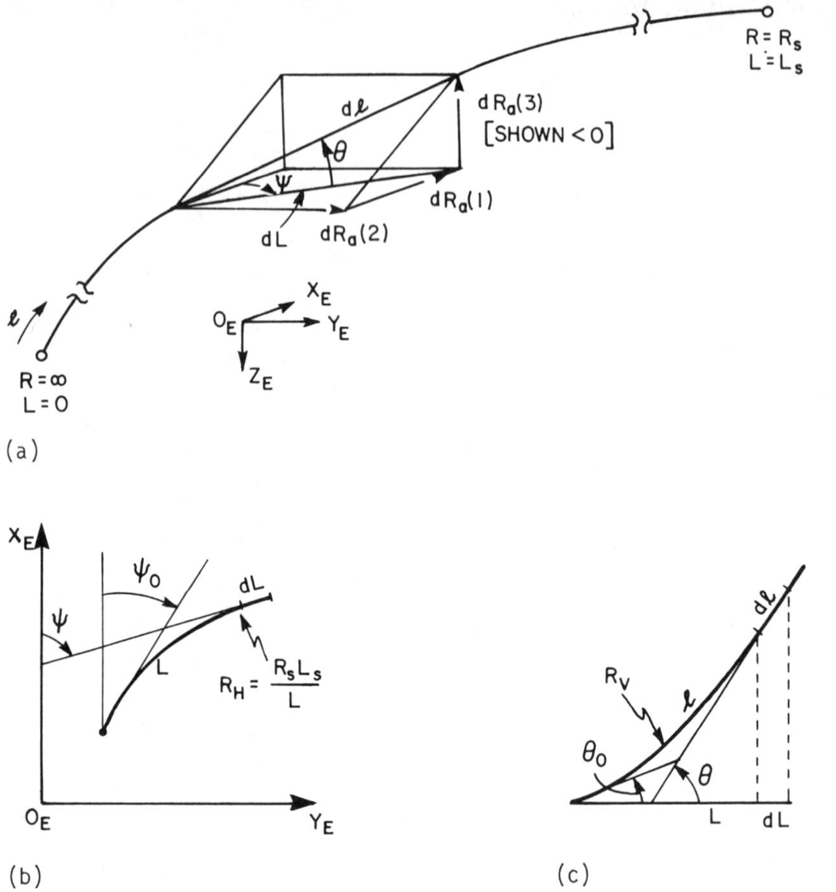

Fig. 7.13 Geometry of a tangent-to-curve spiral combined with a vertical curve: (a) three-dimensional element, (b) projection, and (c) development.

$$
\begin{aligned}
L &= \int_{\theta_o}^{\theta} \cos\theta \, d\ell \\
&= \int_{\theta_o}^{\theta} R_V \cos\theta \, d\theta \\
&= R_V[\sin\theta - \sin\theta_o] \\
&= R_V\left[\sin\left(\theta_o + \frac{\ell}{R_V}\right) - \sin\theta_o\right] \qquad (7.10)
\end{aligned}
$$

or

$$\ell = R_V[\sin^{-1}\left(\frac{L}{R_V} + \sin\theta_o\right) - \theta_o] \tag{7.11}$$

The bank angle along a tangent-to-curve spiral segment is given by

$$\phi = \phi_1 L/L_s \tag{7.12}$$

where ϕ_1 is the bank angle at the end of the segment.

The first and second Euler-angle derivatives for the T→C and VC combination of features in Tables 1 and 2 are obtained by taking the first and second derivatives with respect to ℓ of the expressions in Eqs. (7.6), (7.8) and (7.12). Finally, from Fig. 7.13(a), we obtain:

$$dR_a(1) = d\ell \cos\theta \cos\psi \tag{7.13a}$$

$$dR_a(2) = d\ell \cos\theta \sin\psi \tag{7.13b}$$

$$dR_a(3) = -d\ell \sin\theta \tag{7.13c}$$

Therefore, in general, the first derivative of the alignment axes position with respect to ℓ, expressed in terms of earth axes, is

$$\underset{\sim}{R}_a^{(e)'} = [\cos\theta \cos\psi \quad \cos\theta \sin\psi \quad -\sin\theta]^T \tag{7.14}$$

where superscript T indicates the transpose of the vector.

7.3.3 Actuator Forces

By the application of Newtonian mechanics, it can be shown [1,7] that the translational and rotational equations of motion of a given vehicle rigid body are, respectively,

$$\ddot{\underset{\sim}{x}}^{(m)} = \frac{1}{m}\left(\underset{\sim}{F}_e^{(m)} - \underset{\sim}{F}_t^{(m)} - \underset{\sim}{F}_c^{(m)}\right) \tag{7.15}$$

and

$$\dot{\underset{\sim}{\omega}}_{bm}^{(b)} = [J]^{-1}\left(\underset{\sim}{M}^{(b)} - \underset{\sim}{M}_c^{(b)}\right) \tag{7.16}$$

where

$$\underset{\sim}{F}_t^{(m)} = m\left(\ddot{\underset{\sim}{R}}^{(m)} + \dot{\underset{\sim}{\Omega}}^{(m)} \times \underset{\sim}{x}^{(m)} + \underset{\sim}{\Omega}^{(m)} \times (\underset{\sim}{\Omega}^{(m)} \times \underset{\sim}{x}^{(m)})\right) \tag{7.17}$$

$$\underset{\sim}{F}_c^{(m)} = m\left(2\underset{\sim}{\Omega}^{(m)} \times \dot{\underset{\sim}{x}}^{(m)}\right) \tag{7.18}$$

are the correction forces due to the moving mission axes frame of reference, and

$$\underset{\sim}{M}_c^{(b)} = \underset{\sim}{\omega}^{(b)} \times \left([J]\, \underset{\sim}{\omega}^{(b)}\right)$$
$$+ [J]\left([T_{bm}]\, \dot{\underset{\sim}{\Omega}}^{(m)} - \underset{\sim}{\omega}_{bm}^{(b)} \times \left([T_{bm}]\underset{\sim}{\Omega}^{(m)}\right)\right) \tag{7.19}$$

is the correction moment.

The translational acceleration and angular velocity and acceleration of mission axes required in Eqs. (7.17) to (7.19) are given by:

$$\ddot{\underset{\sim}{R}}^{(m)} = \ddot{\ell}\, \underset{\sim}{E} + \dot{\ell}^2\, \underset{\sim}{F} \tag{7.20}$$

$$\underset{\sim}{\Omega}^{(m)} = \dot{\ell}\, \underset{\sim}{G} \tag{7.21}$$

$$\dot{\underset{\sim}{\Omega}}^{(m)} = \ddot{\ell}\, \underset{\sim}{G} + \dot{\ell}^2\, \underset{\sim}{H} \tag{7.22}$$

where

$$\underset{\sim}{E} = [T_{me}]\, \underset{\sim}{R}_a^{(e)'} + \underset{\sim}{G} \times \underset{\sim}{x}_{am} \tag{7.23}$$

$$\underset{\sim}{F} = [T_{me}]\, \underset{\sim}{R}_a^{(e)''} + \underset{\sim}{H} \times \underset{\sim}{x}_{am} + \underset{\sim}{G} \times (\underset{\sim}{G} \times \underset{\sim}{x}_{am}) \tag{7.24}$$

$$\underset{\sim}{G} = [A_{me}]\underset{\sim}{\alpha}'_{me} \tag{7.25}$$

$$\underset{\sim}{H} = [A_{me}]\underset{\sim}{\alpha}''_{me} + [A'_{me}]\, \underset{\sim}{\alpha}'_{me} \tag{7.26}$$

in which $\underset{\sim}{\alpha}_{me}$, $\underset{\sim}{\alpha}'_{me}$ and $\underset{\sim}{\alpha}''_{me}$ (Eqs. (7.1) to (7.3)) are given by the expressions in Tables 7.1 and 7.2, $\underset{\sim}{R}_a^{(e)'}$ is given by Eq. (7.14) and, taking its derivative with respect to ℓ,

$$\underset{\sim}{R}_a^{(e)''} = \begin{bmatrix} -\psi'_a\cos\theta_a\sin\psi_a - \theta'_a\cos\psi_a\sin\theta_a \\ \psi'_a\cos\theta_a\cos\psi_a - \theta'_a\sin\psi_a\sin\theta_a \\ -\theta'_a\cos\theta_a \end{bmatrix} \tag{7.27}$$

Neglecting subscripts 'me', the earth-to-mission axes transformations in Eqs. (7.23) to (7.26) are given by:

$$[T] = \begin{bmatrix} \cos\theta\cos\psi & \cos\theta\sin\psi & -\sin\theta \\ \sin\theta\sin\phi\cos\psi - \cos\phi\sin\psi & \sin\phi\sin\theta\sin\psi + \cos\phi\cos\psi & \sin\phi\cos\theta \\ \cos\phi\sin\theta\cos\psi + \sin\phi\sin\psi & \cos\phi\sin\theta\sin\psi - \sin\phi\cos\psi & \cos\phi\cos\theta \end{bmatrix} \tag{7.28}$$

$$[A] = \begin{bmatrix} 1 & 0 & -\sin\theta \\ 0 & \cos\phi & \sin\phi\cos\theta \\ 0 & -\sin\phi & \cos\phi\cos\theta \end{bmatrix} \quad (7.29)$$

$$[A'] = \begin{bmatrix} 0 & 0 & -\theta'\cos\theta \\ 0 & -\phi'\sin\phi & -\theta'\sin\phi\sin\theta + \phi'\cos\phi\cos\theta \\ 0 & -\phi'\cos\phi & -\theta'\cos\phi\sin\theta + \phi'\sin\phi\cos\theta \end{bmatrix} \quad (7.30)$$

Ideally, the correction forces and moment (Eqs. (7.17) to (7.19)) need to be supplied by actuators for each rigid body. If the CTS rollers are not superelevated dynamically to conform with the simulated track, Eq. (7.15) becomes

$$\ddot{\tilde{x}}^{(m)} = \frac{1}{m}\left(\tilde{F}_e^{(m)} + \tilde{F}_d^{(m)}\right) \quad (7.31)$$

where

$$\tilde{F}_d^{(m)} = \tilde{F}_{cw}^{(m)} - \tilde{F}_t^{(m)} - \tilde{F}_c^{(m)} \quad (7.32)$$

is the net desired correction force and

$$\tilde{F}_{cw}^{(m)} = mg[0 \quad \sin\phi_a \quad (\cos\phi_a - 1)]^T \quad (7.33)$$

is a correction due to non-superelevation of rollers.

In addition to the net correction force $\tilde{F}_d^{(m)}$ and moment $\tilde{M}_c^{(b)}$, the actuators must incorporate the net external force $\tilde{F}_e^{(m)}$ and moment $\tilde{M}^{(b)}$ arising from aerodynamic and coupler effects, for example, if such effects need to be simulated.

7.3.4 Roller Speeds

Using small angle approximations for circular functions of ϕ_a, the left and right rail radii of curvature are given by:

$$R_\ell = R_H + (K/2)\ \mathrm{sgn}\ \psi_a' \quad (7.34)$$

$$R_r = R_H - (K/2)\ \mathrm{sgn}\ \psi_a' \quad (7.35)$$

where

$$R_H = 1/|\psi_a'| \tag{7.36}$$

and

$$\operatorname{sgn} \psi_a' = \begin{cases} 1, \text{ curving right} \\ 0, \text{ straight track} \\ -1, \text{ curving left} \end{cases} \tag{7.37}$$

Assuming that the nominal roller radii are equal, let

$$r = r_\ell = r_r \tag{7.38}$$

Then, for a given along-track velocity $\dot{\ell}$, the required left and right roller speeds are:

$$\omega_{r\ell} = (R_e/rR_H)\dot{\ell} \tag{7.39}$$

$$\omega_{rr} = (R_r/rR_H)\dot{\ell} \tag{7.40}$$

7.3.5 Orientation of Rollers

The yaw position of a roller stand should ideally be controlled to keep the wheels of the supported wheelset centered on top of the rollers. This could be achieved by feedback control. However, assuming an almost-radial position of wheelsets in curves, the desired yaw positions of roller stands can be calculated by

$$\psi_{rd}(m) = \psi_{adl}(m) - \psi_{ad}, \quad m = 1,\ldots,N_w \tag{7.41}$$

where ψ_{ad} is the desired yaw angle of the alignment axes, $\psi_{adl}(m)$ is the desired track yaw angle at the m'th wheelset, and $\psi_{rd}(m)$ is the desired roller stand yaw position at the m'th wheelset.

Now consider Fig. 7.14, which illustrates the wheel-roller contact surface viewed from the front of a vehicle. The wheel axle, as shown, is parallel to the inertial reference and the roller axle is canted by an angle β. Angle γ is the wheel conicity at the contact surface center 0. The wheel and roller rolling radii at 0 are respectively R and r, whereas, they are R_D and r_D at a distance D away from 0. Relative spin S is given by

$$S = (\Omega_w - \Omega_r)/V \tag{7.42}$$

where V is the forward rolling velocity and

$$\Omega_w = \omega_w(R_D - R)/D \tag{7.43}$$

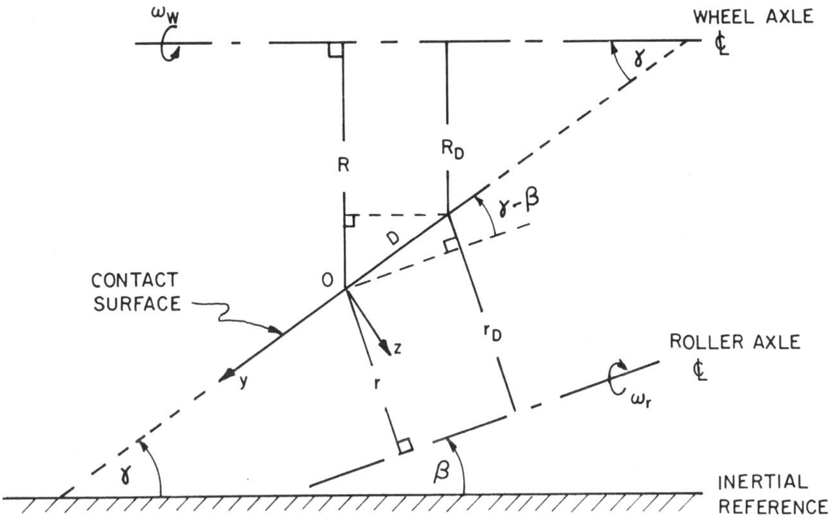

Fig. 7.14 Wheel-roller contact surface viewed from front of vehicle.

and

$$\Omega_r = \omega_r(r_D - r)/D \tag{7.44}$$

are respectively the wheel and roller angular velocities of 0 in the z direction. Since

$$R_D = R - D\sin\gamma \tag{7.45}$$

and

$$r_D = r + D\sin(\gamma - \beta) \tag{7.46}$$

it follows that

$$S = \left(\omega_w \sin\gamma + \omega_r \sin(\gamma - \beta)\right)/V \tag{7.47}$$

Regarding a rail as a non-rotating infinite-diameter roller, the expression for spin in the rail case is obtained by setting $w_r = 0$, i.e.,

$$S_{rail} = (\omega_w \sin\gamma)/V \tag{7.48}$$

Since ω_r is nonzero with rollers, it follows that both spins are equal if and only if the condition

$$\beta = \gamma \tag{7.49}$$

is satisfied. This requires canting of the rollers to match the wheelset conicities.

7.4 CURVED TRACK SIMULATOR SOFTWARE

In this section, we describe software which was developed specifically for the CTS defined in Section 7.1. The interaction between the pertinent programs is shown in Fig. 7.15.

Program CTS1 [6] (Section 7.4.3) accesses five disk data sets prepared off-line and produces one output data set. Detailed data set contents are given in Table 7.3. The input data sets are: track data TRADA, mission data MISDA, vehicle data VEHDA, simulator data CTSDA and run data RUNDA. Of these, the first two are prepar-

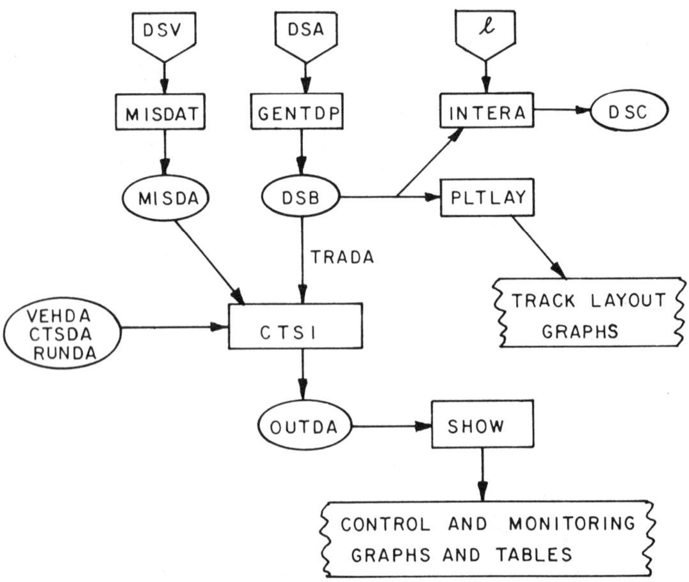

Fig. 7.15 File-flow diagram of CTS software.

ed respectively by means of independent programs GENTDP [2] and MISDAT [3] whereas data for the last three is entered directly by the user. The output CTS control and monitoring data set OUTDA may be used to control and monitor the CTS or these functions may conceivably be carried out directly by executing CTS1 in real time. Tables and graphs produced from OUTDA by program SHOW are useful in predicting operational requirements of CTS control hardware.

Program MISDAT (Section 7.4.1) facilitates the preparation and inspection of mission data consisting of velocity and acceleration profiles. Section 7.4.2 deals with three programs. The first, GENTDP, accepts a simple description of the horizontal and vertical track layouts, generates the appropriate track-related functions and provides means of examining all data and making changes interactively, as required. The generated data corresponds to automatically-determined positions along the track, the total number of which is under the user's control by specifying a curvature proportionality parameter such that the distance between positions is proportional to the radius of curvature. Program INTERA interpolates linearly on this data to obtain corresponding data applicable at any user-specified track position. Scaled plots of the track layouts may be examined by means of program PLTLAY which also relies on data generated by GENTDP.

The above programs were all written in FORTRAN 77 and implemented on a DEC VAX 11/780 computer with a VMS (virtual memory) operating system, although GENTDP, INTERA and PLTLAY were originally developed on an IBM 360/70 time-sharing system. They could readily be adapted to other computer systems by modifying the plotting routines, if necesary.

7.4.1 Mission Data Program

The inertial forces and moments in a railway vehicle's equations of motion are functions of its velocity and acceleration and the track layout (Section 7.3). Since the velocity profile as a function of along-track position is usually given, a program is required to calculate the corresponding acceleration profile. This paper describes program MISDAT which facilitates the preparation of

Table 7.3. Data sets.

Data Set Name	Data Subset Number	Contents
DSV		IDMIS, IDMISP, $\ell_m(r), \dot{\ell}_m(r), \ddot{\ell}_m(r), N_{mT}; r=1,\ldots,N_{mT}$
MISDA		IDMIS, $\ell_m(r), \dot{\ell}_m(r), \ddot{\ell}_m(r), N_{mT}; r=1,\ldots,N_{mT}$
DSA	1	IDTRA
	2	$N_{STH}, \psi_o, N_{SH}, T_{NSH}, \ell_{NSH}$ or $L_{NSH}, R_{NSH}, \phi_{NSH}$
	3	$N_{STV}, \theta_o, N_{SV}, T_{NSV}, \ell_{NSV}, \theta_{NSV}$
	4	$N_m, x_{amk}; k=1,\ldots,N_m$
	5	K_R, K_P
DSB	1	DSA
	2	N_{DT}, ℓ
	3	$\ell_i, \underset{\sim}{E}_k, \underset{\sim}{F}_k, \underset{\sim}{G}_k, \underset{\sim}{H}_k; i=1,\ldots,N_{DT}; k=1,\ldots,N_m$
	4	$\ell_i, \underset{\sim}{\alpha}_i, \underset{\sim}{\alpha}'_i, \underset{\sim}{\alpha}''_i, \underset{\sim}{R}_i, \underset{\sim}{R}'_i, \underset{\sim}{R}''_i; i=1,\ldots,N_{DT}$

DSC	1	$\ell, \underset{\sim}{E}_k, \underset{\sim}{F}_k, \underset{\sim}{G}_k, \underset{\sim}{H}_k; k=1,\ldots,N_m$
	2	$\ell, \underset{\sim}{\alpha}, \underset{\sim}{\alpha}', \underset{\sim}{\alpha}'', \underset{\sim}{R}, \underset{\sim}{R}', \underset{\sim}{R}''$
TRADA		IDTRA, $\ell_i, \phi_{ai}, \psi_{ai}, \psi'_{ai}, \underset{\sim}{R}_{ai}, \underset{\sim}{E}_{ki}, \underset{\sim}{F}_{ki}, \underset{\sim}{G}_{ki}, \underset{\sim}{H}_{ki}, K, N_m, \ell_T,$ N_{DT}; $i=1,\ldots,N_{DT}$; $k=1,\ldots,N_m$
VEHDA		IDVEH, $a(m), N_w, N_b, [J_n], [J_n]^{-1}, m_n$, IDNAM(N); $m=1,\ldots,N_w$; $n=1,\ldots,N_b$
CTSDA		IDCTS, $N_{31}, N_{42}, C_{fr}, G_{tm}, r_r$
RUNDA		IDRUN, $\alpha_t, N_L, N_{LT}, N_{OT}, \ell(o), M(n), \alpha_{bmn}, \omega^{(b)}_{bmn}, x^{(m)}_n, \dot{x}^{(m)}_n$; $n=1,\ldots,N_b$
OUTDA		IDTRA, IDVEH, IDCTS, IDMIS, IDRUN, $N_b, N_w, N_{OT}, N_{LT},$ $\ell_T, \ell(o), [\ell_s, \ell, \dot{\ell}, \dot{\ell}_d, t, C_{Rd}(m), F_{rd}, C'_{fr}(m), \psi_{rd}(m),$ $\underset{\sim}{F}^{(m)}_{tn}, \underset{\sim}{F}^{(m)}_{cwn}, F^{(m)}_{dn}, \underset{\sim}{M}^{(b)}_{cn}, R_{ad}(q), R_{al}(q,m)$; $m=1,\ldots,N_w$; $n=1,\ldots,N_b$; $q=1,2]$ at $\ell_s(j)$, $j=0,\ldots,N_{OT}$ and laps $p=1,\ldots,N_{LT}$

mission data defined by consistent velocity and acceleration profiles.

The program calculates an acceleration profile from an initial velocity profile data description and integrates the former for comparison with the specified velocity profile. If the difference between the two velocity profiles is unacceptable, the user can specify more velocity data and check for resulting improvements interactively and iteratively until the errors are sufficiently small. The mission defined by the latest acceleration and computed velocity profiles is then stored in a disk file for later use by program CTS1. Various tabular and graphical display options are provided for decision-making ease during the iteration phase.

7.4.1.1 Evaluation of acceleration

Given three consecutive data positions ℓ_i, ℓ_j and ℓ_k in an ascending sequence of n positions $\ell_1, \ell_2, \ldots, \ell_n$ and velocities $\dot{\ell}_i$, $\dot{\ell}_j$ and $\dot{\ell}_k$ at those three positions, the acceleration $\ddot{\ell}_j$ at position ℓ_j is given by

$$\ddot{\ell}_j = \dot{\ell}'_j \, \dot{\ell}_j \tag{7.50}$$

where

$$\dot{\ell}'_j = \left.\frac{d\dot{\ell}}{d\ell}\right|_{\ell=\ell_j} \tag{7.51}$$

is the rate of change of velocity $\dot{\ell}$ with respect to along-track distance ℓ, evaluated at $\ell=\ell_j$. In program MISDAT, $\dot{\ell}'_j$ is approximated by

$$\dot{\ell}'_j \simeq \begin{cases} \dot{\ell}'_{jf}, & \ell_j = \ell_1 \\ \left(\dot{\ell}'_{jb}+\dot{\ell}'_{jf}\right)/2, & \ell_1 < \ell_j < \ell_n \\ \dot{\ell}'_{jb}, & \ell_j = \ell_n \end{cases} \tag{7.52}$$

where

$$\dot{\ell}'_{jb} = \left(\dot{\ell}_j-\dot{\ell}_i\right)/\left(\ell_j-\ell_i\right) \tag{7.53}$$

and

$$\dot{\ell}'_{jf} = \left(\dot{\ell}_k-\dot{\ell}_j\right)/\left(\ell_k-\ell_j\right) \tag{7.54}$$

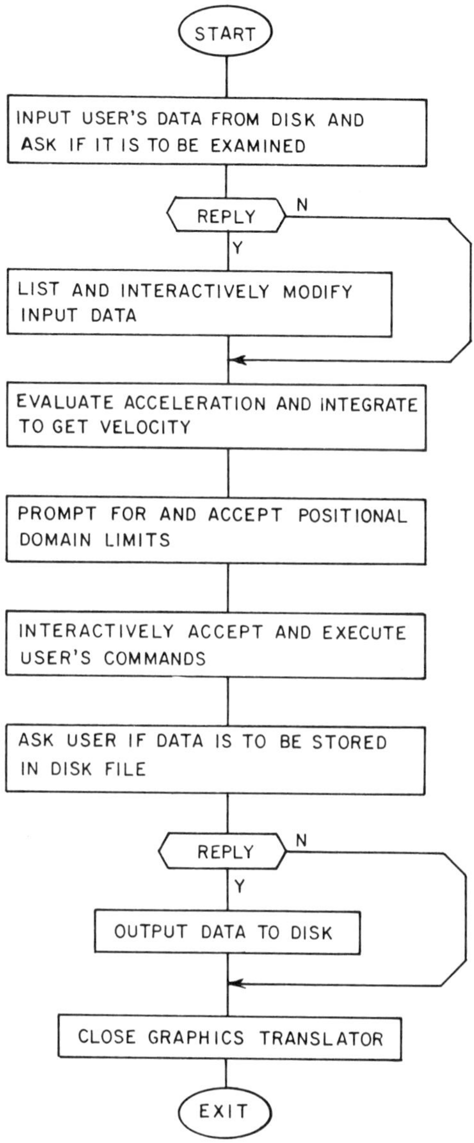

Fig. 7.16 Program MISDAT flowchart.

7.4.1.2 Program description

Before executing MISDAT the user creates a disk file defining: (1) name of disk file into which MISDAT will store the output data, (2) name of disk file to be used for plotting on the printer, (3) total number of data positions, (4) positional values, and (5) corresponding velocity values.

An overview of the program's operation is given in the flow-chart of Fig. 7.16. Upon execution of MISDAT, the user is required to enter the input data disk file name. Once read into memory, the input data can be examined and modified by the change (C), insert (I) and delete (D) commands (Table 7.4). The acceleration profile is then calculated by the above method and a fourth-order Runge-Kutta-Gill integration algorithm is used to compute the corresponding velocity profile.

Table 7.4. Program MISDAT commands.

COMMAND	DESCRIPTION
S,x,z	Select domain from position x to position z
C,x,z	Change velocity at position x to value z
I,x,z	Insert (add) a datum: position x, velocity z
D,x[,z]	Delete datum at position x (z value not required)
PA	Plot the acceleration curve
PV	Plot velocity curves (input and computed)
PC	Plot computed velocity curve
TI	Table printout - Investigation table
TM	Table printout - Mission table
H	Help - Printout of this table
E	Exit - Terminate a program run

At this point the user is asked to specify positional domain limits for investigation purposes. By means of the commands in Table 7.4, he then examines differences between specified and computed velocity profiles observed in tabular displays and graphical displays, such as Figs. 7.17 and 7.18, and modifies the specified data interactively and iteratively until the errors are sufficiently small. When the match is deemed satisfactory, the user indicates whether or not the mission, as defined by the latest acceleration and computed velocity profiles, should be stored in the prescribed disk file. The program responds by performing the required storage before closing the graphics translator and terminating the run.

7.4.1.3 Method of use

The method of use is described under three headings: Preparation for a run, Running the program, and Program outputs.

(i) Preparation for a run. A file is created on disk and data is entered as follows:

(a) Line 1: Name of the disk file into which MISDAT is to store the output data (positions, computed velocities and accelerations).

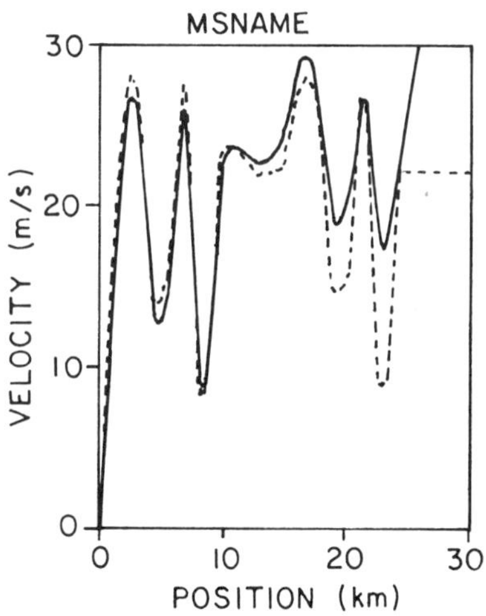

Fig. 7.17 Velocity profile example.

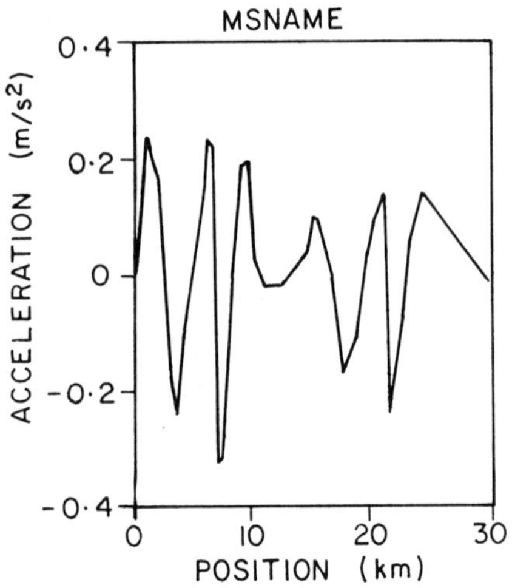

Fig. 7.18 Acceleration profile example.

The name should contain a maximum of 6 characters.

(b) Line 2: Name of the disk file to be used for plotting on the printer. If the program is to be run under a sub-directory, include the directory name in the file name (max. number of characters is 50).

(c) Line 3: An integer value (NP) which defines the total number of data positions (max. 500).

(d) NP positional values, in ascending order.

(e) NP velocity values, corresponding to the NP positions.

The program uses free format for the input of the position and velocity values. There is no specification for the number of values per line, and the number per line need not be consistent. The only restriction is that the first value (position or velocity) must appear on a new line. Within a line, values may be separated by a comma or by one or more spaces and/or tabs. A decimal point need only be included if a value is not a whole number. The FORTRAN floating-point format, where 1.5E5 is equivalent to 150000, may also be used.

To enable graphical displays on the user's terminal, it is necessary to allocate the terminal prior to starting a program run. This is accomplished by the command 'ALLOCATE TT PLOT'. If there will be graphical output to the plotter it is necessary to ready the plotter.

(ii) Running the program. Program MISDAT is executed by the command 'RUN MISDAT'. The user is prompted for all inputs to the program. Most questions require a YES or NO answer. When a table print-out is requested, a prompt "DEVICE:" is given to which a response of P (printer) or T (terminal) is expected. Within the *examine* and *option* phases of the program, a prompt for input of a task consists of ">>".

Initially the user is requested to enter the name (1 to 6 characters) of the disk file which contains the input data. This file is then read and the user is given the option of examining the data and modifying it by the C, D, and I commands (Table 7.4). To exit from this phase of the program, command E (exit) is invoked.

The format for a task input, presented in table form, is displayed on the user's terminal when the program's *option phase* is entered. All available options in Table 7.4 are listed. At any time following the prompt, command H will produce a copy of this table.

For investigation and mission table listings at the terminal, there is a pause following every 20th datum to allow the user to look at the values. Depressing the carriage return releases the pause. When a table is to be printed, it is stored in a file MISTBL.DAT (which may be printed following the program run) and entered into a print queue. The *wait-time* for the printout is dependent upon the work load of the device at that particular time, but program execution continues.

When a plot is requested, following a graphcal display on the user's terminal, an option is provided for a hard-copy output on the printer and/or plotter. Before onput to the plotter, the user must ready the plotter and position the plotting paper.

Command E causes an exit from the option phase. If so requested, MISDAT stores the program data on disk, including total number of points, position, computed velocity and computed acceleration data. Program execution then halts.

(iii) Program outputs. The display options produce the following outputs: (1) specified (dashed line) and computed (solid line) velocity profiles in the positional domain, (2) acceleration profile in the positional domain, (3) computed velocity profile in the positional domain, (4) investigation table which lists position, specified and computed velocity, and acceleration over the full range of data, and (5) mission table which indicates the number of data points and lists position, computed velocity and acceleration.

The graphical displays are output to the terminal screen and optionally to a plotter and/or printer. The tables may be output to either the user's terminal or the line printer. Table listings which are directed to the printer are stored in a file with as many versions as there were table print-outs. Graphical displays which are directed to the printer are temporarily stored on disk in the file whose name was specified by the user in his input data file. When the print-out is completed this file is automatically deleted.

7.4.2 Track Data Programs

Three programs were developed to prepare track data for use in computer modeling of vehicle dynamics: (1) a general track data preparation program (GENTDP), (2) an interpolation program (INTERA), and (3) a program to plot the track layouts (PTLAY). Each one is an independent main program which calls its own set of subroutines. However, both INTERA and PLTLAY require GENTDP's output data set DSB as input (Fig. 7.15).

A track such as the one shown in Figs. 7.19 and 7.20 can be completely specified by data such as in Tables 7.5 and 7.6. The horizontal layout (Table 7.5) requires the horizontal segment number, the type of horizontal segment, its length and, for curve

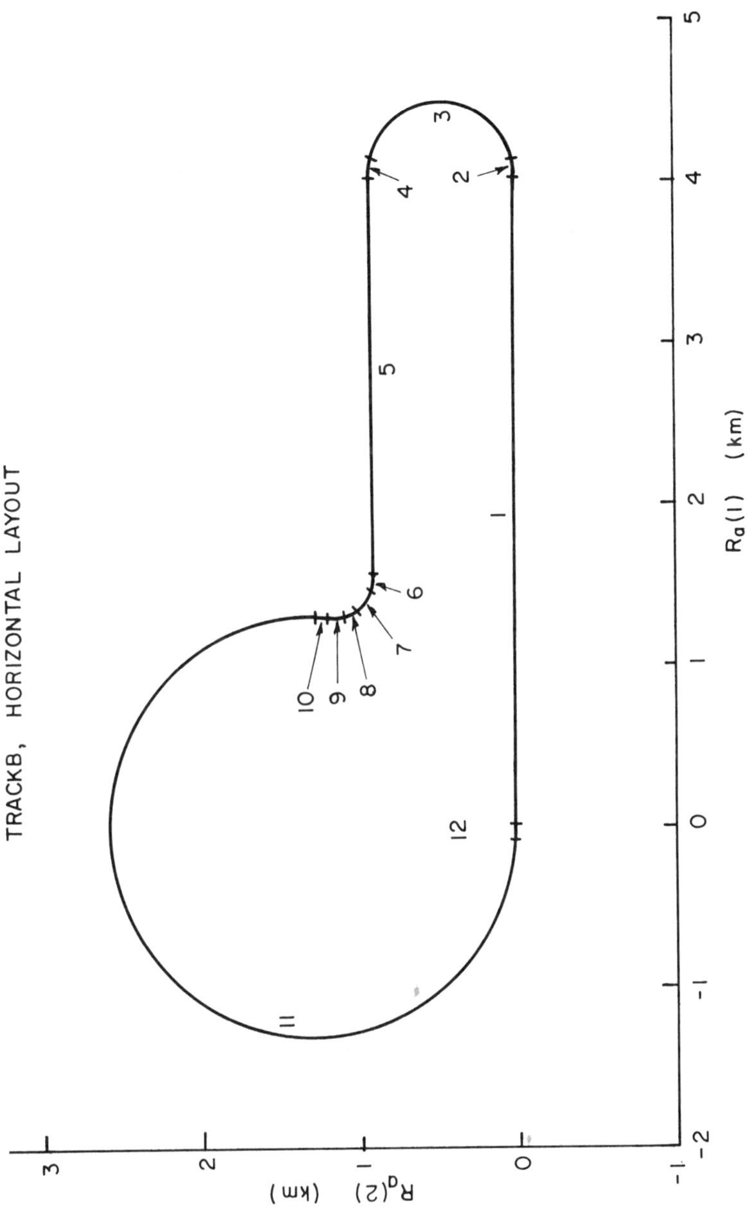

Fig. 7.19 Horizontal layout example.

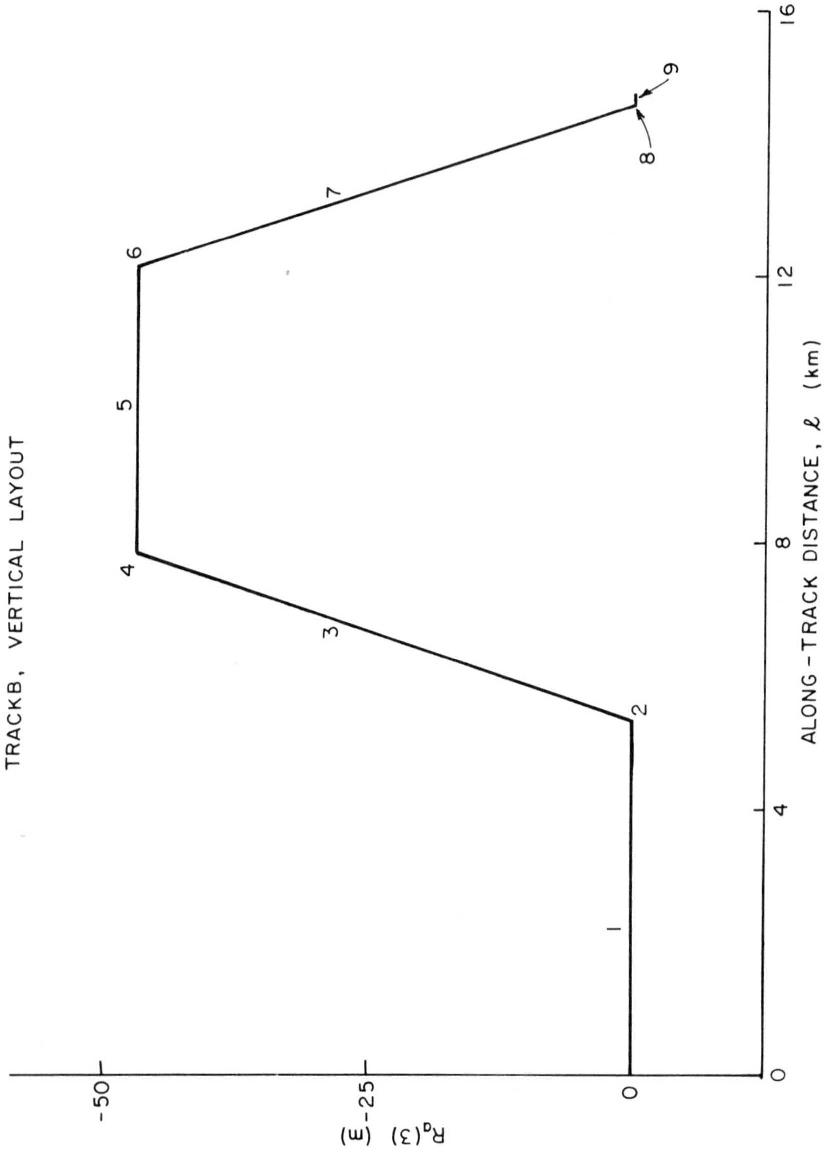

Fig. 7.20 Vertical layout example.

Table 7.5. Horizontal layout for test track example of Fig. 7.19.

Horizontal Segment No.	Type	Length (km)	Radius of Curvature (m)	Bank Angle (deg)
1	T	4.0000		
2	S	0.0893		
3	C	1.3244	-450.000	-4.332
4	S	0.0893		
5	T	2.4366		
6	S	0.1219		
7	C	0.2237	220.000	4.332
8	S	0.1219		
9	T	0.0943		
10	S	0.0409		
11	C	6.0853	-1300.000	-4.332
12	S	0.0409		

segments only, its radius of curvature and bank angle. The vertical layout (Table 7.6) requires the vertical segment number, the type of vertical segment, its length and, for grade segments only, its pitch angle. For program CTS1, only a horizontal layout is required since the CTS is a simulator of only zero-grade track.

The only other information in addition to the above angular position information which is required to determine $\underset{\sim}{E}$, $\underset{\sim}{F}$, $\underset{\sim}{G}$ and $\underset{\sim}{H}$ as functions of along-track position is $\underset{\sim}{x}_{am}$, the position of mission axes relative to alignment axes. Thus, the above mathematical model can be used to compute functions $\underset{\sim}{E}_{ik}$, $\underset{\sim}{F}_{ik}$, $\underset{\sim}{G}_{ik}$ and $\underset{\sim}{H}_{ik}$ corresponding to the i'th along-track positional data point ℓ_i, and the k'th mission frame, as shown in Fig. 7.21. The calculation of data

Table 7.6. Vertical layout for test track example of Fig. 7.20.

Vertical Segment No.	Type	Length (km)	Pitch Angle (deg)
1	G	5.3000	0.000
2	VC	0.0100	
3	G	2.4600	1.085
4	VC	0.0100	
5	G	4.2885	0.000
6	VC	0.0100	
7	G	2.4600	-1.085
8	VC	0.0100	
9	G	0.1000	0.000

points ℓ_i is based on the distance between them being proportional to the alignment's radius of vertical or horizontal curvature, with constant of proportionality K_p as parameter. Specifically, the along-track positional data points ℓ_i, $i=1,\ldots,N_{DT}$ are given by

$$\ell_{i+1} = \ell_i + \Delta\ell_i \tag{7.55}$$

where

$$\ell_1 = 0 \tag{7.56}$$

and the total number of data points is given by $N_{DT} = i$ when $\ell_{i+1} > \ell_T$ and $\ell_i \leq \ell_T$. Having obtained positions ℓ_i, their horizontal projections L_i are computed by the horizontal distance expressions in Table 7.1.

The incremental distance $\Delta\ell_i$ in Eq. (7.55) is determined by considering the vertical and horizontal layouts separately, and

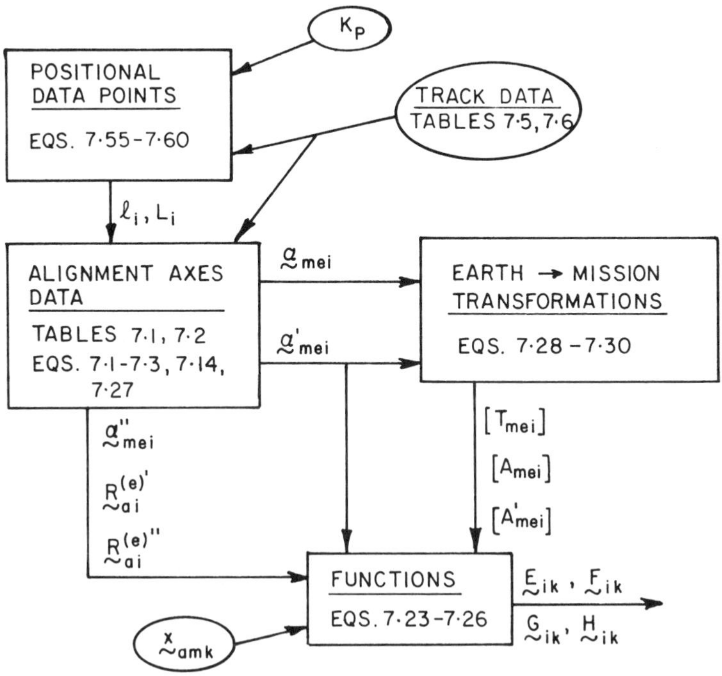

Fig. 7.21 Information flow diagram of track-derived functions.

then taking the minimum without entering the next vertical or horizontal segment.

Thus, for the vertical layout,

$$\Delta\ell_{vi} = \begin{cases} \ell_{NSV}, & T_{NSV} = G \\ K_p |R_V(\ell_i)|, & T_{NSV} \neq G \end{cases} \tag{7.57}$$

and, for the horizontal layout,

$$\Delta L_{Hi} = \begin{cases} L_{NSH}, & T_{NSH} = T \\ K_p |R_H(L_i)|, & T_{NSH} \neq T \end{cases} \tag{7.58}$$

Then, $\Delta\ell_{Hi}$ is computed by means of the along-track distance expressions in Table 7.1. The minimum distance to the next vertical or horizontal segment is computed by

$$D = \min(\ell_{Vo} - \ell_i, \ell_{Ho} - \ell_i) \tag{7.59}$$

and finally,

$$\Delta\ell_i = \min(D, \Delta\ell_{Vi}, \Delta\ell_{Hi}) \tag{7.60}$$

Program GENTDP accepts data set DSA (Table 7.3) interactively from a time-sharing terminal. This data set, partitioned into five subsets, contains a track identification, horizontal and vertical track layout data in the form of Tables 7.5 and 7.6, the total number of mission axes and their positions relative to alignment axes, as well as round-off and curvature proportionality constants used in determining the along-track positional data points. The program stores DSA as the first subset of DSB and applies the conceptual model to determine along-track positional data points, total track length, track-derived functions at mission axes as well as alignment axes positions and positional derivatives at the track data points, all of which is stored in subsets DSB-2,3,4. From a user interaction standpoint, it prompts the user for DSA subsets, gives him print-out options for both DSA and DSB and provides the capability of changing DSA.

Program INTERA performs linear interpolation on subsets DSB-3,4 which apply to specific track locations and produces corresponding data at arbitrary locations specified interactively by the user. Print-out options are also provided.

Program PLTLAY operates on subsets DSB-2,4 to produce two-dimensional plots of horizontal and vertical track layouts such as Figs. 7.19 and 7.20. The user is prompted to specify the desired track layout and the orientation of his sheet of paper on the X-Y plotter. The program then proceeds automatically to scale the problem, draw axes, print the heading and plot the layout.

7.4.3 Control and Monitoring Data Programs

With program CTS1, the user begins by entering data into the input data files on disk. Data set TRADA is accessed as a subset of data set DSB produced by program GENTDP. Data set MISDA is produced by program MISDAT. Vehicle masses and moments of inertia in VEHDA, CTS roller radii and gear ratios in CTSDA, and initial vehicle position in RUNDA are examples of some of the information required in the directly specified data sets (Table 7.3).

The execution of CTS1 provides the option of examining the input data interactively before the computational phase is entered. If the user decides to change the input, he must execute the program anew after making the change. When he is satisfied with the data, the user directs CTS1 to proceed with its computation of OUTDA which contains required actuator forces and moments, roller yaw positions, roller speed control data and other information. Program SHOW provides ten table-printing and seven graph-plotting options interactively.

7.4.3.1 Program CTSI

Figure 7.22 is an information flow diagram of program CTS1 wherein each block corresponds to one of the subroutines described below in order of execution.

(i) Subroutine TRACK. The track-related variable shown as outputs of this module are computed as functions of along-track position ℓ by linear interpolation on track data set TRADA. The desired curvature ratio $C_{Rd}(m)$ at the m'th wheelset is an exception computed from the following general equation:

$$C_R = \omega_d/\omega_m$$
$$= 1 - \left(2N_{31}/N_{42}\right)\left((R/R_r)-1\right) \tag{7.61}$$

where R_r and R are given by Eqs. (7.35) and (7.36) and, in this case, the variables are taken as the desired values applicable at the m'th wheelset. The right and left roller speeds, ω_{rr} and $\omega_{r\ell}$, are related to the main and differential motor speeds, ω_m and ω_d, by gear ratios N_{31}, N_{41} and N_{42}, as follows:

$$\omega_{rr} = N_{31}\omega_m \tag{7.62}$$

$$\omega_{r\ell} = N_{41}\omega_m + N_{42}\omega_d \tag{7.63}$$

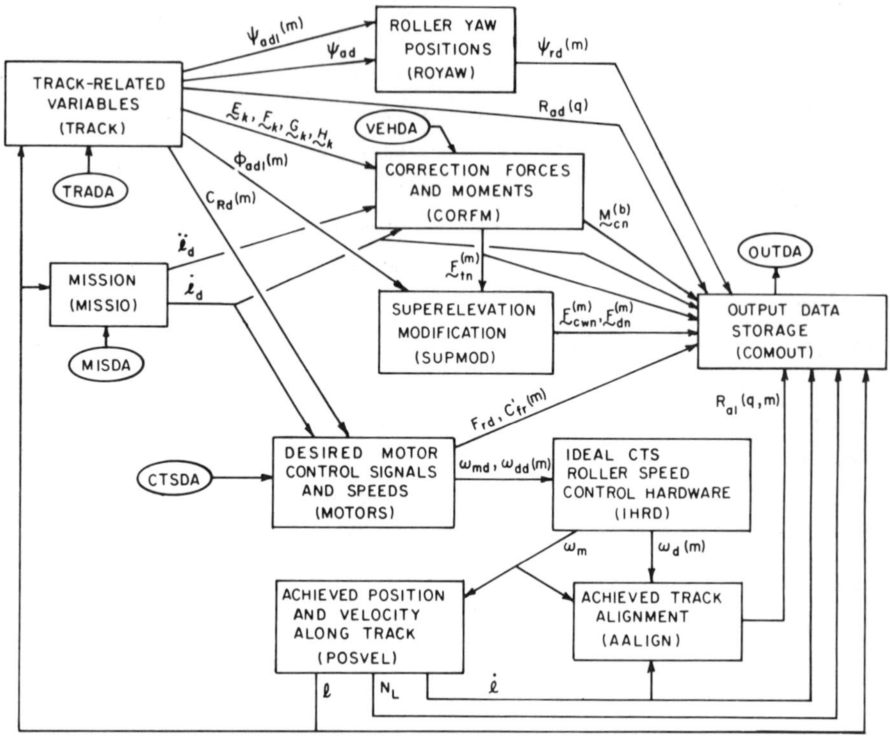

Fig. 7.22 Program CTS1 information flow diagram.

The actual gear ratio values are

$$N_{31} = 0.2865 \tag{7.64}$$

$$N_{41} = 0.2937 \tag{7.65}$$

$$N_{42} = 0.007163 \tag{7.66}$$

so that

$$N_{31} = N_{41} - N_{42} \tag{7.67}$$

and

$$\omega_{r\ell}\begin{cases} < \omega_{rr}, & \omega_d > -\omega_m \\ = \omega_{rr}, & \omega_d = -\omega_m \\ > \omega_{rr}, & \omega_d < -\omega_m \end{cases} \tag{7.68}$$

Equations (7.62) to (7.68) are given as information about the CTS, but are not used in subroutine TRACK.

(ii) Subroutine MISSIO. The desired velocity and acceleration are determined as functions of ℓ by linear interpolation on mission data set MISDA.

(iii) Subroutine ROYAW. The desired CTS roller yaw positions are calculated by Eq. (7.41).

(iv) Subroutine CORFM. Given along-track velocity $\dot{\ell}_d$, acceleration $\ddot{\ell}_d$, track-related functions $\underset{\sim}{E}_k$, $\underset{\sim}{F}_k$, $\underset{\sim}{G}_k$ and $\underset{\sim}{H}_k$, translational and rotational positions and velocities relative to mission coordinates (assumed negligible by specifying them to be zero in run data set RUNDA), and masses and moments of inertia of N_b rigid bodies in vehicle data set VEHDA, this subroutine computes the net transport forces $\underset{\sim}{F}_{tn}^{(m)}$ and inertial moment $\underset{\sim}{M}_{cn}^{(b)}$ for body n. The computations are based on Eqs. (7.17) to (7.19); howver, the Coriolis force $\underset{\sim}{F}_{c}^{(m)}$ is zero because of the above assumptions.

(v) Subroutine SUPMOD. Given the mass m_n (in VEHDA) of each one of the N_b rigid bodies, the net transport force $\underset{\sim}{F}_{tn}^{(m)}$ and the track roll angle $\phi_{adl}^{(m)}$ at the m'th wheelset, this subroutine computes the correction force $\underset{\sim}{F}_{cwn}^{(m)}$ due to non-superelevation of rollers and the net desired correction force $\underset{\sim}{F}_{dn}^{(m)}$ for $n=1,\ldots,N_b$. The calculations are based on the following equations:

$$\phi = \frac{1}{N_w} \sum_{m=1}^{N_w} \phi_{adl}^{(m)} \tag{7.69}$$

$$\underset{\sim}{F}_{cwm}^{(m)} = \begin{matrix} 0 \\ m_n g\phi \\ m_n g(\cos\phi - 1) \end{matrix} \tag{7.70}$$

$$\underset{\sim}{F}_{dn}^{(m)} = \underset{\sim}{F}_{cwn}^{(m)} - \underset{\sim}{F}_{tn}^{(m)} \tag{7.71}$$

where $n=1,\ldots,N_b$ and g is gravitational acceleration.

(vi) Subroutine MOTORS. The desired CTS motor speeds, ω_{md} and ω_{dd}, motor control oscillator frequency F_{rd} and differential ratio control divider variable $C'_{fr}(m)$ are calculated by means of the following equations:

$$\omega_{md} = \dot{\ell}_d/(r_r N_{31}) \tag{7.72}$$

$$\omega_{dd}(m) = \omega_{md} C_{Rd}(m) \tag{7.73}$$

$$F_{rd} = C_{fr} G_{tm} \omega_{md} \tag{7.74}$$

$$C'_{fr}(m) = C_{fr}/C_{Rd}(m) \tag{7.75}$$

where C_{fr} and G_{tm} are constants and $m=1,\ldots,N_w$. The motor speeds are actually controlled by variables F_{rd} and C'_{fr}.

(vii) Subroutine IHRD. This subroutine calculates the main and differential motor speeds, ω_m and $\omega_d(m)$, as functions of their desired values, ω_{md} and $\omega_{dd}(m)$, assuming ideal control hardware:

$$\omega_m = \omega_{md} \tag{7.76}$$

$$\omega_d(m) = \omega_{dd}(m) \tag{7.77}$$

where $m=1,\ldots,N_w$. Its purpose is to provide a stub which may be replaced eventually by real CTS hardware or models thereof such as the one described in Reference 4.

(viii) Subroutine POSVEL. Along-track velocity $\dot{\ell}$ and position ℓ are computed as functions of main motor speed:

$$\dot{\ell} = r_r N_{31} \omega_m \tag{7.78}$$

$$\ell = \ell(0) + r_r N_{31} \int_o^t \omega_m(\tau) d\tau \tag{7.79}$$

The achieved velocity profile $(\ell, \dot{\ell})$ may then be compared with the desired profile $(\ell, \dot{\ell}_d)$ by means of program SHOW.

(ix) Subroutine AALIGN. This subroutine computes the achieved track alignment $R_{a1}(q,m)$ at each wheelset m from motor speeds ω_m and $\omega_d(m)$ and along- track velocity $\dot{\ell}$. Initial positions are obtained by interpolating linearly on ψ_{ai} and $\underset{\sim}{R}_{ai}$ in data set TRADA at $\ell_{wo}(m) = \ell_o + a(m)$ for $m=1,\ldots,N_w$ and $q=1,2$ to obtain $\psi_{a10}(m)$ and $R_{a10}(q,m)$. Achieved distance derivatives of yaw are computed by

$$\delta(m) = \frac{N_{42}}{2N_{31}} \left[1 + \frac{\omega_d(m)}{\omega_m}\right] + 1 \tag{7.80}$$

and

$$\psi'_{a1}(m) = \frac{2[\delta(m)-1]}{K\delta(m)} \tag{7.81}$$

Achieved yaw rates:

$$\dot{\psi}_{a1}(m) = \dot{\ell}\psi'_{a1}(m) \tag{7.82}$$

Achieved yaw positions:

$$\psi_{a1}(m) = \psi_{a10}(m) + \int_0^t \dot{\psi}_{a1}(m)d\tau \tag{7.83}$$

Rates of change of achieved alignment:

$$\begin{bmatrix} \dot{R}_{a1}(1,m) \\ \dot{R}_{a1}(2,m) \end{bmatrix} = \dot{\ell} \begin{bmatrix} \cos \psi_{a1}(m) \\ \sin \psi_{a1}(m) \end{bmatrix} \tag{7.84}$$

Achieved alignment:

$$R_{a1}(q,m) = R_{a10}(q,m) + \int_0^t \dot{R}_{a1}(q,m)d\tau \tag{7.85}$$

The above computations are carried out for $q=1,2$ and $m=1,\ldots,N_W$.

(x) Subroutine COMOUT. This subroutine determines scaled along-track positions $\ell_s(j)$ for output data storage purposes by:

$$\ell_s(j+1) = \ell_s(j) + \Delta\ell_s, \quad j=0,\ldots,N_{OT} \tag{7.86}$$

where

$$\ell_s(0) = \ell(0)/\ell_T \tag{7.87}$$

and

$$\Delta\ell_s = 1/N_{OT} \tag{7.88}$$

and stores the required OUTDA data corresponding to these positions into a disk file allocated to data set OUTDA. The total number of data point positions for output purposes, N_{OT}, is specified by the user in data set RUNDA.

7.4.3.2 Program show

Program SHOW interactively provides ten table-printing and seven graph-plotting options and a selection of output devices including a terminal, plotter, line printer or printer-plotter.

The table-printing or graph-plotting options are exercised respectively by choosing items from the print menu or plot menu.

(i) Print menu.

P1: ℓ_s, ℓ. $\dot{\ell}$, $\dot{\ell}_d$, t

P2: ℓ_s, $C_{Rd}(1),\ldots,C_{Rd}(N_w)$

P3: ℓ_s, F_{rd}, $C'_{fr}(1),\ldots,C'_{fr}(N_w)$

P4: ℓ_s, $\psi_{rd}(1),\ldots,\psi_{rd}(N_w)$

P5: ℓ_s, n, $\underset{\sim}{F}^{(m)}_{tn}$; $n=1,\ldots,N_b$

P6: ℓ_s, n, $\underset{\sim}{F}^{(m)}_{cwn}$; $n=1,\ldots,N_b$

P7: ℓ_s, n, $\underset{\sim}{F}^{(m)}_{dn}$; $n=1,\ldots,N_b$

P8: ℓ_s, n, $\underset{\sim}{M}^{(b)}_{cn}$; $n=1,\ldots,N_b$

P9: ℓ_s, $R_{ad}(q)$; q=1,2

P10: ℓ_s, m, $R_{al}(q,m)$; q=1,2; $m=1,\ldots,N_b$

(ii) Plot menu.

D1: $R_{ad}(1)$ vs $R_{ad}(2)$ or $R_{al}(1,m)$ vs $R_{al}(2,m)$; $m \in [1,N_w]$

D2: $\dot{\ell}_d$ and $\dot{\ell}$ vs ℓ

D3: ℓ_s vs t

D4: $\underset{\sim}{F}^{(m)}_{t}(q,n)$ vs ℓ_s; q=1,2,3; $n \in [1,N_b]$

D5: $\underset{\sim}{F}^{(m)}_{cw}(q,n)$ vs ℓ_s; q=1,2,3; $n \in [1,N_b]$

D6: $\underset{\sim}{F}^{(m)}_{d}(q,n)$ vs ℓ_s; q=1,2,3; $n \in [1,N_b]$

D7: $\underset{\sim}{M}^{(b)}_{c}(q,n)$ vs ℓ_s; q=1,2,3; $n \in [1,N_b]$

(iii) Examples. Figures 7.23 to 7.27 respectively are examples corresponding to items D1, D2, D4, D5 and D6 of the plot menu. Figure 7.23, a specified track layout, exemplifies the first option in D1. The desired and actual velocity profiles (Fig. 7.22) are identical, as expected, in view of the ideal CTS roller speed control hardware (viz. subroutine IHRD). The scales from left to right in Figs. 7.25 to 7.27 are in Newtons and correspond to the curves for q=1,2 and 3, i.e., the solid, dashed and short-dashed

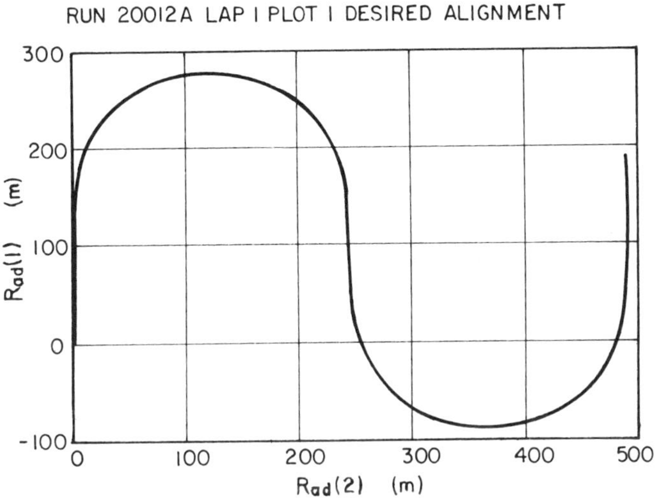

Fig. 7.23 A specified track layout.

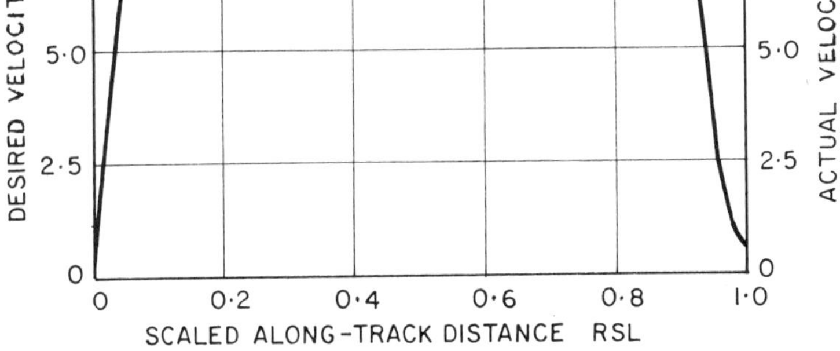

Fig. 7.24 A specified velocity profile.

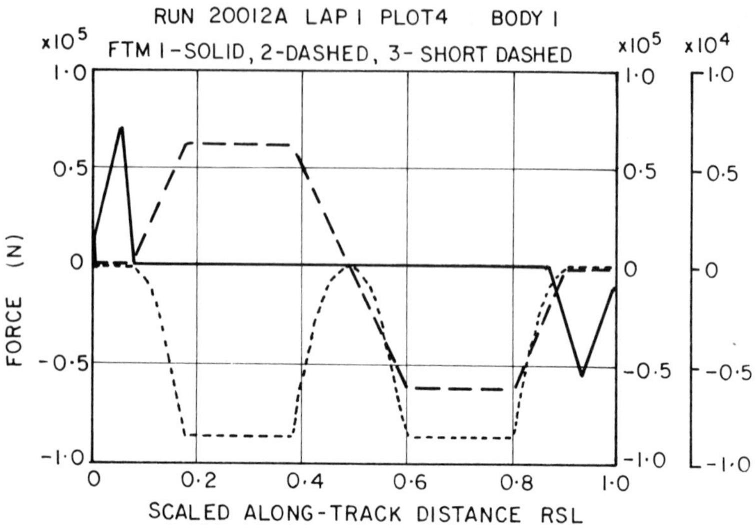

Fig. 7.25 A typical plot of $F_t^{(m)}(q,n)$; q=1,2,3; n=1.

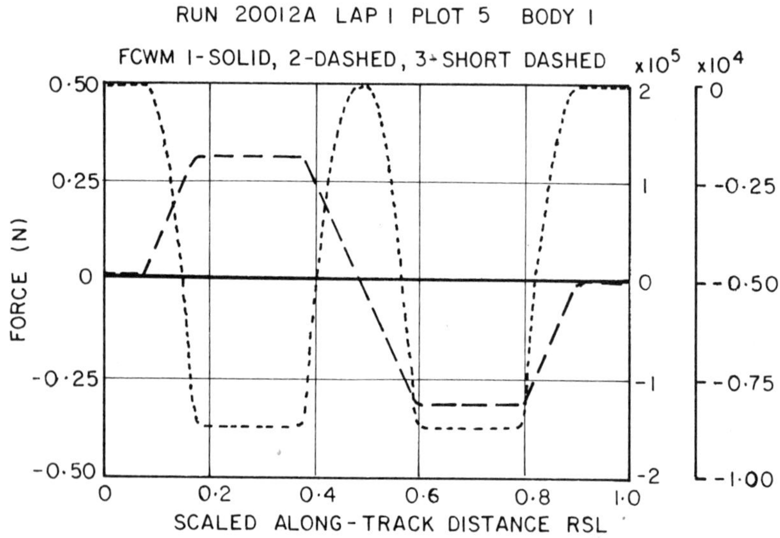

Fig. 7.26 A typical plot of $F_{cw}^{(m)}(q,n)$; q=1,2,3; n=1.

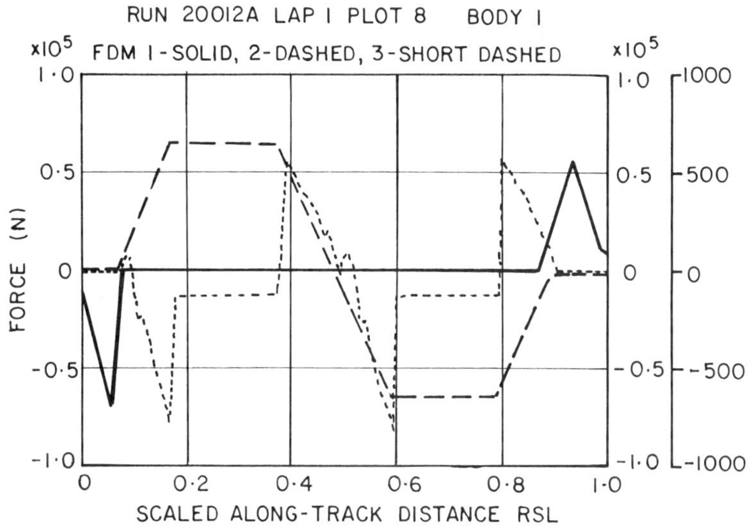

Fig. 7.27 A typical plot of $F_d^{(m)}(q,n)$; q=1,2,3; n=1.

curves, respectively. Body 1 (i.e., n=1) was defined as the main body of the rail vehicle in this example.

Since $\underset{\sim}{F}_{tn}^{(m)}$ is a d'Alembert force, its components are opposite in sense to the required actuator forces, neglecting corrections for non-superelevation of rollers. The curves in Fig. 7.27 are the result of subtracting the curves of Fig. 7.25 from those of Fig. 7.26 according to the equations solved by subroutine SUPMOD. The components of $\underset{\sim}{F}_{dn}^{(m)}$ in Fig. 7.27 are the required actuator force components. On superelevated track, the component of weight acting in the normal track direction is less than for non-superelevated CTS rollers where the whole weight acts vertically. However, the difference goes towards assisting the actuators in providing the normal component of centrifugal force. In this example, it turns out that only a small third component not exceeding ±700N throughout the course remains for the actuator. On the other hand, it is observed that the vehicle was operated at well below balance speed in the curved track region $\ell_s \in [0.18,0.38]$ since $F_t^{(m)}(2,1) < F_{cw}^{(m)}(2,1)$ by a factor of two, approximately.

7.5 VALIDATION

Validation is a necessary condition for the practical interpretation of track simulator results. Their usefulness is dependent upon a knowledge of their relationship to real track performance. Whereas validation can be qualitative or quantitative, extensive quantitative validation gives the most confidence in predicting the performance of a vehicle on real track. The extent of validation determines the degree of credibility of predictions.

7.5.1 Tangent Track Simulators

It was indicated in Section 7.1 that full-scale tangent track simulators can predict critical speeds to within about 10-15 percent. However, Matsudaira et al. [8] showed that the finiteness of roller diameters affects the wavelength of hunting. A kinematic analysis gave the following formula for the hunting wavelength of a wheelset on a roller stand:

$$\lambda = 2\pi(br_w/\gamma_e)^{\frac{1}{2}} \tag{7.89}$$

where

$$\gamma_e = \gamma \left[\frac{1 + (\rho_r/\rho_w)(r_w/r_r)}{1 - (\rho_r/\rho_w)}\right] \tag{7.90}$$

and b is the half-distance between contact points, r_w and r_r respectively are the wheel and roller rolling radii, ρ_w and ρ_r respectively are the wheel and roller lateral radii of curvature at a contact point, γ is the contact angle at the neutral wheelset position, and γ_e is the effective contact angle.

It follows from these equations that the effective slope of a wheel tread on a roller is increased by the factor

$$\rho = 1 + (\rho_r/\rho_w)(r_w/r_r) \tag{7.91}$$

compared with its value on a straight rail ($r_r = \infty$) and, correspondingly, the hunting wavelength of a wheelset on a roller stand is decreased by $\rho^{-\frac{1}{2}}$. If the wheel tread has a straight gradient

($\rho_w = \infty$), such as often exists with new wheels, the hunting wavelength is the same on rollers as on straight rails.

Taking the following values from an actual numerical example,

$r_w = 0.431$ m (1.414 ft) $\qquad r_r = 0.526$ m (1.726 ft)

$\rho_w = 0.120$ m (0.394 ft) $\qquad \rho_r = 0.050$ m (0.164 ft)

$b = 0.549$ m (1.801 ft) $\qquad \gamma = 0.097$

we have

$\rho = 1.34 \qquad \rho^{-\frac{1}{2}} = 0.84$

$\gamma_e = 0.223 \qquad \lambda = 6.47$ m (21.23 ft)

Thus, the effective wheel tread slope and hunting wavelength are respectively 34% larger and 16% smaller than their corresponding values for the wheelset on straight rails. It can be concluded, from this example and the above analysis, that the hunting wavelength of a worn wheelset can be significantly less and the hunting behaviour different on a roller stand than on straight rails; but, if the wheel tread is a straight gradient, the effect of finiteness of roller radii may be neglected.

The finiteness of roller radii also diminishes the contact area which, in turn, changes the creep coefficient values. However, Matsudaira et al. stated that this effect would not usually be so large as to produce important changes in hunting behaviour.

In summary, it can be said that tangent track simulators can be used to predict critical speeds to within approximately 10-15 percent. However, the degree of wear of wheel treads must be taken into consideration when a more detailed investigation of hunting behaviour is required. Given this knowledge regarding the effects of roller-diameter finiteness, it is meaningful to proceed with the problem of validation of curved track simulators.

7.5.2 Curved Track Simulators

Since there is a lack of literature on the quantitative validation of physical simulators for curving track, research results on this subject would be welcomed. Some of the impediments

to successful validation may include: (1) wheels not on top-dead-center of rollers in curve simulations, (2) differences in lateral roller stiffness compared with lateral track stiffness, (3) differences in torsional natural frequencies of roller drive systems compared with corresponding track natural frequencies, (4) inaccurate application of inertial forces and moments in curve simulations, taking non-superelevation of rollers into account, (5) an incomplete set of roller stands, (6) inaccurate roller yaw positioning, (7) inaccurate or non-dynamic roller cant angle settings, and (8) inaccurate control of transient and steady-state roller speeds.

Validation may be accomplished by comparing curved track simulator results with: (i) track results using the same vehicle, or (ii) validated mathematical simulation results using correct vehicle and track parameter values. The following variables are among several which could be monitored for validation purposes: (a) wheelset and roller speeds, (b) lateral and yaw displacements of wheelsets and rollers, and (c) inertial and creepage forces and moments. From these, it would be possible to determine control errors, critical speeds, hunting wavelengths, and L/V ratios. Further instrumentation would be required to determine carbody and truck motions.

Despite the apparent lack of strict quantitative validation of curved track simulators, their utilization in various successful applications constitutes a certain degree of qualitative validation. Some of these applications are described in the following section.

7.6 TRACK SIMULATORS SURVEY

A recent survey has revealed the existence of six physical track simulators located respectively in: (i) Pueblo, U.S.A., (ii) Kunitachi, Japan, (iii) Berlin, East Germany, (iv) Munich, West Germany, (v) Vitry, France, and (vi) Derby, England. In addition to the facility under development in Ottawa, Canada, a new curved-track simulator being constructed in Kunitachi, Japan, is to

be commissioned partially in 1987. A synopsis of the capabilities and limitations of the above facilities appears in Tables 7.8 to 7.10, using parameters defined in Table 7.7. Installations (i) to (iii) are straight-track simulators, whereas (iv) to (vi) are curved-track simulators; however, the Derby installation's capabilities for curving track are very limited. Of course, the latter three can also be used to simulate straight track. In Sections 7.6.1 to 7.6.3, we describe some applications of the Kunitachi, Vitry and Munich plants to illustrate the range of uses of track simulators.

Table 7.7. Track simulator parameters.

Parameter	Description
(a)	Ideal level and straight track
(b)	Adjustable wheelbase
(c)	Curved track simulated by yaw of rollers about vertical z-axis
(d)	Superelevation simulated by cant of roller stands and/or application of equivalent vehicle forces
(e)	Different gauges simulated by adjusting the lateral distance between rollers (mm)
(f)	Tilted rail simulated by canting rollers about their longitudinal x-axis
(g)	Faulty rail level simulated by vertical motion of rollers
(h)	Faulty rail alignment simulated by yaw of rollers, or displacing both rollers laterally
(i)	Wheel load (kN)
(j)	Simulated speed (km/h)
(k)	Driving power/axle (kW)

Table 7.8. Existing straight-track simulators.

Parameter	Pueblo USA	Kunitachi Japan	Berlin E. Germany
(a)	Y	Y	Y
(b)	Y	Y	Y
(e)	N	1000-1676	1000-1676
(i)	225	90	100
(j)	463	300	250
(k)	450	300	110

Y: Yes, N: No.

Table 7.9. Existing curved-track simulators.

Parameter	Munich W. Germany	Vitry France	Derby England
(a)	Y	Y	Y
(b)	Y	Y	?
(c)	Y	N	Y
(d)	Y	Y	N
(e)	1435-1700	1430-1680	N
(f)	Y	N	?
(g)	Y	Y	N
(h)	Y	Y	N
(i)	125	100	125
(j)	500	300	320
(k)	600	220	55

Table 7.10. Planned curved-track simulators.

Parameter	Kunitachi Japan	Ottawa Canada
(a)	Y	Y
(b)	Y	Y
(c)	Y	Y
(d)	Y	Y
(e)	1000-1676	1432-1524
(f)	Y	Y
(g)	Y	N
(h)	Y	N
(i)	100	165
(j)	450	240
(k)	500	150

7.6.1 Kunitachi Facility

The straight-track simulator at Kunitachi has contributed much to the understanding of the running characteristics of a recently-developed bolsterless truck, the development of a high-speed truck, and clarification of freight-car derailment phenomena. The new plant is expected to be of considerable assistance in the development of high-speed trucks for trains running at 300 km/h.

7.6.2 Vitry Facility

The Vitry-sur-Seine installation was operational in 1967. During the first ten years, seventy-seven vehicles were tested, including passenger cars, freight cars and locomotives. Of these, fifty-one were production vehicles; the others were prototypes or the first of a new series. Since then, many more have been tested.

The theoretical and practical experience gained at Vitry has demonstrated that the facility is not suitable for simulating wheel-rail hunting instability phenomena. The main reasons given are: (1) wheel-roller and wheel-rail kinematic differences, (2) high roller-stand rigidity compared with track rigidity, (3) different roller- and rail-material properties, and (4) lack of rigour in roller synchronization. The facility is used mainly for ride quality studies using sinusoidal or stochastic excitations. In the latter case, the applied vertical excitations are such as to produce wheelset-acceleration power spectral densities which are similar to those measured in the field. Strictly-speaking, this is the only simulation of real track calimed to be performed at the Vitry site. Nevertheless, the plant has been very useful. In addition to ride quality studies, structural vibration properties and constraints of vehicles and trucks have been determined.

7.6.3 Munich Facility

The Munich curved-track simulator plant, which began operations in the late seventies, has been used for many varied tests including the following.

(i) Verification of a computer model of a railway vehicle. Physical simulator tests to verify the computer model were carried out in three stages: (1) single-wheel-set tests, (2) truck tests, and (3) complete vehicle tests. A very close qualitative and quantitative correlation between physical simulation and computer simulation data was achieved, thereby verifying the accuracy of the computer model as a railway vehicle representation.

(ii) Tests with a newly-developed wheelset. A non-rigid wheelset prototype with controllable torque transfer between wheels was built to increase the running stability on straight track, but still retaining the necessary guidance ability in curves. Tests were conducted on the simulator for a variety of parameter adjustments. The intended function of this new wheelset was obtained for speeds up to 503 km/h.

(iii) Tests with a newly-developed truck. An active torsional-damping system with torque retardation was designed into a truck instead of a conventional passive system. Simulation tests showed that the truck was stable at all simlulated speeds whereas, without torque retardation, the critical speed was 160 km/h.

(iv) Investigation of an experimental high-speed truck. An experimental high-speed truck (bogie) was intended for use at speeds reaching 350 km/h. Simulator tests were conducted to determine: (1) the response to given track irregularities, (2) behaviour in the marginal stability region, (3) dynamic rigidity in the primary and secondary levels, (4) braking behaviour during unstable running, (5) effects of axle-bearing transverse play, and (6) effects of primary-suspension rigidity variations.

(v) Investigation of the electric-motor drive of a locomotive. The performance of an electric-motor drive installed in a redesigned locomotive was investigated. Tests were conducted on both wheelsets of the first bogie. To control the vehicle, an external driver's cab was set up in the measuring and control centre of the test facility. Simulation tests covered the entire driving and braking range. Included were: (1) driving and braking traction-characteristics tests, (2) speed tests, (3) endurance tests creating limiting temperatures in the converters and drive motors, (4) load capacity tests of the clutch between the drive system and wheelsets, and (5) tests to determine the influence of wheel-diameter variations on the distribution of motor currents and torques. The simulator was well-suited to this last test in that wheel-diameter variations could be simulated by varying roller speeds.

(vi) Tests with rubber-ring sprung wheels. Tests were conducted to determine the effect of rubber-ring sprung wheels on the reduction of noise and vibration levels in the environment. By comparison of results with wheels devoid of rubber rings, it was possible to determine the frequency range in which an improvement was obtained because of the intermediate rubber layer.

(vii) Trials with a high-speed locomotive. A four-axle diesel-electric locomotive was equipped with newly-developed bogies for high-speed operation up to 350 km/h. Comprehensive tests were carried out on the simulator to determine: (1) the rigidity and kinematics of the wheelset coupling, (2) the effect of torque retardation variations, and (3) the locomotive's drive performance. The results of these tests and real track tests confirmed the suitability of the newly-developed bogie configuration for the next generation of high-speed rail vehicles.

(viii) Investigation of the safety of transported goods. The safety of transported goods can be determined by simulating measured track spectra and investigating the load's behaviour regarding lateral displacement and rolling. Centrifugal force effects can be simulated by tilting the roller stands.

7.7 NOMENCLATURE

A dot or double dot above a variable denotes the first or second derivative with respect to time whereas a prime or double prime superscript denotes the first or second derivative with respect to along-track distance ℓ, respectively.

A vector or matrix is denoted by bold-faced type (or an underlining tilde in diagrams) or square brackets and its transpose is indicated by superscript T outside of the brackets.

Subscripts A,B,E or M ascribed to X,Y,Z axes denote alignment, body, earth or mission axes, respectively.

Subscripts 0 and 1 denote values at the beginning and end of a track or segment, respectively, and subscripts NSH and NSV denote values pertaining to horizontal segment N_{SH} and vertical segment N_{SV}, respectively.

The following subscripts apply to angular position, velocity and acceleration variables:

a or ae: variable of alignment axes relative to earth axes
me : variable of mission axes relative to earth axes
bm : variable of body axes relative to mission axes

Superscript (a), (b), (e) or (m) added to a vector denotes that it is expressed in terms of alignment, body, earth or mission axes components, respectively.

The addition of subscript i to a variable associates it with the i'th along-track positional data point whereas subscript k associates it with the k'th mission axes. Subscript n associates it with the n'th rigid body.

The symbols are defined below without the above modifiers in most cases.

Finally, lower-case ℓ and upper-case L, regardless of subscripts, refer to along-track distances and their horizontal projections, respectively.

7.7.1 Roman

a(m)	Longitudinal distance of m'th wheelset center of gravity from alignment frame origin, positive towards front of vehicle
b	Half-distance between contact points of a wheelset
$[A_{me}]$	Matrix relating Euler angular rates to mission axes angular velocity relative to earth axes expressed in terms of mission axes
C_{fr}	Main ratio control divider constant
$C'_{fr}(m)$	Differential ratio control divider variable at m'th wheelset
C_R	Curvature ratio
$C_{Rd}(m)$	Desired curvature ratio at m'th wheelset
CTSDA	CTS data set
D	Distance from wheel-roller contact center
$\underset{\sim}{E}$	Track-derived function for use in determining correction forces and moments
$\underset{\sim}{F}$	Track-derived function for use in determining correction forces and moments
$\underset{\sim}{F}_c$	Coriolis acceleration force
$\underset{\sim}{F}_{cw}$	Correction force due to non-superelevation of rollers
$F_{cw}(q,n)$	Q'th component of n'th body correction force due to non-superelevation of rollers
$\underset{\sim}{F}_d$	Net desired correction force
$F_d(q,n)$	Q'th component of net desired correction force at n'th rigid body

$\underset{\sim}{F}_e$	External force applied to rigid body
$\underset{\sim}{F}_n$	Net external force applied to n'th rigid body
F_{rd}	Desired reference oscillator frequency
$\underset{\sim}{F}_t$	Transport acceleration force
$\underset{\sim}{G}$	Track-derived function for use in determining correction forces and moments
G_{tm}	Main tachometer gain
$\underset{\sim}{H}$	Track-derived function for use in determining correction forces and moments
IDCTS	CTS data name
IDMIS	Mission data name
IDMISP	Name of disk file for plotting on a printer
IDNAM(N)	Name of n'th rigid body
IDOUT	Output data name
IDRUN	Run data name
IDTRA	Track data name
IDVEH	Vehicle data name
J	Polar moment of inertia
$[J_n]$	Principal moments of inertia matrix of n'th rigid body
$[J_n]^{-1}$	Inverse of principle moments of inertia matrix of n'th rigid body
K	Track gauge
K_P	Proportionality constant
K_R	Along-track positional data point round-off constant
ℓ	Distance along track alignment
$\ell(o)$	Initial along-track position
$\ell(j)$	J'th track data point position for output data
L	Horizontal projection distance along track alignment
$\dot{\ell}_d$	Desired vehicle velocity at actual position
$\ell_m(r)$	R'th track data point position for mission data
$\dot{\ell}_m(r)$	R'th mission data velocity
ℓ_s	Scaled along-track position
$\ell_s(j)$	J'th scaled track data point position generated by subroutine COMOUT

L_s	Alignment length of a spiral segment
ℓ_T	Total length of track
m	Rigid body mass
$\underset{\sim}{M}$	External moment applied to a rigid body
$\underset{\sim}{M}_c$	Net inertial moment
$\underset{\sim}{M}_c(q,n)$	Q'th component of net inertial moment for body 'n'
$\underset{\sim}{M}_{cb}$	Inertial moment due to relative body motion only
$\underset{\sim}{M}_{cbm}$	Inertial moment due to the interaction of relative body motion and mission axes motion
$\underset{\sim}{M}_{cm}$	Inertial moment due to mission axes motion only
M(n)	Mission axes number associated with n'th rigid body
MISDA	Mission data set
N_b	Total number of rigid bodies
N_{DT}	Total number of track data point positions generated by program GENTDP
N_L	Lap number
N_{LT}	Total number of laps for output purposes
N_m	Total number of mission reference frames
N_{mT}	Total number of mission data point positions
N_{OT}	Total number of output positional increments per lap
N_{SH}	Horizontal segment number
N_{STH}	Total number of horizontal segments
N_{STV}	Total number of vertical segments
N_{SV}	Vertical segment number
N_w	Total number of wheelsets
N_{31}	Roller stand gear ratio
N_{41}	Roller stand gear ratio
N_{42}	Roller stand gear ratio
OUTDA	CTS control and monitoring data
R,R_D	Wheel rolling radii at contact center and distance D, respectively
r,r_D	Roller rolling radii at contact center and distance D, respectively

$\underset{\sim}{R}_a$	Position of alignment axes relative to earth axes
$R_a(q)$	Q'th component of track alignment
$R_{ad}(q)$	Q'th component of desired alignment
$R_{ado}(q)$	Q'th component of desired initial alignment
$R_{a1}(q,m)$	Q'th component of achieved alignment at m'th wheelset
$R_{a10}(q,m)$	Q'th component of initial alignment at m'th wheelset
R_c	Radius of curvature of curve segment
R_H	Horizontal radius of curvature
R_ℓ, R_r	Left and right rail radii of curvature
r_ℓ, r_r	Left and right nominal rolling radii of a roller stand
R_I	Inner rail radius of curvature
R_O	Outer rail radius of curvature
r_w	Wheel rolling radius
RUNDA	Run data set
S	Relative spin
T	Track segment type
t	Time
$[T_{bm}]$	Transformation from mission coordinates to body coordinates
$[T_{me}]$	Transformation from earth coordinates to mission coordinates
TRADA	Track data set
V	Forward rolling velocity
$\underset{\sim}{x}$	Position of body axes relative to mission axes
$\underset{\sim}{x}_{am}$	Position of mission axes relative to alignment axes
$\underset{\sim}{x}_n(o)$	Initial position of body 'n' relative to mission axes

7.7.2 Greek

$\underset{\sim}{\alpha}_a$	Angular position of alignment axes relative to earth axes expressed in Euler angles
$\underset{\sim}{\alpha}_{bm}$	Angular position of a body relative to mission axes, expressed in Euler angles
α_t	Small angle approximation threshold angle
β	Roller cant angle
γ	Wheel conicity

γ_e	Effective wheel conicity
λ	Hunting wavelength of a wheelset
$\underset{\sim}{\omega}_{bm}$	Angular velocity of a body relative to mission axes
$\omega_d(m)$	Differential motor speed at m'th wheelset
$\omega_{dd}(m)$	Desired differential motor speed at m'th wheelset
ω_m	Main motor speed
ω_{md}	Desired main motor speed
$\omega_{r\ell},\omega_{rr}$	Left and right angular rolling velocities of a roller stand
Ω_w,Ω_r	Spin velocities of wheel and roller, respectively
ω_w,ω_r	Angular rolling velocities of wheel and roller, respectively
ϕ_a	Bank angle of alignment axes relative to earth axes
$\phi_{adl}(m)$	Desired track yaw angle at m'th wheelset
ψ_a	Yaw angle of alignment axes relative to earth axes
ψ_{ad}	Desired yaw angle of alignment axes
$\psi_{adl}(m)$	Desired track yaw angle at m'th wheelset
$\psi_{al}(m)$	Yaw position of achieved alignment at m'th wheelset
$\psi_{rd}(m)$	Desired roller stand yaw position at m'th wheelset
ρ_r,ρ_w	Roller and wheel radii of curvature at a contact point

7.8 SUMMARY

The use of rollers to simulate real track was discussed in this chapter. Thereby, the concept of physical simulators was introduced. The presentation of general aspects of physical track simulators was accomplished through the medium of a specific curved track simulator (CTS).

The architecture of roller stands and control systems was described first, followed by the mathematical foundation of track simulators. Here, we defined earth, alignment, mission and body coordinate systems and the mathematical representation of track layouts.

The speeds and orientations of rollers were respectively shown to be functions of the simulated forward velocity and the track curvature. It was found necessary to cant the rollers in order to minimize spin-creep errors.

Various CTS computer programs were described. Their generated data could be used to control and monitor the CTS in real time or to predict operational requirements of CTS control hardware.

The problem of validating physical track simulators was discussed. It was found that tangent track simulators can be used to predict critical speeds to within approximately 10-15 percent, whereas a large part of the validation of curved track simulators remains in the research domain.

Results of a recent survey on physical track simulators were presented, including three existing straight-track simulators, three existing curved-track simulators, and two of the latter type which are under development. Various parameters were used to compare these facilities and a range of actual applications was described.

REFERENCES

1. AMYOT J.R. and MUFTI I.H. (1975): Track Model for Computer Studies of Railway Vehicle Dynamics; in: *Proceedings of the Third International Conference on Vehicle System Dynamics*, H.K. Sachs (Ed.), 54-66, Swets and Zeitlinger, Amsterdam.
2. AMYOT J.R. and CAMIRÉ G.A. (1981): Track Data Preparation Programs for Railway Vehicle Dyanmics; *Proceedings of the Summer Computer Simulation Conference*, Washington, D.C., U.S.A., 710-714, 16-17 July 1981.
3. AMYOT J.R. and STOCK F.T. (1982): Mission Data Preparation Program for Railways Vehicle Dynamics; *Proceedings of the Summer Computer Simulation Conference*, Denver, Colorado, U.S.A., 224-226, 19-21 July 1982.
4. AMYOT J.R. (1982): Computer Model of Roller Drive Systems in a Railway Curved Track Simulator; *Proceedings of the 10th IMACS World Congress on System Simulation and Scientific Computation*, Vol. 3, Montreal, Canada, 159-162, 8-13 August 1982.

5. AMYOT J.R. (1982): Computer Simulation in Design of a Railway Curved Track Simulator; Computers in Engineering 1982, Vol. 1, ASME; *Proceedings of the Second International Computer Engineering Conference*, San Diego, California, U.S.A., 243-244, 15-19 August 1982.

6. AMYOT J.R. (1983): Analytical Program for a Curved Track Simulator; *Proceedings of the Summer Computer Simulation Conference*, Vol. 1, Vancouver, Canada, 319-326, 11-13 July 1983.

7. HO A., APPARAO T.A.P.S. and BILLING J.R. (1975): Definition of a Model of Study of the Dynamics of a MAGLEV System; Report of the Transit Systems Research and Development Branch, Research and Development Division, Ontario Ministry of Transportation and Communications, Toronto, Canada, September 1975.

8. MATSUDAIRA T., MATSUI N., ARAI S. and YOKOSE K. (1969): Problems on Hunting of Railway Vehicle on Test Stand; *Transactions of the ASME, Journal of Engineering for Industry*, 879-890, August 1969.

9. SMITH C.A.M. and BOWLER E.H. (1981): A Curved Track Simulator; *Proceedings of the XV Panamerican Railway Congress*, Mexico, D.F., 465-483, November 1981.

10. SMITH C.A.M. and DUKKIPATI R.V. (1983): The NRC Curved Track Simulator and its Potential Use; *Post-Conference Proceedings of the Conference on the Economics and Performance of Freight Car Trucks*, Montreal, Canada, 255-262, 25-27 October 1983.

INDEX